Artificial Intelligence Tools and Technologies for Smart Farming and Agriculture Practices

Rajeev Kumar Gupta
Pandit Deendayal Energy University, India

Arti Jain
Jaypee Institute of Information Technology, India

John Wang
Montclair State University, USA

Santosh Kumar Bharti
Pandit Deendayal Energy University, India

Samir Patel
Pandit Deendayal Energy University, India

A volume in the Practice, Progress, and Proficiency in Sustainability (PPPS) Book Series

IGI Global
PUBLISHER of TIMELY KNOWLEDGE

Published in the United States of America by
IGI Global
Engineering Science Reference (an imprint of IGI Global)
701 E. Chocolate Avenue
Hershey PA, USA 17033
Tel: 717-533-8845
Fax: 717-533-8661
E-mail: cust@igi-global.com
Web site: http://www.igi-global.com

Library of Congress Cataloging-in-Publication Data

Names: Gupta, Rajeev Kumar, 1984- editor. | Jain, Arti, 1977- editor. | Wang, John, 1955- editor. | Bharti, Santosh Kumar, editor. | Patel, Samir, editor.
Title: Artificial intelligence tools and technologies for smart farming and agriculture practices / edited by Rajeev Kumar Gupta, Arti Jain, John Wang, Santosh Kumar Bharti, and Samir Patel.
Description: Hershey, PA : Engineering Science Reference, [2023] | Includes bibliographical references and index. | Summary: Artificial Intelligence Tools and Technologies for Smart Farming and Agriculture Practices discusses various tools and technologies that can be used in smart farming and agriculture practice and explores the role of different emerging technologies like the internet of things, big data, machine learning, deep learning, and AI from agricultural prospects. Covering key topics such as farming, pests, soil, and weeds, this premier reference source is ideal for environmentalists, farmers, agriculturalists, industry professionals, researchers, academicians, scholars, practitioners, instructors, and students.
Identifiers: LCCN 2023004836 (print) | LCCN 2023004837 (ebook) | ISBN 9781668485163 (hardcover) | ISBN 9781668485170 (paperback) | ISBN 9781668485187 (ebook)
Subjects: LCSH: Agricultural informatics. | Alternative agriculture. | Artificial intelligence--Agricultural applications.
Classification: LCC S494.5.A65 A78 2023 (print) | LCC S494.5.A65 (ebook) | DDC 630.2/085--dc23/eng/20230421
LC record available at https://lccn.loc.gov/2023004836
LC ebook record available at https://lccn.loc.gov/2023004837

This book is published in the IGI Global book series Practice, Progress, and Proficiency in Sustainability (PPPS) (ISSN: 2330-3271; eISSN: 2330-328X)

British Cataloguing in Publication Data
A Cataloguing in Publication record for this book is available from the British Library.

All work contributed to this book is new, previously-unpublished material. The views expressed in this book are those of the authors, but not necessarily of the publisher.

For electronic access to this publication, please contact: eresources@igi-global.com.

Practice, Progress, and Proficiency in Sustainability (PPPS) Book Series

Ayman Batisha
International Sustainability Institute, Egypt

ISSN:2330-3271
EISSN:2330-328X

Mission

In a world where traditional business practices are reconsidered and economic activity is performed in a global context, new areas of economic developments are recognized as the key enablers of wealth and income production. This knowledge of information technologies provides infrastructures, systems, and services towards sustainable development.

The **Practices, Progress, and Proficiency in Sustainability (PPPS) Book Series** focuses on the local and global challenges, business opportunities, and societal needs surrounding international collaboration and sustainable development of technology. This series brings together academics, researchers, entrepreneurs, policy makers and government officers aiming to contribute to the progress and proficiency in sustainability.

Coverage

- Innovation Networks
- Strategic Management of IT
- Sustainable Development
- Technological learning
- E-Development
- Green Technology
- Knowledge clusters
- Global Business
- Environmental informatics
- Outsourcing

IGI Global is currently accepting manuscripts for publication within this series. To submit a proposal for a volume in this series, please contact our Acquisition Editors at Acquisitions@igi-global.com or visit: http://www.igi-global.com/publish/.

The Practice, Progress, and Proficiency in Sustainability (PPPS) Book Series (ISSN 2330-3271) is published by IGI Global, 701 E. Chocolate Avenue, Hershey, PA 17033-1240, USA, www.igi-global.com. This series is composed of titles available for purchase individually; each title is edited to be contextually exclusive from any other title within the series. For pricing and ordering information please visit http://www.igi-global.com/book-series/practice-progress-proficiency-sustainability/73810. Postmaster: Send all address changes to above address.

Titles in this Series

For a list of additional titles in this series, please visit: www.igi-global.com/book-series

Financing Regions Toward Sustainability in the Midst of Climate Change Risks and Uncertainty
Beata Zofia Filipiak (University of Szczecin, Poland) Dominika Kordela (University of Szczecin, Poland) and Izabela Nawrolska (University of Szczecin, Poland)
Information Science Reference • © 2023 • 330pp • H/C (ISBN: 9781668476208) • US $225.00

Pragmatic Legal and Policy Implications of Environmental Lawmaking
Hussein Movahedian (Islamic Azad University, UAE & Department of Private Law, Islamic Studies and Law Faculty, Imam Sadiq University, Tehran, Iran) and Nima Norouzi (Islamic Azad University, UAE & Law and Political Science Department, University of Tehran, Tehran, Iran)
Information Science Reference • © 2023 • 269pp • H/C (ISBN: 9781668441589) • US $215.00

Leadership and Governance for Sustainability
Neeta Baporikar (Namibia University of Science and Technology, Namibia & SP Pune University, India)
Information Science Reference • © 2023 • 350pp • H/C (ISBN: 9781668497111) • US $250.00

Implications of the COVID-19 Pandemic and the Russia-Ukraine Crisis on the Agricultural Sector
Vítor João Pereira Domingues Martinho (ESAV, CERNAS-IPV Research Centre, Polytechnic Institute of Viseu, Portugal)
Engineering Science Reference • © 2023 • 515pp • H/C (ISBN: 9781668489239) • US $215.00

Innovations in Environmental Legislation and Justice Environmental and Water-Energy-Food Nexus Laws
Nima Norouzi (Islamic Azad University, UAE & Law and Political Science Department, University of Tehran, Tehran, Iran) and Hussein Movahedian (Islamic Azad University, UAE & Department of Private Law, Islamic Studies and Law Faculty, Imam Sadiq University, Tehran, Iran)
Information Science Reference • © 2023 • 268pp • H/C (ISBN: 9781668471883) • US $215.00

Sustainable Approaches and Strategies for E-Waste Management and Utilization
A. M. Rawani (National Institute of Technology, Raipur, India) Mithilesh Kumar Sahu (O.P. Jindal University, India) Siddharth S. Chakarabarti (O.P. Jindal University, India) and Ajit Kumar Singh (GLA University, India)
Engineering Science Reference • © 2023 • 300pp • H/C (ISBN: 9781668475737) • US $240.00

Handbook of Research on Solving Societal Challenges Through Sustainability-Oriented Innovation
Luísa Cagica Carvalho (Instituto Politécnico de Setúbal, Portugal & CEFAGE, University of Évora, Portugal) Paulo Bogas (Instituto Politécnico de Setúbal, Portugal) Jordana Kneipp (Federal University of Santa Maria, Brazil) Lucas Avila (Federal University of Santa Maria, Brazil) and Elis Ossmane (Universidade Aberta, Portugal)
Engineering Science Reference • © 2023 • 419pp • H/C (ISBN: 9781668461235) • US $295.00

IGI Global
PUBLISHER of TIMELY KNOWLEDGE

701 East Chocolate Avenue, Hershey, PA 17033, USA
Tel: 717-533-8845 x100 • Fax: 717-533-8661
E-Mail: cust@igi-global.com • www.igi-global.com

Editorial Advisory Board

Table of Contents

Detailed Table of Contents

The world's population is predicted to grow by nearly 10 billion in the year 2050, which increases the food and agricultural demands. Hence, farmers worldwide are under continuous pressure to increase crop yield to satisfy the increasing food demand. To support the farmers in food production, crop monitoring and soil management are essential tasks. The traditional human observation-based crop and soil monitoring require more time and manpower. Hence, there is a need for smart crop and soil monitoring systems. This chapter is focused on different digital imaging-based Artificial Intelligence (AI) approaches such as machine learning, and deep learning that are used for smart crop and soil monitoring. There has been limited research work on the usage of smart intelligent techniques in crop and soil monitoring. This chapter attempts to identify the research gap in smart crops and soil monitoring. Future research trends would focus on integrating hybrid AI and reinforcement learning techniques for agricultural monitoring and management.

Indian agriculture industry gives a significant rise to GDP and provides food and livelihood to a large portion of the whole global population. This demand is increasing rapidly by population growth; the world's population is projected to reach 9.7 billion by 2050. It shoots up pressure on farmers to uplift production and grow crops that are better suited to their local weather and soil conditions. Traditional methods of crop recommendation are often based on subjective assessments and can be inaccurate. However, to overcome these challenges, new technologies like IoT can help farmers to make better decisions and growth of the crop towards holistic rural development. IoT makes it possible to take better decisions on run time and interactively by combining the previous data, past experience of farmers, scientific methodologies, feedback, soil health, weather forecasting, etc. So, in this chapter the use of the IoT and sensors have been discussed with practicality to adopt IoT easily and effectively in the agriculture field.

being adopted to ensure future food security. With the help of CV, AI, and related methods, such as Machine Learning (ML) together with Deep Learning (DL), key stakeholders can gain invaluable insights into the performance of agricultural and farm initiatives, enabling them to make data-driven decisions without the need for direct interaction. This chapter presents a comprehensive overview of the requirements, techniques, applications, and future directions for smart farming and agriculture. Different vital stakeholders, researchers, and students who have a keen interest in this field would find the discussions in this chapter insightful.

Air quality forecasting is an important way to safeguard public health since it provides ample notice of harmful air pollutants in a given location. Exhaust emissions from gas-guzzling vehicles, industrial pollution, and wildfires obfuscate some of the country's most dramatic vistas and can pose a significant hazard. If forecasts are trustworthy and accurate enough, they can be a crucial component of an air quality control system that complements more conventional emissions-based strategies. This study uses multivariate LSTM Time series forecasting techniques to understand and apply multiple variables together to contribute more accuracy to forecasting due to its proven track record of success with time-series data. The factors that are found to have a major effect on the air quality are dew point, wind speed, temperature, snowfall and rain conditions. In particular, this model focuses on predicting and forecasting the concentrations of PM2.5 (Atmospheric Particulate Matter with a diameter of less than 2.5mm) as this pollutant has a major effect on human health worldwide.

The resource that the global economy most heavily depends on is agricultural productivity. Nowadays, drones are widely used in a variety of civic applications, including agriculture. However, the aim of this work is to provide an overview of how unmanned aerial vehicles are used in agriculture fields. A development in drone technology for agricultural events can restructure the farm sector to become dynamic and intelligent rather than static and manual. Precision agriculture can be made more effective with the deployment of drones integrated with the proper camera, sensors, and integrating modules. The two main forces behind agriculture automation are precision agriculture and wireless sensor network. Drones, automation, and spectral cameras are all combined to increase the possibilities of managing crop yield and weed issues more effectively. The best design concept and analysis of an autonomous agricultural crop and weed management UAV are discussed. In order to meet the increasing need for food over the past few decades, farming practices have undergone a significant evolution.

The food security of any country may be jeopardized due to improper management of agricultural insect pests. Accurate pest detection and efficient pest control strategies must be employed in time to grow healthy crops for achieving food security in a country and worldwide. Hence, developing efficient and robust techniques to detect agricultural insect pests using computer vision approaches is one of the essential steps for timely managing insect pests. This chapter presents a short survey of deep learning-based object detection techniques focusing on insect pest detection and associated insect pest image datasets. Subsequently, a transfer learning-based custom You Only Look Once (YOLOv5) model is developed using the publicly available dataset IP102 for detecting agricultural insect pests with the help of computer vision approaches. The hyperparameters of the proposed insect pest detector are optimized using the genetic algorithm-based hyperparameter evolution method. The performance metrics of the proposed insect pest detector are found to be promising.

Agriculture is inextricably linked to the environment. Climate change has an effect directly on agricultural activities. India gets severely impacted if there is a loss of yields, which affects human lives. Hence, monitoring climate and its impact on the agricultural field is essential for a country like India. This chapter proposes a novel deep learning-based hybrid vector autoregressive–gated recurrent unit model (VAR-GRU model) for weather forecasting involving the four important weather parameters such as temperature, pressure, humidity, and wind speed for the cities of Bengaluru and temperature, pressure, dew point, and wind speed for the cities of Dongsi. The effectiveness of the proposed VAR-GRU model is proven by comparing its performance metrics (MAE, MSE, RMSE, and R2 Score) with that of other baseline models such as LSTM, VAR, GRU, and another hybrid VAR-LSTM model. The outcomes of this research work can help in increasing crop yields by utilizing the weather forecasting results in smart farming applications.

Agriculture is the foremost factor which is important for the survival of human beings. Farming contributes to a very big part of GDP; still, several areas exist where improvements are required. One of those is crop recommendation. Crop productivity is boosted as a result of accurate crop prediction. As crop production has already started to suffer from climate change, improving crop output is consequently desirable because agronomists are impotent to select the appropriate crop(s) depending on environmental and soil parameters, and the mechanism of forecasting the selection of the appropriate crops manually has failed. Factors like soil characteristics, soil types, climate characteristics, temperature, rainfall, area, humidity, geographic location etc. affect crop forecast. This chapter focuses mainly on building a recommendation system, i.e., suggesting the kind of the crop by applying various machine learning and deep learning techniques depending upon several parameters. The system would help the farmers for the appropriate decision to be taken regarding the crop type.

Agriculture and farming have gotten smarter as a result of the use of current technology such as Wireless Sensor Networks (WSN) and the Internet of Things (IoT). Smart farming is an enhanced agriculture system that offers data such as temperature, soil moisture, and so on, to assist in the growth of plants and cattle. It integrates wireless sensors and the internet to collect and communicate information with farmers. The priority event-based energy efficient algorithm developed in this study is utilized for accurate and efficient information transmission regarding power consumption and node priority. The major goal of the IoT-sensor network in this chapter is to increase farm productivity and extend its lifespan by applying intelligent algorithms such as Artificial Neural Network (ANN) to recognize environmental conditions and improve total production. Priority event-based energy efficient method reduces energy usage and increases the lifetime using Dijkstra's algorithm.

Preface

Rapid developments in Artificial Intelligence (AI) and related technologies are having a profound impact on the agricultural sector around the world. These innovations have revolutionized conventional agricultural methods and prepared the path for a new era of smart farming. AI supported tools and technologies have emerged as effective allies in optimizing production, improving resource efficiency, and assuring sustainable agriculture across a wide range of applications such as precision farming, crop monitoring, livestock management, predictive analysis of crops.

The authors of the book *Artificial Intelligence Tools and Technologies for Smart Farming and Agriculture Practices* discuss the promising applications of AI and the Internet of Things in the modern agriculture. In this book, readers would have an in-depth analysis of how innovative technologies are revolutionizing agriculture, as well as helpful guidance would be provided on how to put into practice AI-driven solutions in today's farming arena. We hope that by publishing this book, we would be able to bridge the gap between scientific developments in AI and their applications in agriculture in the real-world. We have brought together a group of experienced contributors, including researchers, and technologists who have offered their knowledge, expertise, and experiences to present a comprehensive guide for leveraging AI tools and technologies in the field of smart farming.

The book is organized into Thirteen Chapters, each addressing a crucial aspect of AI and IoT in agriculture. Beginning with an introduction to the fundamental concepts of AI and its various sub-fields, like machine learning, deep learning and computer vision. The remaining chapters examine roles of AI in precision agriculture, covering topics such as soil monitoring, crop yield prediction, and automated irrigation.

The topics in this book includes AI and IoT based on crop and soil monitoring, crop recommendation, IoT based crop protection system from animal, role of computer vision in smart agriculture, impact of air pollution in agriculture, crop and weeds monitoring using unmanned aerial, pest detection, role of weather pattern in smart agriculture, plant disease detection and classification, and many more.

ORGANIZATION OF THE BOOK

The book is organized into 13 chapters which includes the role of AI and IoT in agriculture and their applications. A brief description of each chapter are as follows:

Chapter 1 provides a detailed introduction of different AI techniques for smart crop and soil monitoring. The world's population is predicted to grow by nearly 10 billion in the year 2050, which increases the food and agricultural demands. Hence, farmers worldwide are under continuous pressure to increase crop yield to satisfy the increasing food demand. To support the farmers in food production, crop monitoring and soil management are essential tasks. The traditional human observation-based crop and soil monitoring require more time and manpower. Hence, there is a need for smart crop and soil monitoring systems. This chapter is focused on different digital imaging-based Artificial Intelligence (AI) approaches such as machine learning, and deep learning that are used for smart crop and soil monitoring. There has been limited research work on the usage of smart intelligent techniques in the crop and soil monitoring. This chapter attempts to identify the research gap in smart crop and soil monitoring. Future research trends would focus to integrate hybrid AI and reinforcement learning techniques for agricultural monitoring and management.

Chapter 2 discusses different sensors that are used in the smart farming. Indian agriculture industry has a significant role in the Gross Domestic Product (GDP) and provides food and livelihood to a large portion of the whole global population. This demand is increasing rapidly by population growth and the world's population is projected to reach 9.7 billion by 2050. It increases pressure on farmers to increase production and grow crops better suited to their local weather and soil conditions. Traditional methods of crop recommendation are often based on subjective assessments and can be inaccurate. However, to overcome these challenges, new technologies like IoT can help farmers to take better decisions and substantial growth of crops, which in turn would turn holistic towards rural development. IoT makes it possible to take better decisions on run time and interactively combines the previous data, past experiences of farmers, scientific methodologies, feedbacks, soil health, weather forecasting, etc. The use of IoT and related sensors are discussed in this chapter, with practicality to adopt IoT easily and effectively in the agriculture field.

Chapter 3 illustrates how IoT and machine learning can help the farmers to protect their crops from animals. No one can imagine a world without agriculture. Due to overpopulation and deforestation animal interference in residential areas is increasing day by day which affects human life and property causing human-animal conflict. At present time, farmers in India and other countries are also facing a serious threat from wildlife animals which causes less production in crop yielding and harms to the crops. It is not possible to stay in the field for 24 hours to guard the crops against animals. In spite of good production, the crops on farms are many times ravaged by animals and leads to huge losses for the farmers, in terms of production, effort, quality, and money. To overcome this problem, IoT-based crop protection system is being proposed in the present work. The system uses a microprocessor and a set of sensors to detect wild animals approaching near the field. The sensor data is analyzed through a machine learning algorithm and helps the microprocessor to take action. This system helps farmers to protect their crops from animals.

Chapter 4 focuses on different AI techniques to monitor and forecast the growth of plants. Smart agriculture has become crucial in meeting the increasing dietary needs of a growing population, particularly in countries where agriculture has significant economic impact. Deep learning techniques, such as convolutional neural networks and recurrent neural networks are extensively researched and applied in the agriculture during recent years. In this study, recent research articles on deep learning in agriculture over the past five years are analyzed to identify key contributions and challenges. This study also explores agriculture parameters that are monitored by the internet of things, and use them to train the deep learning algorithm for analysis. The study compares various factors across different studies, including the agriculture area of focus, dataset used, deep learning model and framework, data preprocessing and augmentation methods, and accuracy of results.

Chapter 5 covers various computer vision techniques for the smart farming. The advent of cutting-edge techniques such as Computer Vision (CV) and Artificial Intelligence (AI) have sparked a revolution in the agricultural industry, with applications ranging from crop and livestock monitoring to yield optimization, crop grading and sorting, pest and disease identification, and pesticide spraying among others. By leveraging these innovative techniques, sustainable farming practices are being adopted to ensure future food security. With the help of CV, AI, and related methods, such as Machine Learning (ML) together with Deep Learning (DL), key stakeholders can gain invaluable insights into the performance of agricultural and farm initiatives, enabling them to make data-driven decisions without the need for direct interaction. This chapter presents a comprehensive overview of requirements, techniques, applications, and future directions for smart farming and agriculture. Different vital stakeholders, researchers, and students who have a keen interest in this field would find the discussions in this chapter insightful.

Chapter 6 addresses AI methods for air quality forecasting. Air quality plays a significant role in agriculture and has a substantial impact on plant growth and productivity. As plants require oxygen for their respiration, polluted air leads to reduction in the oxygen concentration in the air which has a negative impact on the plant growth. Air quality forecasting is an important way to safeguard plant health since it provides ample notice of harmful air pollutants in a given location. Exhaust emissions from gas-guzzling vehicles, industrial pollution, and wildfires obfuscate some of the country's most dramatic vistas and can pose a significant hazard. If forecasts are trustworthy and accurate enough, they can be a crucial component of an air quality control system that complements more conventional emissions-based strategies. This study uses multivariate Long-Short Term Memory (LSTM) time series forecasting techniques, to understand and apply multiple variables together, to contribute more accuracy to forecasting due to its proven track record of success with time-series data. The factors that are found to have a major effect on air quality are dew point, wind speed, temperature, and snowfall and rain conditions. In particular, focus is on predicting and forecasting the concentrations of PM2.5 (Atmospheric Particulate Matter with a diameter of less than 2.5mm), as this pollutant has severe effects on plant health worldwide.

Chapter 7 proposes a dynamic application of Unmanned Aerial Vehicle (UAV) for analyzing the growth of crop and weed for precision agriculture. The resource that the global economy most heavily depends on is agricultural productivity. Nowadays, drones are widely used in a variety of civic applications, including agriculture. This work provides an overview of how UAVs are used in agriculture fields. A development in drone technology for agricultural events can restructure the farm sector to become dynamic and intelligent rather than static and manual. Precision agriculture would be made more effective with the deployment of drones integrated with the proper camera, sensors, and integrating modules. The two main forces behind agriculture automation are- precision agriculture and wireless sensor network. Drones, automation, and spectral cameras, are all combined to increase the possibilities of managing crop yield and weed issues more effectively. The best design concept and analysis of an autonomous agricultural crop and weed management through UAV are discussed.

Chapter 8 exemplifies a deep learning based solution to detect insect pests for plant using YOLO. The food security of any country may be jeopardized due to improper management of agricultural insect pests. Accurate pest detection and efficient pest control strategies must be employed in time to grow healthy crops for achieving food security in the country and worldwide. Hence, developing efficient and robust techniques to detect agricultural insect pests using computer vision approaches is one of the essential steps for timely managing insect pests. This chapter presents a short survey of deep learning-based object detection techniques focusing on insect pest detection and associated insect pest image datasets. Subsequently, a transfer learning-based custom You Only Look Once version5 (YOLOv5) model is developed using the publicly available dataset (IP102) for detecting agricultural insect pests with the help of computer vision approaches. The hyperparameters of the proposed insect pest detector are optimized using the Genetic Algorithm-based hyperparameter evolution method. The performance metrics of the proposed insect pest detector are found to be promising.

Chapter 9 illustrates a hybrid method for Long-Term Forecasting of Weather Parameters for Smart Farm. Agriculture is inextricably linked to the environment. Climate change has an effect directly on agricultural activities. India gets severely impacted if there is a loss of yields, which affects human lives. Hence, monitoring climate and its impact on the agricultural field is essential for a country like India. This chapter proposes a novel deep learning-based hybrid Vector Autoregressive–Gated Recurrent Unit (VAR-GRU) model for weather forecasting. The four important weather parameters such as temperature, pressure, humidity, and wind speed for the cities of Bengaluru; and temperature, pressure, dew point, and wind speed for the cities of Dongsi are considered. The effectiveness of the VAR-GRU model is proven by comparing its performance metrics (MAE, MSE, RMSE, and R2 Score) with that of other baseline models such as LSTM, VAR, GRU, and another hybrid VAR-LSTM model. The outcomes of this research work would help in increasing crop yields by utilizing the weather forecasting results in smart farming applications.

Chapter 10 talks about crop recommendation system based on soil and weather patterns using deep learning. Agriculture is the foremost factor which is important for survival of human beings. Farming contributes to a very big part of GDP, still several areas exist where improvements are required. One of those is the Crop Recommendation. Crop productivity is boosted as a result of accurate crop prediction. As, crop production has already started to suffer from climate change, improving crop output is consequently desirable because agronomists are impotent to select the appropriate crop(s) depending on environmental and soil parameters, and mechanism of forecasting the selection of appropriate crops manually has failed. Factors like soil characteristics, soil types, climate characteristics, temperature, rainfall, area, humidity, geographic location etc. affect crop forecast. This chapter focuses mainly on building a recommendation system i.e. suggesting a kind of crop by applying various machine learning and deep learning techniques depending upon several parameters. The system would help farmers for appropriate decision to be taken regarding the crop type.

Chapter 11 integrates Internet of Things (IoT) and Wireless Sensor Networks (WSN) system for Smart Agriculture. The smart farming is an enhanced agriculture system that offers data such as temperature, soil moisture, and so on to assist in the growth of plants and cattle. It integrates wireless sensors and internet to collect and communicate information with farmers. The Priority Event-based Energy Efficient Algorithm is developed in this study, that is utilized for accurate and efficient information transmission regarding power consumption and node priority. The major goal of the IoT sensor network is to increase farm productivity and extend its lifespan by applying intelligent algorithms such as Artificial Neural Network (ANN) to recognize environmental conditions and to improve the total production. Priority Event-based Energy Efficient method reduces energy usage and increases the lifetime using Dijkstra's algorithm.

Chapter 12 presents AI-based plant disease detection and classification using pre-trained models. Plants growth are crucial to an agricultural industry, and to the economy of a nation. Consequently, taking care of plants is essential. Like humans, plants are susceptible to several bacterial, fungal, and viral diseases. To avoid plants from being destroyed, prompt disease detection and treatment are crucial. The goal of this chapter is to introduce plant disease detection and classification using deep learning and transfer learning based models. Using images of leaves, the models can identify several plant ailments.

The models, namely- Convolutional Neural Network (CNN), MobileNet, and VGG16 are used for plant disease identification. To train them, the PlantVillage dataset is used. To enhance the sample size, the dataset is augmented. For experimentation purpose, images of both healthy and damaged plants are taken. Experiment results revel that VGG16 has outperformed CNN and MobileNet models for the detection of tomato, potato and apple plant diseases. Accuracy of the VGG16 model is 0.89, 0.92, and 0.95 for the tomato, potato and apple plant diseases respectively.

Chapter 13 discusses different AI techniques for plant disease detection and classification. Any developing economy can be thought of as having its foundation in the agricultural sector. One industry where automation and IoT have a significant impact is agriculture and modern farming. Due to the country's expanding population and rising food demands, agriculture is important in India. As a result, agricultural yield needs to be increased. Diseases brought on by bacteria, fungus, and viruses are one of these significant factors in decreased crop yields. By using methods for plant disease detection, this could be avoided and managed. Farmers must have access to the latest technologies and practices in order to get the most yield possible from their crops. A maximum agricultural output must be maintained, which means retaining healthy plants and keeping an eye on their surroundings to spot or detect diseases. Artificial intelligence, machine learning, and deep learning have all been proven to be incredibly useful in the field of modern agriculture as a means of sophisticated image analysis.

Rajeev Kumar Gupta
Pandit Deendayal Energy University, India

Arti Jain
Jaypee Institute of Information Technology, India

John Wang
Montclair State University, USA

Santosh Kumar Bharti
Pandit Deendayal Energy University, India

Samir Patel
Pandit Deendayal Energy University, India

Acknowledgment

Editors of the book entitled, *Artificial Intelligence Tools and Technologies for Smart Farming and Agriculture Practices*, would place on record special gratitude to their respective University/Institute Pandit Deendayal Energy University (Gujarat, India), Jaypee Institute of Information Technology (Uttar Pradesh, India) and Montclair State University (New Jersey, USA) respectively, for providing the motivation and positive research environment to start this work. A special thanks to all the contributors who have submitted their chapters from which 13 quality chapters have been selected post review for the publication in the book. We also would like to extend our special gratitude to all the esteemed advisory members and reviewers throughout the globe for their valuable suggestions to improve and enhance the quality of the book.

Dr. Rajeev Kumar Gupta would dedicate this book to his parents Shri Kedarnath Gupta and Smt. Malti Gupta. A special thanks to the backbone of his family especially his daughters Miss. Mehar & Mishika Gupta, son Atharv Gupta and wife Neetu Gupta for allowing and foregoing quality time to be dedicated to this book. A special acknowledgement of all the near and dear ones. Last but not the least secial thanks to book co-editors Dr. Arti Jain and Prof. John Wang who have been constantly the driving force behind this project.

Dr. Arti Jain would like to dedicate this book to her parents, mother Smt. Luxmi Gupta and late father Shri Rajender Kumar Gupta. A very special thanks to her son Master Chirayu Jain who always stand by her side. She would also like to acknowledge her friends and colleagues for motivating her throughout this journey. She would like to thank authors, reviewers, advisory board members, and co-editors, especially Dr. Rajeev Kumar Gupta and Prof. John Wang without their support this project would not be completed. Last but not the least, her heartfelt thanks to the Almighty for burning a ray of light in her and showing her the right path to move on.

Dr. Santosh Bharti would like to dedicate this book to his parents – father Shri Ram Bilas Sah and mother Asha Devi. A special thanks to his daughter Miss Nainshi Gupta and wife Laxmi for their cooperation throughout this journey. Last but not the least special thanks to co-editors Dr. Rajeev Kumar Gupta, Dr. Arti Jain and Dr. John Wang who have been the key driving force behind this project.

Dr. Samir Patel would like to dedicate this book to his late parents – Shri Bhupendrabhai and Smt. Muktaben. A special thanks to his wife Heena and his sons Nihar and Kirtan for their cooperation throughout this journey. Last but not the least, special thanks to co-editors Dr. Rajeev Kumar Gupta, Dr. Arti Jain, Prof. John Wang, and Dr. Santosh Bharti who have worked day and night to complete this project. It is my great pleasure to be part of this journey.

A very special thanks to IGI Global for the continuous support throughout this project, starting from commencement stage to the final publication stage in a timely manner. Wishing to work on more upcoming projects with IGI Global.

Chapter 1
Computational Intelligent Systems for Crop and Soil Monitoring Through Digital Imaging:
A Survey

Mahesh Kumar S. V.
Amrita College of Engineering and Technology, India

Sreya U. Parvathi
College of Agriculture, Vellayani, India

ABSTRACT

The world's population is predicted to grow by nearly 10 billion in the year 2050, which increases the food and agricultural demands. Hence, farmers worldwide are under continuous pressure to increase crop yield to satisfy the increasing food demand. To support the farmers in food production, crop monitoring and soil management are essential tasks. The traditional human observation-based crop and soil monitoring require more time and manpower. Hence, there is a need for smart crop and soil monitoring systems. This chapter is focused on different digital imaging-based Artificial Intelligence (AI) approaches such as machine learning, and deep learning that are used for smart crop and soil monitoring. There has been limited research work on the usage of smart intelligent techniques in crop and soil monitoring. This chapter attempts to identify the research gap in smart crops and soil monitoring. Future research trends would focus on integrating hybrid AI and reinforcement learning techniques for agricultural monitoring and management.

DOI: 10.4018/978-1-6684-8516-3.ch001

INTRODUCTION

Agriculture is a practice of farming, which includes cultivating the soil, crops, and domesticating animals. The commonly used traditional agricultural practices are agroforestry, crop rotation, intercropping, cover crops, organic composting, shifting cultivation, and integrated farming. These conventional farming practices are preferred still today because they were developed based on indigenous knowledge and vast experience accumulated over many years. Even though there are many advantages due to these conventional farming practices, such as enhanced soil fertility, resource usage, carbon sequestration, biodiversity preservation, sustainability and environmental protection, a few agricultural practices, like shifting agriculture have drawbacks (Hamadani *et al.*, 2021).

Crop production is an important agricultural practice followed by farmers all over the world, aimed at growing and producing crops for domestic and commercial purposes. In terms of crop production, India ranks among the top three countries in the world. Wheat and rice are the major two crops farmed in India's traditional seasonal agriculture. Nowadays, High-Yielding Variety (HYV) seeds were also sowed with the help of increased usage of fertilisers, insecticides, and pesticides to attain better crop productivity. However, farmers still struggle to choose the most profitable crop for their land due to the wide variety of soil types found worldwide and the limited technical facilities available for crop selection. Hence, the agricultural industry is transitioning from conventional to contemporary practices (Nisha and Megala, 2015).

Soil testing and monitoring are vital requirements for crop production. Generally, the soil comprises primary nutrients such as nitrogen, phosphorous, and potassium (commonly known as NPK), which are essential for plant growth. Hence, the testing of soil is a very essential procedure for farmers. To test the soil, the farmers must take the soil samples to any testing laboratory. The conventional soil testing process is time-consuming and the cost of soil testing is also high. Hence, many farmers avoid the soil testing process. To reduce the complexities in existing soil testing and classification, a few cost-effective theoretical frameworks were developed. These soil classification frameworks were guided by empirical data and practical circumstances. Moreover, these soil classification frameworks establish a connection between soil samples and different types of natural entities on the land surfaces (Eswaran H, Ahrens R, Rice T. J., 2002).

Issues in Traditional Crop and Soil Monitoring

The traditional crop and soil monitoring methods have few limitations related to the increasing food demand. According to a study conducted by Van Dijk *et al.* (2021), the demand for food is increasing due to the increase in the global population. Hence, there is a vital need to improve crop production also. A productive soil with the proper nutrients, such as nitrogen, phosphorus, and potassium, is essential for improved crop production. Poor quality crops may result if such nutrients are not adequately presented in the soil. Soil quality identification is a challenging task in traditional soil monitoring. Moreover, climatic factors such as temperature, rainfall, and humidity also play a vital role in traditional cropping. Due to these climatic factors, farmers are struggling to prepare their soil and select the proper plant seeds. Hence, traditional crop and soil monitoring heavily depend on expert's opinion and labour-intensive procedures. Due to this limitation, the traditional crop and soil monitoring methods are inadequate to meet the increasing food demand.

Need for Intelligent Systems in Crop and Soil Monitoring

In 2050, the world population is predicted to reach 10 billion, which requires a 56% increase in agricultural productivity to meet the food demand (Ranganathan *et al.*, 2018). Hence, there is a vital need for intelligent systems in smart farming, to attain significant agricultural productivity without increasing agricultural land. However, in comparison to other fields, agriculture is still less digitalized. Hence, the agriculture industry needs to adopt digital transformation techniques, to deal with the factors poised to destabilise the agricultural field. To improve the agricultural-related tasks throughout the food supply chain, the agricultural industry need to transform into AI-based smart farming and monitoring technologies. These advanced technologies can significantly reduce labour requirements in the agricultural field.

AI-based smart farming technologies can help farmers to monitor the soil, produce healthier crops, and organise the data required for farming. Continuous monitoring of crops using various sensors along with AI techniques can help the farmers to closely observe the crop's growth and decide the harvest time, and also help to attain a high yield. Recently, AI-based agricultural monitoring systems are used along with advanced Internet of Things (IoT) technologies (Pandithurai *et al.*, 2017). Using these AI and IoT techniques, farmers can be aware of the change in weather conditions, and also closely observe the soil moisture or dryness conditions continuously. Hence, AI and IoT-based soil and crop monitoring techniques can be used for increasing the crop production significantly.

Organization of the Chapter

The overview of intelligent crop and soil monitoring systems discussed in this Chapter is illustrated in Figure 1. The remainder of this chapter is organized as follows: The various processing steps of the AI-based crop, and soil monitoring system is described first. After that, the technical details of various imaging modalities used in agricultural monitoring systems are discussed. Then, the state-of-the-art AI-based crop monitoring systems are briefed. After that, AI-based soil monitoring systems are focused. Then, detailed information about various database sources used for the development of smart agricultural monitoring systems are provided, and the performance of various soil and crop monitoring intelligent systems are also compared. Finally, based on the comparative analysis, future research direction is highlighted in the conclusion Section of this Chapter.

AI-BASED CROP AND SOIL MONITORING SYSTEM

The process flow diagram of AI-based crop, and soil monitoring system is shown in Fig. 2. The AI-based crop, and soil monitoring systems use the data collected from satellites, drones, soil sensors, and manual data, to create a database from which classifiers can make decisions about the crop, and soil management. The sensors placed in agricultural field are used to collect data about different parameters like temperature, soil moisture levels, and soil nutrients. After data collection, the ML algorithms, and deep structures analysed the collected data, and predict the crop, and soil conditions based on the historical data. The IoT-based techniques can also be integrated along with the ML algorithms, and deep structures for efficient predictive analysis, and accurate decision-making.

Figure 1. Overview of intelligent crop and soil monitoring systems

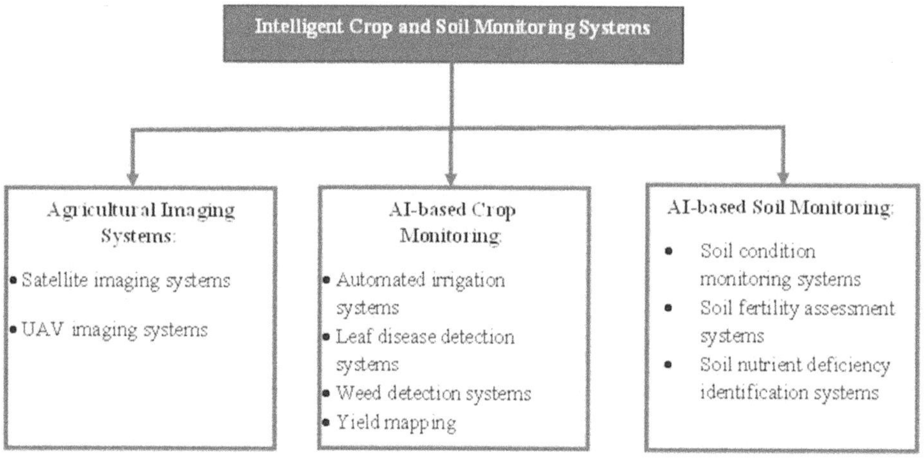

In ML-based crop, and soil monitoring systems linear regression models are used for prediction. The ML-based model's prediction outcome $y(x,w)$ is represented by equation (1).

$$y(x) = w^T x + w_0 \tag{1}$$

where w is denoted as the weight vector, and w_0 is the bias function. In this model, the input x is assigned as class 1, if $y(x)\geq 0$, and it is assigned to class 2, otherwise. In this way, this model classifies the input vector. Multiclass classification framework also largely used in agricultural monitoring systems. The multiclass model is represented by equation (2).

$$y_K(x) = W_K^T x + w_{k0} \tag{2}$$

where k indicates k-class classification. Recently, deep structures are largely used for crop, and soil monitoring systems, to attain accurate prediction results by using automated feature analysis. The Convolutional Neural Network (CNN) is a commonly used DL-based method used for agricultural monitoring systems. In CNN-based models, the image acquired from the agricultural field is formed as a small matrix, which is defined as a kernel. The feature map is computed from these matrices using equation (3).

$$G[m,n] = (f * h)[m,n] = \sum_j \sum_k h[j,k] f[m-j, n-k] \tag{3}$$

where f is the input image, and h is the kernel. The rows and columns of the matrix is represented as m, and n. The feature map is used as the input for predictive analysis.

Figure 2. Process flow diagram of AI-based crop and soil monitoring system

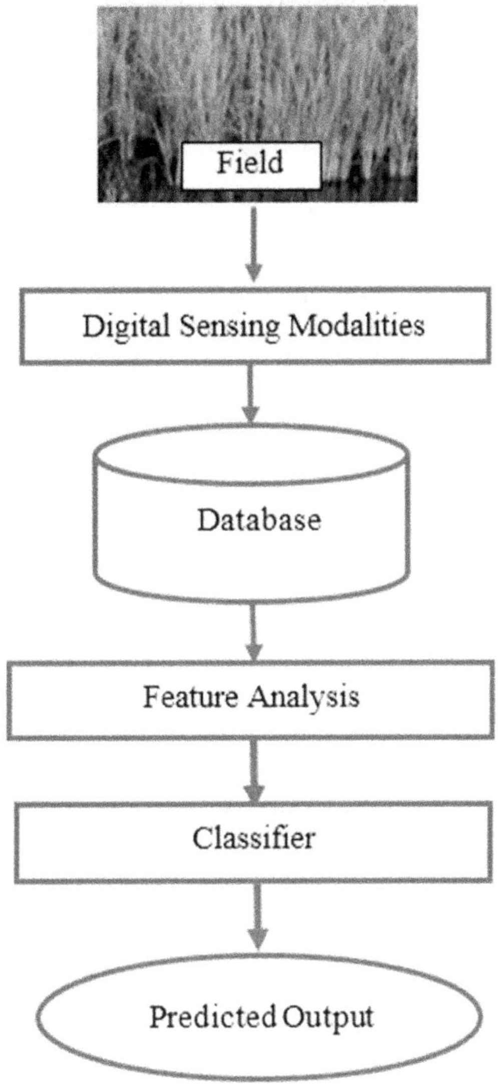

OVERVIEW OF AGRICULTURAL IMAGING SYSTEMS

Image Analysis-Based Agricultural Monitoring

A vital crop monitoring method is the analysis of vegetation spectrum response with the help of optical sensors for non-destructive remote crop monitoring. Various imaging sensors and cameras are also used for crop monitoring and soil assessment. A general block diagram of an image processing-based crop monitoring system is shown in Fig. 3. The camera sensors capture the crop images in Red (R), Green(G), Blue (B), and Near-Infrared (NIR) wavelengths. After that, the acquired images are transmitted to the remote analysis system with the help of wireless communication modules. Finally, a software-based image analysis module is used to interpret the data from the captured crop images. This interpretation is used to predict crop health conditions. Recently, imaging sensors and cameras are attached to drones for continuous crop monitoring.

Figure 3. General block diagram of image processing-based crop monitoring system

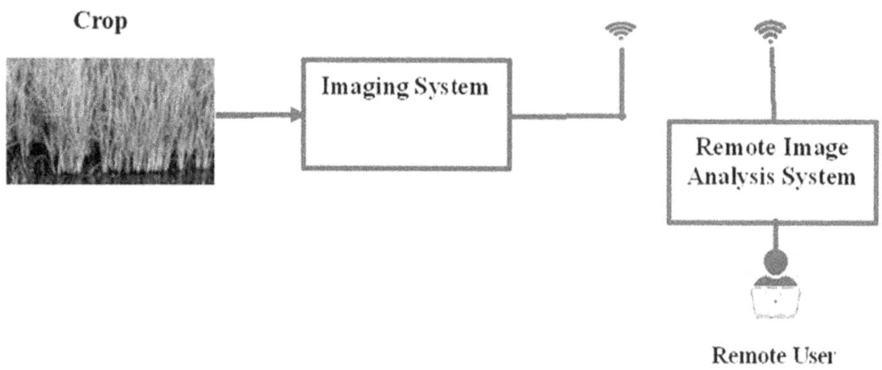

Satellite Imaging Systems

Precision in farming is attained with the help of satellite data-based monitoring systems. The conceptual block diagram of a satellite data-based agricultural monitoring system is illustrated in Fig. 4. In these systems, high-performance field sensors with high spectral, spatial, and temporal resolutions are used for data acquisition. In the data acquisition module, the physical layer is used for sensing and collecting information related to changes in the agricultural field conditions such as soil moisture, and crop growth status. The sensed information is remotely stored in cloud storage with the help of satellite communication techniques. The data stored in the cloud storage can be accessed and analysed by a remote user at any time. These satellite data-based monitoring systems have a vital role in crop and soil condition monitoring, which is essential for effective farm management (Jackson, 1984).

The remote-sensing data acquired from an agricultural field is used for smart farm planning. Commonly, this data is acquired with the help of different high-frequency satellites. The satellite data-based systems are appropriate for continuous crop health monitoring and are used to maximize the yielding also. In satellite imaging-based systems, remote sensing techniques were used to process the high-resolution digital images acquired from various agricultural resources. These high-resolution satellite images clearly indicate the differences in the organic matter and the drainage patterns. Using these high-resolution images, it is possible to differentiate the soils with higher levels of organic matter and lighter, sandier soil with lower levels of organic matter. Hence, satellite imaging data can be used for the real-time analysis of larger areas of growing crops. The data acquired from the satellite imaging systems can also be combined with other data types for effective visualization and further analysis, with the help of a Geographic Information System (GIS) (Romeijn, 2008). The sample satellite imagery data of the Xinavane – Mozambique sugar cane plantation (Image source: https://www.satimagingcorp.com/) is shown in Fig. 5.

Figure 4. Conceptual diagram of a satellite-based agricultural monitoring system

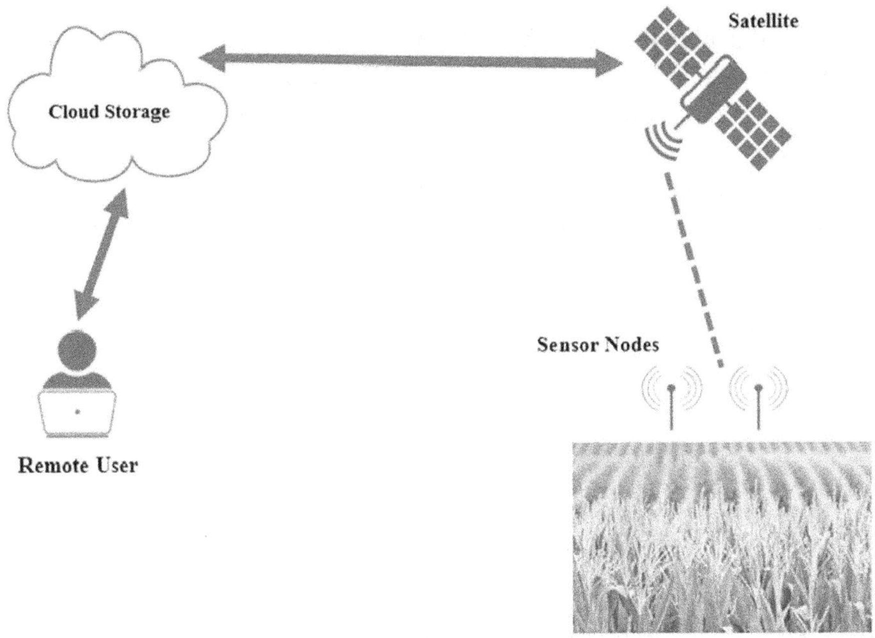

Figure 5. Sample satellite imagery data of Xinavane -Mozambique sugar cane plantation
(Image source: https://www.satimagingcorp.com/)

UAV Imaging Systems

Aerial images of an agricultural field are broadly used to spot the issues related to crop monitoring and also used to improve yielding. The aerial imaging systems have high-quality cameras and sensors, which are used to acquire a high-resolution image of a crop area to rectify the field issues such as soil and field variations, irregular irrigation, and pest infection. Commonly, these aerial images are acquired by satellites or low-altitude aircraft like planes or drones. The sample aerial imaging of a vegetation field collected from a public domain web source (https://pxhere.com/en/photo/1547017) is shown in Fig. 6. Recently, Unmanned Aerial Vehicles (UAVs) are also used in various crop management applications, to acquire images with high spatial and temporal resolution (Tsouros *et al.*, 2019). UAV technologies are largely used in different countries of the world to monitor different crop types.

Figure 6. Aerial imaging of a vegetation field
(Image Source: https://pxhere.com/en/photo/1547017)

UAV-based monitoring is mainly used for crop-related applications such as crop growth monitoring, weed mapping, and crop health monitoring. Crop growth monitoring is useful for yield estimation also. Continuous monitoring of crops using UAVs is useful to accurately observe crop growth and it is also useful for recording variations in several field parameters. In weed mapping, the UAVs acquired the images of an agricultural field to make a weed map, which shows the regions where chemicals are required. Another major application of UAVs is crop health monitoring. Regular crop health monitoring is essential to identify crop diseases in its earlier stage. Monitoring systems based on UAVs use the features extracted from the crop images to detect variations in crop biomass and crop health. Using this approach, crop diseases can be easily identified in the early stages. Early diagnosis of crop diseases can be useful for farmers to reduce the losses in cultivation.

REVIEW OF AI-BASED CROP MONITORING SYSTEMS

Automated Irrigation Systems

Automated irrigation systems can estimate the optimal quantity of water required for a specific crop, by analysing the factors like the type of the plant, soil nature, moisture condition, and climatic condition. Generally, AI-based automated irrigation systems have many sensors for the measurement of various parameters such as atmosphere temperature, soil moisture, and humidity (Khalifeh *et al.*, 2021). Using these automated systems, human errors in cultivation can be significantly eliminated. However, most of the earlier stage AI-based models normally run on a cloud environment and require a fixed communication infrastructure. Due to this limitation, AI-based models are not suitable for greenhouses, which has unstable mobile connectivity issue. To overcome this communication-related limitation, edge nodes are used in the farm field itself along with small computers (Kawai and Mineno, 2020).

IoT techniques are also used along with AI techniques in automated irrigation system development. Aggarwal and Kumar (2019) developed an automated irrigation system using Artificial Neural Network (ANN) technique along with an IoT device. The above research work was focused on the development of a low-priced smart device to collect the soil moisture data, pH, and soil temperature of the field. The ANN was used to analyse the collected data, and based on the outcome of the analysis, the compact device triggers the valve to start or stop the water flow automatically. Many low-cost embedded devices such as Arduino Uno, and Raspberry Pi were also used in automated irrigation systems along with machine learning (ML) approaches (Shekhar *et al.*, 2017). A decision support system (DSS) developed by Khan *et al.* (2018) effectively assists farmers in automated irrigation practices. Using this DSS, water-deficient regions in the field were identified with the help of soil moisture sensors. This method shows the effectiveness of Wireless Sensor Networks (WSN) in irrigation management.

Leaf Disease Detection Systems

Early detection of crop diseases is vital to satisfy the food supply demands, which one is closely related to the increasing population. Diagnosis of plant diseases using AI-based approaches reduces the big-scale monitoring work in the larger fields of crops. These AI-based diagnosis systems can easily identify the disease symptom, when it appears on the plant leaves itself. Numerous soft computing methods are used in AI-based leaf disease diagnosis systems. The Support Vector Machine (SVM) is one of the mainly used classifier in leaf recognition systems (Salman *et al.*, 2017; Zhang *et al.*, 2012). The SVM has generalization potential, moreover, it is exceptionally robust also. However, it has speed and feature size limitations, which is closely related to the classifier's training and testing phases.

Computer vision-based systems are another type of automated system, which completely ignores the image background, to speed up the leaf recognition process, and makes the diagnosis system appropriate for extremely complex leaf samples also (Azfar Firdaus Azlah *et al.*, 2019). Deep Learning (DL)-based models are also broadly used in leaf disease diagnosis. In DL models, convolutional layers are used for the automated identification of plant features such as the colours and textures of the lesions. Compared with other classifiers, the CNN can extract and recognize the image features concurrently. Hence, DL-based approaches are considered as a rapid and promising tool for leaf disease diagnosis. Ahmad *et al.* (2023) presented a detailed review of various DL methods used for leaf disease detection. This review included 70 DL-based application studies and discussed their use in leaf disease detection and management. Moreover, this study addressed the research gaps in the existing leaf disease diagnosis systems and guided future research trends in leaf disease diagnosis.

Weed Detection Systems

Weed detection on the agricultural farm has become a challenging task, because of the randomly spaced crops on the farm. The traditional method of weed detection is done only with the naked eye. Therefore, it becomes a complicated task, because of a wide variety of weeds like narrow, wide leaves, and broad-leaf existing in the plantings. Hence, there is a vital need for automated weed detection systems. Commonly, these automated systems use image processing and AI techniques for weed detection purposes. The DL-based methods are largely used for weed detection. Umamaheswari *et al.* (2018) proposed an automated weed detection system for farm crops by using image processing techniques. In this work, the automated system was trained with crop and weed images using a DL-based framework. The convolutional network-based systems developed by Lottes *et al.* (2018) and Sa *et al.* (2018) achieved significant performance in automated weed detection process. Moreover, these systems attained significant accuracy in the crop-weed classification problem without retraining the model.

The computational vision-based method developed by Irias Tejeda and Castro Castro (2019) is also used in the weed detection process. This method improved the overall crop management process's performance, reduced the time required to search the weeds all over the agricultural field, and focused on the weed removal process in specific regions for effective weed control. Drone-aided system developed by Koshelev *et al.* (2023) is also appropriate for automated weed detection. In this system, the RGB and multispectral cameras fixed on the drone were used for the aerial acquisition of images from the field. The transfer learning approach was used to identify the weeds in the field from the acquired aerial images. To support the farmers, a low-cost weed detection system was also developed by Liang *et al.* (2019) using RGB images acquired by drones. This system accurately detected the weeds with significant accuracy of 98.8%. This method attained similar detection accuracy compared with other high-cost automated weed detection methods, at a low cost.

Yield Mapping

Yield mapping, also known as yield monitoring, is an agricultural technique that uses Global Positioning System (GPS) information to analyse the variables like crop yielding and moisture content in an agricultural field. Dr. Graeme Cox created the world's first yield map in 1996, using the data collected from the mass flow sensor of the harvester and the GPS data. The first yield map collected from Davco farming (http://www.davcofarming.com/) is shown in Fig.7. In developed countries, most of the farmers use GPS-linked yield monitors in their harvesting equipment. For example, in the United States of America (USA) 70% of farmers have GPS-linked yield monitors, and 18% of farmers have yield monitors without any GPS-linked harvesters (Franzen *et al.*, 2018). Recently, ML techniques are also largely used to predict the yielding of an agricultural field. Rashid *et al.* (2021) conducted a detailed review of the role of ML techniques in crop yield prediction. This study addresses the advantages and difficulties in ML-based yield prediction, and provides potential solutions for existing problems in yield prediction.

Figure 7. World's first sugar cane yield map of Davco farming
(Image source: http://www.davcofarming.com/)

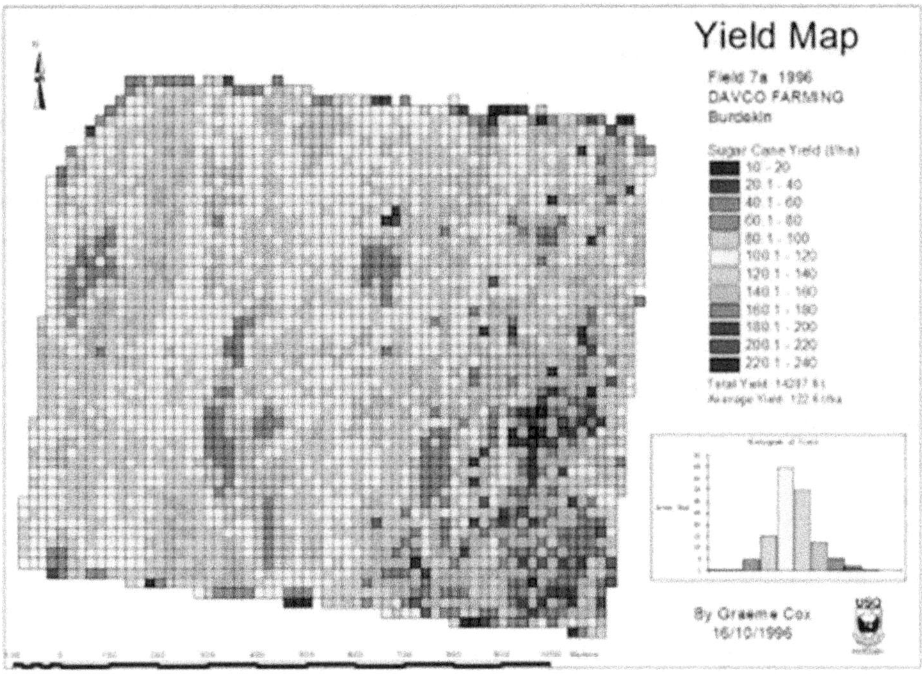

REVIEW OF AI-BASED SOIL MONITORING SYSTEMS

Soil Condition Monitoring Systems

Soil is one of the vital resource of worldwide agriculture, which is also the foundation of food production. Hence, soil monitoring is considered as an important task in agricultural practice. AI-based automated systems are broadly used in soil monitoring by using various algorithms. These systems used supervised or unsupervised approaches to perform the data analysis for providing effective solutions to soil monitoring-related problems. AI-based soil monitoring systems are used to perform various soil monitoring tasks such as soil colour monitoring, and soil moisture measurement. In a soil colour analysis system developed by Al-Naji *et al.* (2021), RGB images and the ANN are used for the field application of smart irrigation. This system used the soil colour data as an input to the ANN model to take a binary decision, irrigating the soil or not.

Soil moisture monitoring is essential to manage the water supply in the agricultural field and is used to predict natural disasters also. An IoT-based system proposed by Anand *et al.* (2019) identified the soil moisture and other atmosphere components using ML techniques. In this system, moisture sensors, temperature sensors, and rain sensors were used to acquire information on various farm parameters. An open-source platform ThingSpeak Cloud was used for data storage. An automated system proposed by Santi *et al.* (2018) estimates the soil moisture using C-Band, X-Band SAR, and two ML techniques Support Vector Regression (SVR), and ANN. This system shows that the combination of X-band, and C-band images give accurate information required for soil moisture retrieval. Another ANN-based automated system proposed by Tyagi *et al.* (2021), retrieves the soil moisture using the CYGNSS data. The efficiency of this system was verified using spatial, and temporal correlation analysis of the CYGNSS data.

Soil Fertility Assessment Systems

Soil fertility assessment is an essential process for the development of a productive and sustainable agroecosystem. The conventional soil fertility assessment is expensive and time-consuming also. The fertility of any soil sample is related to different soil parameters such as temperature, moisture, NPK contents, and pH level. Sensor-based measurement of these soil parameters is simple, but it is focused only on the specific region of the agricultural field. Hence, hyperspectral-imaging systems are preferred to cover the overall agricultural field. Recently, Phute *et al.* (2021) developed an ML-based soil fertility assessment model using hyperspectral-image analysis. In this model, the correlation between the hyperspectral images, temperature and the moisture of the soil was investigated using linear regression, and random forest approaches.

Web GIS is also used for soil fertility assessment. Agrawal *et al.* (2021) developed a web-GIS-based DSS for soil fertility assessment. This system attained 85.2% accuracy. Recently, IoT devices have also been used along with AI techniques for assessing soil fertility. Aliparo *et al.* (2022) developed a system, which assessed and monitored various soil parameters such as moisture level, acidity level, NPK content, and humidity using IoT devices. Based on the assessment outcome, this system can suggest the crop type, which is suitable for the specific soil type. In a recent study conducted by Misbah *et al.* (2021), the performance of various ML-based approaches such as random forest, SVR, partial least square regression, principal component analysis (PCA), and ANN in soil NPK content detection using hyperspectral imagery was investigated. This study shows that the performance of the above ML-based approaches is significant in view of agricultural field application.

Soil Nutrient Deficiency Identification Systems

Soil with a sufficient quantity of nutrients is essential for higher crop yield, which is vital to meet the increasing food demand. For higher crop yield, the plantations must be nourished with various elements such as NPK. Electrochemical and optical sensing methods were commonly used to identify and analyse the nutrients existing in the soil. Electrochemical-based methods used ion-selective electrodes for soil nutrient identification. Optical sensing-based methods used colorimetric or spectroscopic techniques for soil nutrient measurement (Potdar *et al.*, 2021). According to a study conducted by Tagad *et al.* (2013), optical sensing techniques are preferred in cost-effective, portable, and small-sized soil nutrient sensor development. Moreover, optical-based sensing methods have many advantages compared with electrical-based methods, like high selectivity, sensitivity, repeatability, and resistance to electromagnetic interference. In optical sensors, the reflected energy from the soil surface is used to measure the soil nutrient (Stenberg *et al.*, 2010). However, the sensor-based soil nutrient deficiency detection systems are complex and not appropriate for continuous remote monitoring. To overcome these limitations, image-based diagnosis systems are used for soil nutrient deficiency detection.

Digital image processing (DIP) and AI techniques are broadly used in soil nutrient monitoring systems. A diagnosis system developed by Puno *et al.* (2017) detects the soil nutrient and pH levels using DIP and ANN techniques. The usage of ANN significantly improved image processing-based soil nutrient diagnosis accuracy. Various ML-based algorithms were also used in soil nutrient measurement. Morellos *et al.* (2016) developed an ML-based prediction model to measure the total nitrogen (TN), organic carbon (OC), and moisture content (MC) of the soil using the NIR spectroscopy. This study compared the soil nutrient prediction ability of linear multivariate approaches such as the principal component regression (PCR), partial least squares regression (PLSR), and ML-based models such as least squares-SVM (LS-SVM), and cubist. Data mining techniques were also used in soil nutrient analysis (Chiranjeevi and Nadagoudar, 2018). These techniques were commonly used for the analysis of large-sized soil nutrient databases. The image and data processing-based methods are appropriate for non-contact and remote soil monitoring.

PERFORMANCE COMPARISON AND DISCUSSIONS

Agricultural Databases

For developing computational intelligence-based agricultural monitoring systems using images, many representative databases are required. The PlantVillage database (Hughes and Salathé, 2015) is the most commonly used database for the development of AI-based crop disease diagnosis systems. This database contains 54,309 images of 38 different disease categories, which affect 14 crop varieties. These images were captured in a controlled laboratory environment with a uniform background. The rice disease database (Prajapati *et al.*, 2017) is another commonly used crop database for crop disease diagnosis system development. This database contains 3355 images of four categories such as healthy, hispa, brown spot, and leaf blast. In this database also, the images were captured in the controlled laboratory environment with a uniform background. The Digipathos (*Repositório Digipathos: Home*, n.d.) is another large-sized plant disease database which contains 46,513 images of 171 disease categories affecting 21 crop types. This database was used in various studies focused on crop disease diagnosis. However, over 95% of the images in this database did not represent any plant disease category caused in real-agricultural field conditions (Ahmad *et al.*, 2023). Hence, there is a vital need for a crop image database with disease types caused in real-agricultural field conditions.

In the PlantDoc database (Singh *et al.*, 2020), most of the crop images were acquired in field conditions. This database contains 2598 images of 17 disease categories affecting 13 crop varieties. Few crop-specific databases are also available for intelligent crop disease detection system development. The Northern Leaf Blight (NLB) database (Wiesner-Hanks *et al.*, 2018) contains 18,222 infected cornfield images captured using different imaging techniques such as aerial imaging, handheld imaging, and mounting the camera on the boom. The NLB dataset contains many real-agricultural field images. However, this database only contains images of corn leaves with a single disease category. Hence, it is not appropriate to develop the automated diagnosis systems for multiple disease categories. The RoCoLe-coffee disease database (Parraga-Alava *et al.,* 2019) is another crop-specific database that comprises images of coffee diseases along with severity data. This database contains 1560 images captured by a smartphone camera in a field with 390 coffee plants in Ecuador. The comparison of various crop image databases is provided in Table 1.

Commonly, soil monitoring systems used various sensors to analyse the soil content. Hyperspectral imagery is also used for remote analysis of the soil contents (Misbah *et al.*, 2021). Recently, colour images of the soil are also be used for smart farming purposes. Al-Naji *et al.* (2021) developed a colour soil image database using a compact digital camera (Nikon D5300 model). In this database, the soil images were collected from the agricultural nursery in Baghdad, Iraq. The different soil surfaces were analysed based on the brightness levels of RGB bands. The Indian Regions Soil Image Database (IRSID) developed by Pandiri *et al.*(n.d.) is another country-specific database representing various soil categories of India. This database contains colour images of the clay, sand, loam, sandy loam, and loam sandy soil types. These soil images were captured using a mobile phone camera. The GFSAD30 developed by Oliphant *et al.* (2022) is a high-resolution global cropland database developed with the support of the NASA fund. This database used multiple sensor remote sensing data, secondary data, and field-plot data to document the cropland variations.

Table 1. Comparison of different crop image databases

Name of the Database	**Number of crop types available**	**Number of categories focused**	**Number of images available**
PlantVillage database (Hughes and Salathé, 2015)	14	38	54,309
Rice disease database (Prajapati *et al.*, 2017)	Single crop type (Rice)	04	3355
Digipathos (*Repositório Digipathos: Home*, n.d.)	21	171	46,513
PlantDoc database (Singh *et al.*, 2020)	13	17	2598
NLB database (Wiesner-Hanks *et al.*, 2018)	Single crop type (Corn)	Northern Leaf Blight infection disease	18,222
RoCoLe- coffee disease database (Parraga-Alava *et al.*, 2019)	Single crop type (Coffee)	06	1560

Performance Measures for Computational Intelligence Systems

The classification performance measures attained from the confusion matrix were used to assess the computational intelligence-based agricultural monitoring systems. This specific performance measure allows to visualize the performance of an ML-based algorithm in crop, and soil monitoring applications. The confusion matrix compares the computational intelligence system's diagnosis outcome with the human observation result (ground truth) to evaluate the agricultural monitoring system's performance. The confusion matrix of a two-class crop disease diagnosis system is illustrated in Figure 8.

Figure 8. Confusion matrix used in computational intelligence-based crop disease diagnosis

Human observation	Computational intelligence system's diagnosis outcome	
	Unhealthy Crop	*Healthy Crop*
Unhealthy Crop	True Positive (TP)	False Negative (FN)
Healthy Crop	False Positive (FP)	True Negative (TN)

Using this confusion matrix illustrated in Fig.7, four classes of possible outcomes can be attained. These four potential outcomes are stated as follows:

TP: This performance measure is focused on unhealthy crops. If the predicted value is unhealthy, the outcome is TP for an unhealthy crop image.
FN: This performance measure is also focused on the unhealthy crops. For the class of unhealthy crop image, if the predicted value is healthy, the outcome is FN.
TN: This performance measure is focused on the healthy crops. For the class of a healthy crop image, if the predicted value is healthy, the outcome is TN.
FP: This performance measure is also focused on the healthy crops. If the predicted value is unhealthy, the outcome is FP for the class of a healthy crop image.

From these TP, FP, TN, and FN results, three performance evaluation metrics can also be derived using Equations (4)-(6).

$$\text{Sensitivity} = \frac{\text{TP}}{\text{TP} + \text{FN}} \tag{4}$$

$$\text{Specificity} = \frac{\text{TN}}{\text{TN} + \text{FP}} \tag{5}$$

$$\text{Accuracy} = \frac{\text{TP} + \text{TN}}{\text{TP} + \text{FP} + \text{FN} + \text{TN}} \tag{6}$$

Classification accuracy is the commonly used performance evaluation metric for ML and DL-based intelligent agricultural monitoring systems. However, other performance metrics such as the mean square error (MSE), root mean square error (RMSE), and coefficient of determination (R^2) were also used to evaluate the performance of intelligent agricultural monitoring systems.

The MSE is a performance measure, which shows the average squared difference between the agricultural intelligent system's prediction outcome and the ground truth values in the given database. The RMSE is a metric that shows the average distance between the diagnosis model's predicted outcome and the actual outcome in the database. The lower value of RMSE indicates that the diagnosis model is well-fitted with a database value. The R^2 value indicates the prediction fitness of the diagnosis model. The R^2 value ranges between 0 and 1. If the R^2 value is 1 means, it indicates that the predicted outcome is similar to the actual value.

Comparative Analysis and Discussions

The comparison of recent research works in crop and soil-related computational intelligent systems development is given in Table 2. In crop monitoring, leaf disease detection and weed detection are essential tasks. The leaf can be effectively recognized with the help of image processing techniques and ML approaches such as the Canny edge detector and SVM (Salman *et al.*, 2017). Drone imaging plays a significant role in weed detection. The drone imaging-based monitoring systems are low-cost (Liang *et al.*, 2019) and also eliminated the need for multispectral hardware in weed detection (Koshelev *et al.*, 2023). Drone imaging along with CNN was also be used for accurate weed detection (Liang *et al.*, 2019). The fully CNN-based system developed by Lottes *et al.* (2018) attained a significant precision, up to 99.1% (for weed) and 98.3% (for crop). The multispectral imaging-based CNN model developed by Sa *et al.* (2018) was focused on dense semantic weed classification. In this model, a micro aerial vehicle was used for image acquisition. Computer vision techniques were also used for efficient weed monitoring (Irias Tejeda and Castro Castro, 2019). Computer vision techniques increased crop management performance and reduced the time required for weed searching.

Table 2 also shows that the ANN plays a vital role in automated soil fertility assessment and soil nutrient detection. The ANN-based systems can help the farmers to decide whether they need to irrigate the soil or not, based on soil colour analysis outcome. The ANN-based system developed by Al-Naji *et al.* (2021) attained low MSE values in the training, testing, and validation phases of the classifier. Moreover, the ANN-based system developed by Puno *et al.* (2017) indicates the soil nutrient and pH levels qualitatively. The IoT techniques were also be used for soil fertility assessment. The IoT-based system developed by Aliparo *et al.* (2022) effectively detected the NPK content of the soil and also supports healthy and fast crop growth. The IoT system is used along with ML algorithms also (Anand *et al.*, 2019). This combined system effectively tested the soil quality and also suggested the crop type for the plantation. The ML algorithm also uses C- and X-band SAR data for soil moisture content (SMC) estimation (Santi *et al.*, 2018). The above review shows the importance of AI-based intelligence systems in soil fertility assessment and soil nutrient detection.

CONCLUSION AND FUTURE RESEARCH DIRECTIONS

Increasing food demand due to the population is a major issue all over the world. Worldwide farmers are under pressure to increase crop yield to satisfy the increasing food demand. Soil and crop monitoring is a vital step required to increase crop yield. The traditional human observation-based crop and soil monitoring is time-consuming and required manpower also. Hence, researchers all over the world are focused on the development of different smart soil and crop monitoring systems. Digital imaging and computational intelligence techniques were broadly used for the development of smart crop and soil monitoring systems. Numerous intelligent systems used for crop and soil monitoring were discussed in this Chapter. The intelligent agricultural monitoring systems were categorized as soil monitoring, crop monitoring systems, and various imaging, computational intelligence techniques used in these monitoring systems were also briefed in this chapter. Moreover, this chapter also addressed the various image databases used for the automated agricultural monitoring system development.

The comparative analysis of this chapter shows that the digital imaging and computational intelligence-based systems performed significantly in automated crop and soil monitoring. However, the outcome of these intelligence systems is heavily dependent on the imaging models, training database, and method used for data analysis. Future research works may include the crop and soil monitoring using multimodal imaging techniques. The huge-scale training data-related technical problems can be solved by developing computationally intensive databases. Moreover, remote sensing-based intelligent models can also be used to improve agricultural spatial analysis efficiency. Future research trends may also be focused on the integration of hybrid AI and reinforcement learning techniques for efficient agricultural monitoring and management.

Table 2. Comparison of crop and soil-related computational intelligence systems

Authors	Technique used	Issues addressed	Performance measures and Observations
Salman *et al.* (2017)	Canny edge detector, SVM	Leaf detection	· This system extracted 15 leaf features using the Canny method. · This system can classify 22 plant types using an SVM classifier. · The overall accuracy of this system is 85% to 87% for worst cases to the best cases respectively.
Liang *et al.*(2019)	Drone imaging and CNN	Weed detection	· This method is a low-cost method. · This method attained 98.8% accuracy compared with other high-cost weed detection methods.
Koshelev *et al.* (2023)	Drone imaging and Embedded image processing	Weed detection	· This method can be used to eliminate multispectral hardware in the drone-based weed detection system. · This method attained 1.1% increase in terms of mean intersection over the union metric, when evaluated using RGB camera stream data, and 1.4% increase for orthophoto data.
Lottes *et al.*(2018)	CNN	Weed detection and precision farming	· This method labels the images pixel-wise as crop and weed. · This method attained a precision value of up to 99.1% (for weed) and 98.3% (for crop).
Sa *et al.*(2018)	Multispectral imaging and CNN	Weed classification	· The multispectral images were acquired by the micro aerial vehicle. · This method was focused on dense semantic weed classification. · This method attained 0.78 area under the curve (AUC) classification metric.
Irias Tejeda and Castro Castro (2019)	Computer Vision	Weed detection	· This method established accurate weed monitoring. · This method increased crop management performance and reduced the time required for weed searching. · The sensitivity of this algorithm is 97%, and the specificity is 79%.
Agrawal *et al.* (2021)	Web-GIS-based DSS	Soil fertility assessment	· This method recommended fertilizer levels required according to soil fertility levels available. · This method attained 85.2% accuracy in soil fertility assessment.
Al-Naji *et al.* (2021)	ANN	Soil colour analysis for smart irrigation	· This method can be used to take the decision, whether the need to irrigate the soil or not, based on soil colour analysis. · This method attained an MSE value of 1.616×10^{-6} (in training), 1.004×10^{-5} (in testing) and 1.809×10^{-5} (in validation).
Puno *et al.*(2017)	Image processing and ANN	Soil nutrients detection	· This system indicated the qualitative soil nutrient and pH levels. · The MSE is less than $1x10^{-10}$.
Aliparo *et al.* (2022)	IoT	Soil fertility assessment	· This method monitored the NPK content of the soil. · The experimental results show that crop growth is healthier and faster while using this system.
Anand *et al.* (2019)	IoT and ML	Soil moisture detection	· This system tested the soil quality and suggested the crop type for the plantation. · Using this system farmers can get the message useful for crop production. · This method was not quantitatively evaluated.
Santi *et al.* (2018)	C- and X-band SAR data, ML	SMC estimation	· This method attained RMSE between 2%, and 4% of SMC, and the correlation coefficient R between 0.85 and 0.97, for various combination of inputs. · Experimental results show that the addition of X- and C-band images provided valuable information for SMC retrieval.

REFERENCES

Aggarwal, S., & Kumar, A. (2019). A Smart Irrigation System to Automate Irrigation Process Using IOT and Artificial Neural Network. *2nd International Conference on Signal Processing and Communication, ICSPC 2019 - Proceedings*, (pp. 310–314). IEEE. 10.1109/ICSPC46172.2019.8976631

Agrawal, N., Verma, K., & Kumar, T. (2021). Soil Fertility Assessment using Web-based Spatial Decision Support System. *2021 IEEE International Conference on Emerging Trends in Industry 4.0, ETI 4.0 2021*. IEEE. 10.1109/ETI4.051663.2021.9619240

Ahmad, A., Saraswat, D., & El Gamal, A. (2023). A survey on using deep learning techniques for plant disease diagnosis and recommendations for development of appropriate tools. *Smart Agricultural Technology*, *3*, 100083. doi:10.1016/j.atech.2022.100083

Al-Naji, A., Fakhri, A. B., Gharghan, S. K., & Chahl, J. (2021). Soil color analysis based on a RGB camera and an artificial neural network towards smart irrigation: A pilot study. *Heliyon*, *7*(1), e06078. doi:10.1016/j.heliyon.2021.e06078 PMID:33537493

Aliparo, J., Santos, C. D., Sese, C. M., Ver Trinidad, J., Beano, M. G., Capuno, M. E. A., Medina, O., Sigue, A. L., & Soriano, A. (2022). IoT-based Assessment and Monitoring of NPK Content and Fertility Condition of Soil. *IEEE Region 10 Annual International Conference, Proceedings/TENCON, 2022-Novem*. IEEE. 10.1109/TENCON55691.2022.9978040

Anand, R., Sethi, D., Sharma, K., & Gambhir, P. (2019). Soil Moisture and Atmosphere Components Detection System Using IoT and Machine Learning. *Proceedings of the 2nd International Conference on Smart Systems and Inventive Technology, ICSSIT 2019*, (pp. 842–847). IEEE. 10.1109/ICSSIT46314.2019.8987754

Azfar Firdaus Azlah, M., Suan Chua, L., Razan Rahmad, F., Izana Abdullah, F., & Rafidah Wan Alwi, S. (2019). Review on Techniques for Plant Leaf Classification and Recognition. *Computers*, *8*(77), 77. doi:10.3390/computers8040077

Chiranjeevi, & Nadagoudar, R. B. (2018). Analysis of Soil Nutrients using Data Mining Techniques. International Journal of Recent Trends in Engineering and Research, 4(7), 103–107. doi:10.23883/IJRTER.2018.4363.PDT1C

Eswaran, H., Ahrens, R., & Rice, T. J. S. B. A. (2002). Soil classification: a global desk reference. CRC Press. doi:10.1201/9781420040364

Franzen, D. W., Casey, F. X. M., & Derby, N. (2018, May). Yield Mapping and Use of Yield Map Data. *Ndsu Extension*. https://extensionpublications.unl.edu/assets/pdf/ec2004.pdf

Hamadani, H., Rashid, S. M., Parrah, J. D., Khan, A. A., Dar, K. A., Ganie, A. A., Gazal, A., Dar, R. A., & Ali, A. (2021). Traditional Farming Practices and Its Consequences. *Microbiota and Biofertilizers*, *2*, 119–128. doi:10.1007/978-3-030-61010-4_6

Hughes, D. P., & Salathé, M. (2015). An open access repository of images on plant health to enable the development of mobile disease diagnostics through machine learning and crowdsourcing. In ArXiv.

Irias Tejeda, A. J., & Castro Castro, R. (2019). Algorithm of weed detection in crops by computational vision. *CONIELECOMP 2019 - 2019 International Conference on Electronics, Communications and Computers*, (pp. 124–128). IEEE. 10.1109/CONIELECOMP.2019.8673182

Jackson, R. D. (1984). *Remote Sensing of Vegetation Characteristics for Farm Management.* SPIE. doi:10.1117/12.966243

Kawai, T., & Mineno, H. (2020). Evaluation environment using edge computing for artificial intelligence-based irrigation system. *Proceedings - 2020 16th International Conference on Mobility, Sensing and Networking, MSN 2020.* IEEE. 10.1109/MSN50589.2020.00046

Khalifeh, A., Al-Qammaz, A., Darabkh, K. A., Abualigah, L., Khasawneh, A. M., & Zinonos, Z. (2021). An AI Based Irrigation and Weather Forecasting System utilizing LoRaWAN and Cloud Computing Technologies. *Proceedings of the 2021 IEEE Conference of Russian Young Researchers in Electrical and Electronic Engineering, ElConRus 2021*, (pp. 443–448). IEEE. 10.1109/ElConRus51938.2021.9396431

Khan, R., Ali, I., Zakarya, M., Ahmad, M., Imran, M., & Shoaib, M. (2018). Technology-Assisted Decision Support System for Efficient Water Utilization: A Real-Time Testbed for Irrigation Using Wireless Sensor Networks. *IEEE Access : Practical Innovations, Open Solutions*, *6*, 25686–25697. doi:10.1109/ACCESS.2018.2836185

Koshelev, I., Savinov, M., Menshchikov, A., & Somov, A. (2023). Drone-Aided Detection of Weeds: Transfer Learning for Embedded Image Processing. *IEEE Journal of Selected Topics in Applied Earth Observations and Remote Sensing*, *16*, 102–111. doi:10.1109/JSTARS.2022.3224657

Liang, W. C., Yang, Y. J., & Chao, C. M. (2019). Low-cost weed identification system using drones. *Proceedings - 2019 7th International Symposium on Computing and Networking Workshops, CANDARW 2019*, (pp. 260–263). IEEE. 10.1109/CANDARW.2019.00052

Lottes, P., Behley, J., Milioto, A., & Stachniss, C. (2018). Fully convolutional networks with sequential information for robust crop and weed detection in precision farming. *IEEE Robotics and Automation Letters*, *3*(4), 2870–2877. doi:10.1109/LRA.2018.2846289

Misbah, K., Laamrani, A., Chehbouni, A., Dhiba, D., & Ezzahar, J. (2021). Use of Hyperspectral Prisma Level-1 Data and ISDA Soil Fertility Map for Soil Macronutrient Availability Quantification In A Moroccan Agricultural Land. *International Geoscience and Remote Sensing Symposium (IGARSS)*, (pp. 7051–7054). 10.1109/IGARSS47720.2021.9553190

Morellos, A., Pantazi, X. E., Moshou, D., Alexandridis, T., Whetton, R., Tziotzios, G., Wiebensohn, J., Bill, R., & Mouazen, A. M. (2016). Machine learning based prediction of soil total nitrogen, organic carbon and moisture content by using VIS-NIR spectroscopy. *Biosystems Engineering*, *152*, 104–116. doi:10.1016/j.biosystemseng.2016.04.018

Nisha, G., & Megala, J. (2015). Wireless sensor Network based automated irrigation and crop field monitoring system. *Sixth International Conference on Advanced Computing, ICoAC 2014*, (pp. 189–194). IEEE. 10.1109/ICoAC.2014.7229707

Oliphant, A., Thenkabail, P., & Teluguntla, P. (2022). Global food-security-support-analysis data at 30-m resolution (GFSAD30) cropland-extent products—Download Analysis. In Open-File Report. US Geological Survey. doi:10.3133/ofr20221001

Pandiri, D. N. K., Murugan, R., & Goel, T. (n.d.). *Indian Regions Soil Image Database (IRSID): A dataset for classification of Indian soils.* IEEE Dataport. https://ieee-dataport.org/documents/indian-regions-soil-image-database-irsid-dataset-classification-indian-soils

Pandithurai, O., Aishwarya, S., Aparna, B., & Kavitha, K. (2017). Agro-tech: A digital model for monitoring soil and crops using internet of things (IoT). *ICONSTEM 2017 - Proceedings: 3rd IEEE International Conference on Science Technology, Engineering and Management.* IEEE. 10.1109/ICONSTEM.2017.8261306

Parraga-Alava, J., Cusme, K., Loor, A., & Santander, E. (2019). RoCoLe: A robusta coffee leaf images dataset for evaluation of machine learning based methods in plant diseases recognition. *Data in Brief, 25*, 104414. doi:10.1016/j.dib.2019.104414 PMID:31516934

Phute, M., Madake, J., & Bhatlawande, S. (2021). Precision Farming Based Soil Fertility Assessment Techniques. *ACCESS 2021 - Proceedings of 2021 2nd International Conference on Advances in Computing, Communication, Embedded and Secure Systems*, (pp. 170–172). IEEE. 10.1109/ACCESS51619.2021.9563284

Potdar, R. P., Shirolkar, M. M., Verma, A. J., More, P. S., & Kulkarni, A. (2021). *Determination of soil nutrients (NPK) using optical methods: a mini review, 44*(12), 1826–1839. Taylor & Francis. doi:10.1080/01904167.2021.1884702

Prajapati, H. B., Shah, J. P., & Dabhi, V. K. (2017). Detection and classification of rice plant diseases. *Intelligent Decision Technologies, 11*(3), 357–373. doi:10.3233/IDT-170301

Puno, J. C., Sybingco, E., Dadios, E., Valenzuela, I., & Cuello, J. (2017). Determination of soil nutrients and pH level using image processing and artificial neural network. *HNICEM 2017 - 9th International Conference on Humanoid, Nanotechnology, Information Technology, Communication and Control, Environment and Management.* IEEE. 10.1109/HNICEM.2017.8269472

Ranganathan, J., Waite, R., & Searchinger, T. (2018). *How to Sustainably Feed 10 Billion People by 2050, in 21 Charts.* WRI. https://www.wri.org/insights/how-sustainably-feed-10-billion-people-2050-21-charts

Rashid, M., Bari, B. S., Yusup, Y., Kamaruddin, M. A., & Khan, N. (2021). A Comprehensive Review of Crop Yield Prediction Using Machine Learning Approaches with Special Emphasis on Palm Oil Yield Prediction. *IEEE Access : Practical Innovations, Open Solutions, 9*, 63406–63439. doi:10.1109/ACCESS.2021.3075159

Repositório Digipathos: Home. (n.d.). DigiPathos. https://www.digipathos-rep.cnptia.embrapa.br/

Romeijn, L. J. (2008). Agriculture Crop Management and Production Improved by Satellite Remote Sensing Technology and Geographic Information Systems (GIS). *Satellite Imaging Corporation, Houston, Texas.* https://www.prweb.com/releases/2008/06/prweb1044924.htm

Sa, I., Chen, Z., Popovic, M., Khanna, R., Liebisch, F., Nieto, J., & Siegwart, R. (2018). WeedNet: Dense Semantic Weed Classification Using Multispectral Images and MAV for Smart Farming. *IEEE Robotics and Automation Letters, 3*(1), 588–595. doi:10.1109/LRA.2017.2774979

Salman, A., Semwal, A., Bhatt, U., & Thakkar, V. M. (2017, October 13). Leaf classification and identification using Canny Edge Detector and SVM classifier. *Proceedings of the International Conference on Inventive Systems and Control, ICISC 2017*. IEEE. 10.1109/ICISC.2017.8068597

Santi, E., Pettinato, S., Paloscia, S., Dabboor, M., Notarnicola, C., Padovano, A., Greifeneder, F., & Cuozzo, G. (2018). Estimating soil moisture from C and X band SAR using machine learning algorithms and compact polarimetry. *International Geoscience and Remote Sensing Symposium (IGARSS)*. IEEE. 10.1109/IGARSS.2018.8518469

Shekhar, Y., Dagur, E., & Mishra, S. (2017). Intelligent IoT Based Automated Irrigation System. *International Journal of Applied Engineering Research: IJAER*, *12*(18), 7306–7320.

Singh, D., Jain, N., Jain, P., Kayal, P., Kumawat, S., & Batra, N. (2020). PlantDoc: A dataset for visual plant disease detection. *ACM International Conference Proceeding Series*, (pp. 249–253). ACM. 10.1145/3371158.3371196

Stenberg, B., Viscarra Rossel, R. A., Mouazen, A. M., & Wetterlind, J. (2010). Visible and Near Infrared Spectroscopy in Soil Science. In Advances in Agronomy, Vol. 107, (pp. 163–215). Academic Press. doi:10.1016/S0065-2113(10)07005-7

Tagad, C. K., Kim, H. U., Aiyer, R. C., More, P., Kim, T., Moh, S. H., Kulkarni, A., & Sabharwal, S. G. (2013). A sensitive hydrogen peroxide optical sensor based on polysaccharide stabilized silver nanoparticles. *RSC Advances*, *3*(45), 22940–22943. doi:10.1039/c3ra44547j

Tsouros, D. C., Bibi, S., & Sarigiannidis, P. G. (2019). A Review on UAV-Based Applications for Precision Agriculture. *Information, 10*(11), 349. doi:10.3390/info10110349

Tyagi, S., Pandey, D. K., Putrevu, D., Srivastava, P. K., & Misra, A. (2021). Machine Learning Based Soil Moisture Retrieval Algorithm and Validation at Selected Agricultural Sites Over India Using CYGNSS Data. *International Geoscience and Remote Sensing Symposium (IGARSS)*, (pp. 6335–6338). IEEE. 10.1109/IGARSS47720.2021.9555095

Umamaheswari, S., Arjun, R., & Meganathan, D. (2018, July 2). Weed Detection in Farm Crops using Parallel Image Processing. *2018 Conference on Information and Communication Technology, CICT 2018*. IEEE. 10.1109/INFOCOMTECH.2018.8722369

Van Dijk, M., Morley, T., Rau, M. L., & Saghai, Y. (2021). A meta-analysis of projected global food demand and population at risk of hunger for the period 2010–2050. *Nature Food 2021 2:7, 2*(7), 494–501. doi:10.1038/s43016-021-00322-9

Wiesner-Hanks, T., Stewart, E. L., Kaczmar, N., Dechant, C., Wu, H., Nelson, R. J., Lipson, H., & Gore, M. A. (2018). Image set for deep learning: Field images of maize annotated with disease symptoms. *BMC Research Notes*, *11*(1), 1–3. doi:10.118613104-018-3548-6 PMID:29970178

Zhang, H., Yanne, P., & Liang, S. (2012). Plant species classification using leaf shape and texture. *Proceedings of the 2012 International Conference on Industrial Control and Electronics Engineering, ICICEE 2012*, (pp. 2025–2028). IEEE. 10.1109/ICICEE.2012.538

Chapter 2
IoT System for Crop Recommendation Based on Soil and Weather Patterns

Harsh Tiwari
Samrat Ashok Technological Institute, India

Divya Rishi Sahu
https://orcid.org/0000-0002-2446-3507
Samrat Ashok Technological Institute, India

Rajeev Kumar Gupta
https://orcid.org/0000-0002-5317-9919
Pandit Deendayal Energy University, India

ABSTRACT

Indian agriculture industry gives a significant rise to GDP and provides food and livelihood to a large portion of the whole global population. This demand is increasing rapidly by population growth; the world's population is projected to reach 9.7 billion by 2050. It shoots up pressure on farmers to uplift production and grow crops that are better suited to their local weather and soil conditions. Traditional methods of crop recommendation are often based on subjective assessments and can be inaccurate. However, to overcome these challenges, new technologies like IoT can help farmers to make better decisions and growth of the crop towards holistic rural development. IoT makes it possible to take better decisions on run time and interactively by combining the previous data, past experience of farmers, scientific methodologies, feedback, soil health, weather forecasting, etc. So, in this chapter the use of the IoT and sensors have been discussed with practicality to adopt IoT easily and effectively in the agriculture field.

DOI: 10.4018/978-1-6684-8516-3.ch002

1. INTRODUCTION

The agricultural sector has seen a significant transformation with the utilization of Internet of Things (IoT), Artificial Intelligence and Machine Learning (AIML), etc. technologies. There are integrated multi-disciplinary approaches using IoT to predict soil conditioning, Disease identification, Time predication, smart storage, etc. Few of the integrated technological solutions are shown in the figure given below-

Figure 1. Technology integration for smart farming practices
(TIH, 2023)

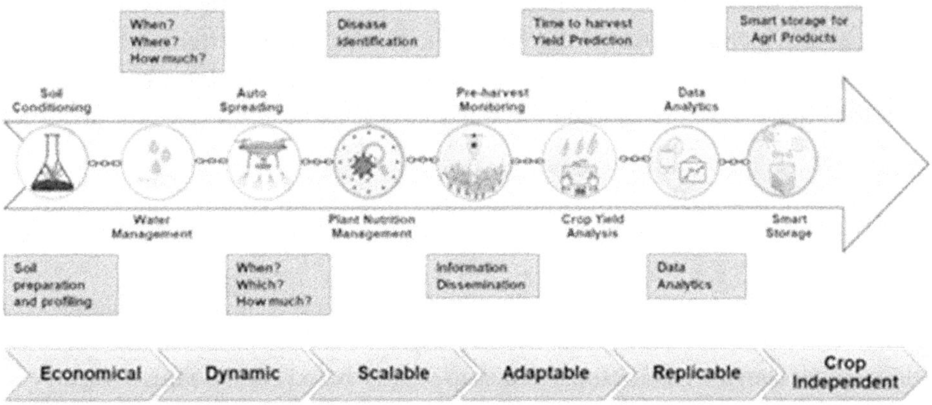

One of the key applications of IoT technology in agriculture is crop recommendation, which involves suggesting the most suitable crop to grow on a particular farmland based on its soil and weather pattern. This process can be challenging, requiring accurate and timely data on soil and weather conditions. However, with the use of IoT devices such as soil and weather sensors, farmers can collect real-time data and make informed decisions on crop selection and planting dates.

In this chapter, we will explore the features and benefits of using IoT technology for crop recommendation, including soil sensors, weather sensors, and other IoT devices that can significantly improve the efficiency and productivity of farming operations.

Since we all know that agriculture is the most department and crucial sector for every aspect of living. Agriculture provides food and material which help in physical, social, and economic development. After agriculture, the second most important is the

1.1 IoT in Agriculture

The Internet of Things (IoT) refers to a network of physical devices, vehicles, home appliances, and other objects embedded with electronics, software, and sensors that enable the exchange of data. IoT has become increasingly popular in agriculture, and farmers are using IoT devices to collect data on soil and weather conditions and to monitor the growth and health of their crops. The use of IoT in agriculture offers several advantages, including-

- Real-time data collection: IoT devices can collect data in real-time, providing farmers with up-to-date information about their fields.
- Improved decision-making: IoT devices can help farmers make informed decisions about irrigation, fertilization, and pest control, leading to increased crop yields and improved efficiency.
- Reduced costs: IoT devices can help farmers reduce their costs by reducing the amount of water, fertilizer, and pesticides they need.

1.2 AIML in Agriculture

Artificial Intelligence (AI) with Machine Learning (ML) refers to the use of AIML, where we enable machines to learn from their surroundings and predict unseen future results. Involvement in AI in agriculture not only helps in increasing the quality of production but also in better predicting results. It helps in the personalization of the method of farming according to the surroundings and needs of the farmer.

Apart from this, the use of AIML in agriculture has several more advantages in which few are as follows:

- **Disease Detection**: Using it, we can analyse the condition of plants and crops and, in advance, detect the prone diseases.
- **Recommendation**: it can better study environmental change and suggest changes in the method of farming.
- **Selection**: Better suggestions for tools and crop variety based on soil nutrients, water availability, and possible methods of farming, it can suggest suitable tools and crop varieties.
- **Accuracy**: Overall, leading to better and more fail-proof farming with the implementation of these in agriculture.

2. LITERATURE SURVEY

An yield prediction system (Pande, et Al., 2021) for the farmers on the basis of type of soil and its mobile application for the help of farmers connectivity and location with GPS. Applied and compared the machine learning methods including Support Vector Machine (SVM), Artificial Neural Network (ANN), Random Forest (RF), Multivariate Linear Regression (MLR), and K-Nearest Neighbour (KNN) to predict the crop yields.

A system (Seru, 2022) to recommend about crops using classifier machine learning algorithm basically focused the parameters including soil type, Rainfall, Groundwater level, Temperature, Fertilizers, Pesticides and season according to the south Indian states. Crop recommendation is based on factors of the surrounding environment, using machine learning algorithm, enable a better understanding of the environment and farmers can better predicts of the combination of crop and agriculture inputs

The use of better technology in farming and production can yield better output and can save from error or failure (Geetha, 2015). For the use of technology better, data should be very accurate, with data mining techniques, one can get better information on the detail of this production.

A crop recommendation system (Hossain et. Al., 2023) using IoT and machine learning methods to measure the nutrients of soil. Author measures the moisture, temperature, humidity, rainfall, nitrogen, phosphorous, potassium and pH of the soil as parameters to identify the soil fertility. Hossain consider it as the most important factor for crop selection because less nutrient soil eventually leads to bad production.

Parameters including water level and moisture to improve the quality of crop as well as to minimize the human efforts (Mishra, et Al., 2018). Water and moist are the essential element involved in the agriculture production. Proper irrigation is a must and effectively done so that water and moisture can be reached evenly and in the required area. This can be done by using IOT-based equipment and a technological approach, which will also help in reducing human labor.

A system for planning of farming including water level and its time of requirement focused on concept that agriculture is a field of work where it required most attention in every aspect of the production stage (Saleh, et Al., 2022). Traditional methods sometimes become hard to monitor it all the time. Using IoT monitor system, it can be achieved very significantly and with accurate reading and details. It will improve production and will produce a better yield.

Use of NPK sensor to identify the condition and maintain the health of soil (B. Cheruv et Al., 2023). Soil plays an important role in production, health, and nutrient-rich soil produces good and nutrient products. Proper health care of the soil and its nutrients like Nitrogen, phosphorus, and potassium. With the help of IoT sensor - NPK sensor, proper monitoring of the soil can be achieved accurately and give proper a better idea about the condition and step which should be taken to maintain it in its potential required form.

Convolution neural network (CNN) with image processing is proposed (Sagarika et Al., 2020) to examine the plant and identify the Plant Disease on rice. Rice production is a long process that involves many resources like time, labor, natural resources like water, etc. Therefore, it proper care of paddy plant is required to safeguard disease or any condition which can lead to crop failure or poor yield. This can be prevented by examine the plant and most efficiently with image processing and convolution neural network (CNN) which help in detecting and prevent any difficult causes and getting the most out of it

Since, we know that IoT based system can improve the production and quality of production author proposed a smart gateway system for connected farming (Ghodeswar et Al. 2022). But implementation of IoT can be challenging for some farming and people related to it, this could be eliminated by Green IoT, A smart IoT gateway for connected farming which will help in implementing IoT systems in the agriculture environment and executing agricultural operations in an efficient and smart way.

With the increase in implementation of IoT in farming and agricultural activities, this creates a requirement of search engine where query related IoT can be processed like time and better predict can be made based on the case studies and database processed by the search engine. A special search engine (Hatcher et Al., 2021) to query about the farming is developed that combine several realistic data, a clear view of query can be obtained by user.

As discussed in this section, the researchers proposed and developed different tools and techniques in the filed of IoT and machine learning to improve the crop production. Few of the method discussed in this section includes IoT based system using NPK sensor, machine learning approach including CNN, Support Vector Machine (SVM), Artificial Neural Network (ANN), Random Forest (RF), Multivariate Linear Regression (MLR), K-Nearest Neighbor (KNN), etc. By applying these technologies researchers are predicting the best output to maintain the temperature, moisture, water level, humidity, rainfall, nitrogen, phosphorous, potassium and pH, etc. towards accurate growth of the crop.

In the next section we will discuss one of the hands-on of crop recommendation system to improve the health of soil according to the favorable conditions for crop, season and geography.

3. CROP RECOMMENDATION SYSTEM

A crop recommendation system is a combination of software and hardware application that helps farmers make informed decisions about which crops to grow based on local soil and weather conditions. These system uses data from various sources, including weather sensors, soil sensors, and historical weather patterns, to provide farmers with accurate information about their fields. With this gathered information, farmers can make informed decisions about which crops will be most suitable for their soil and weather conditions, leading to higher yields and lower costs. Hence, improving productivity.

Few of the Crop Recommendation Software or Apps (DA&CFW, DMA, 2023) are-

- Kisan Suvidha
- Pusa Krishi
- Shetkari Masik Android App
- Bhuvan Hailstorm App
- Agrimarket

Few of the Crop Recommendation Systems Hardware (DA&CFW, DMA, 2023) are-

- Weather Station
- Remote Sensing Device

3.1 Factors Involved in the Selection of Crop

Common factors that are involved in the selection of crop are identification of soil type, climate condition, water level and condition, temperature, fertilizers, etc. Few of the factors are described as follows.

F1. Soil Type: Soil is the most crucial factor that determines the selection of crops. The soil type in the field will determine the type of crops that can be grown. For example, certain crops grow better in sandy soil, while others prefer loamy soil.

F2. Climate Condition: Climate plays a vital role in the growth of crops. The selection of crops should be based on the climate of the region. For instance, crops that require a lot of water may not be suitable for dry areas. We can get suitable environmental conditions for any crop you want to produce from the national government food and research website of the country (DA&FW, Weather Details, 2023 & SFAC, 2023).

F3. Water Availability: The availability of water in the field is an essential factor to consider when selecting crops. Crops that require a lot of water may not be suitable for areas with limited water availability.

F4. Market Demand: It is essential to consider the market demand for crops. Growing crops that are in high demand can ensure a good profit for the farmer.

F5. Farmer's Preference: The farmer's preference also plays a role in the selection of crops. Some farmers may prefer to grow crops they are familiar with or have experience growing.

F6. Crop Rotation: It is recommended to rotate crops to maintain soil fertility and reduce the risk of pests and diseases.

F7. Temperature: It is the most crucial factor for the cultivation of crops, each crop has its specific range of survival temperature and this has to take into account for selecting a crop

F8. Fertilizer: For the healthy growth and field of crops, crop needs essential nutrients such as nitrogen, phosphorus, etc. Fertilizer acts as supplements and provides an essential resource.

F9. Pests and Diseases: Pests and Diseases are the great enemies of the crop and their production. Those crops are chosen which have resistance to pests and disease prone to that region.

Let's understand these factors with a real-life example, for the sample crop let's say Rice. Requirements for the production of rice on agricultural land are given in the table 1 given below (NFSM, 2023).

Table 1. Requirements for the production of rice on agricultural land (NFSM, 2023)

Requirements	Description
Soil Type	Rice can be grown in different soil types, but it prefers fertile, well-drained soil with a pH between 5.0 and 7.5.
Climate Condition	Analysis of climate conditions or weather forecasting to identify the requirements of water and protection from exceed weather conditioning. Rice production needs hot and humid climate
Water Availability	Rice requires a consistent supply of water, either from rainfall or irrigation. The amount of water needed depends on the stage of growth and the type of rice being grown.
Market Demand	In year 2011-2012 105.31 million tons of rice produced for the market
Farmer's Preferences	Farmer choose rice which they have plenty of water and sunshine
Temperature	Rice is a warm-season crop and prefers temperatures between 20 and 35 degrees Celsius.
Fertilizer	Rice requires nutrients such as nitrogen, phosphorus, and potassium, which can be supplied through organic or chemical fertilizers. The amount and timing of fertilizer application depend on soil nutrient levels and the stage of growth.
Pests and Diseases	Rice is susceptible to a range of pests and diseases, such as blasts, sheath blight, and stem borers. Integrated pest management strategies can help control these issues.

3.2 Integration of IoT With Crop Recommendation System

Nowadays Internet of Things plays a vital role in every field and aspect of life. Integration of IoT technology with crop recommendation systems to improve agriculture is one of the fields. In this chapter, we focus on one of the practical to understand crop recommendation systems with their implementation using IoT.

The data collected from different soil and weather sensors health measuring sensors can be integrated into a crop recommendation system to provide farmers with a complete picture of their fields. This information can be used to make informed decisions about crop selection, soil management, and weather management, leading to higher yields and lower costs.

In smart irrigation system, Researchers are mainly using 10 parameters as showing in figure 2. Among them most of the researchers are focusing on soil moisture.

IoT can work to identify the different factors as we discussed in an earlier section. For better understanding and practice, we will see the integration of soil and weather sensors for the recommendation of factors for the desired crop. It needs the understanding of technologies and its integration as discussed in a next section.

Figure 2. Parameters measured in no of papers for smart farming
(García et Al., 2020)

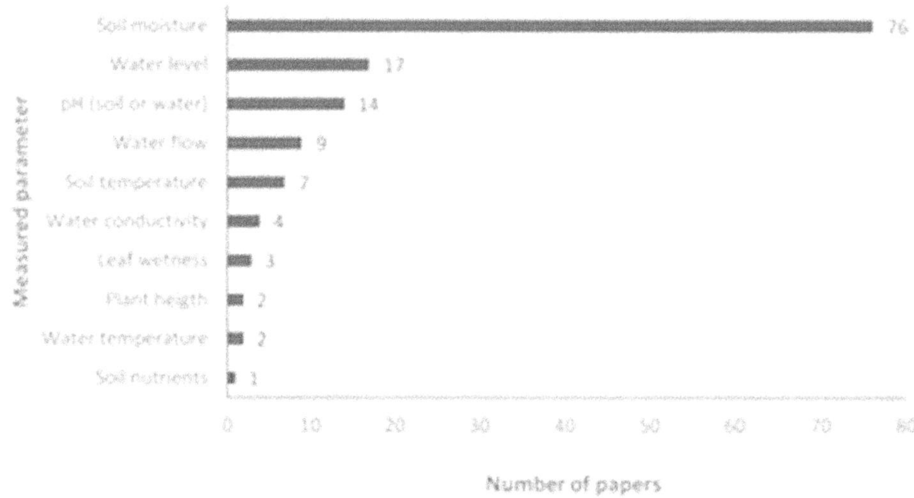

3.2.1 Technology Bucket

The point of integration of IoT in agriculture is get the best possible yield from production by the use of today's modern technology which provides and can predicate accurate results. This is reducing the risk of failure as its surveillances using tough like drought or cold, saving mankind's efforts.

It is possible by integrating various IoT, AI and ML technologies including soil sensors, weather sensor, humid or any of the actuators which are connected to development boards via network and process the data received from these. In this chapter we are using focusing of sensors to identify the heath of soil moisture and weather conditions for practical hands-on experiences.

3.2.1.1 Sensors Used to Identify the Health of Soil

Different sensors may be used to measure physical and chemical properties of soil, including temperature, moisture, pH, and nutrient levels. These sensors can be placed in and across the soil, and they collect data that can be used to make informed decisions about crop selection and soil management. For example, soil sensors can measure moisture levels in the soil and provide farmers with information about when to irrigate their fields and provide better care.

Types of Sensors to Measure Soil Health

There are several types of sensors available in the market that can be used to measure the health of the soil. Each sensor is designed to capture specific data points. The most common types of soil sensors that can be used to identify the health of soil includes-

- **Moisture sensors**: These sensors measure the soil's moisture content at various depths. It measures the volumetric water content in soil They are designed to provide real-time data on soil moisture levels, allowing farmers to optimize irrigation practices and reduce water waste.

- **Temperature sensors**: These sensors measure the soil's temperature at different depths. They are useful for monitoring soil temperature changes and can help farmers determine the optimal time to plant crops.
- **Nutrient sensors**: These sensors measure the levels of essential nutrients in the soil, including nitrogen, phosphorus, and potassium. They help farmers determine the appropriate amount of fertilization required to maximize crop yields.
- **Soil pH sensor**: These sensors measure the pH level of the soil, which gives the idea of the acidity of the soil, and certain crops which require a certain pH level can be selected for farming.

In this chapter we are focusing on the temperature sensor and humidity sensor for the practically identification of soil health for crop recommendation system.

3.2.1.2 Temperature Sensors

It measures the heat in the surrounding environment and converting to the electrical signals. It basically used to identify the ambient condition of soil required for the growth of crop in crop recommendation system. The most utilized temperature sensors as shown in figure 3 given below.

Figure 3. Different temperature sensors used by no. of researchers
(García et Al., 2020)

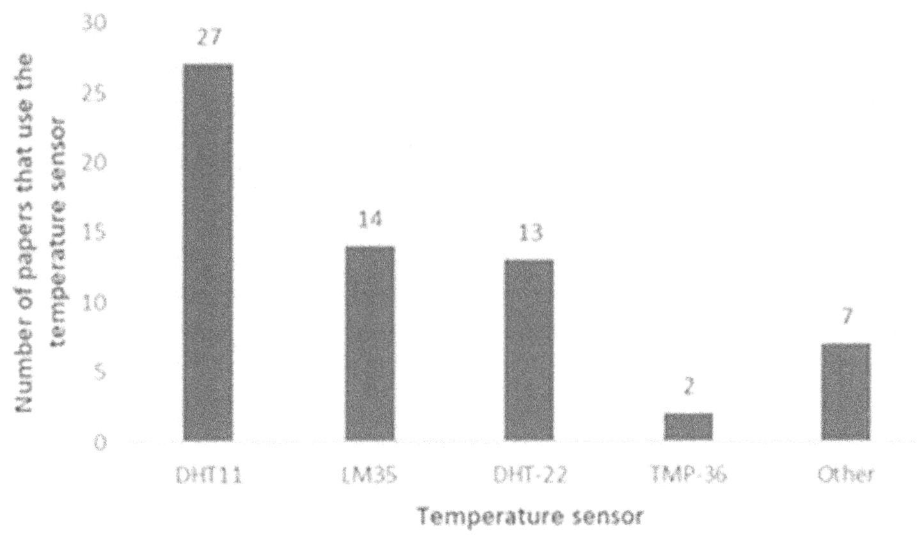

3.2.1.2 Humidity Sensors

It basically measures the relative humidity in surrounding environment. Relative Humidity (RH) is the ratio of the actual water vapour pressure to the saturation water vapour pressure. It basically used to identify the ambient condition of soil for the growth of crop in crop recommendation system. The most utilized humidity sensors as shown in figure 4 given below.

Figure 4. Different humidity sensors used by no. of researchers
(García et Al., 2020)

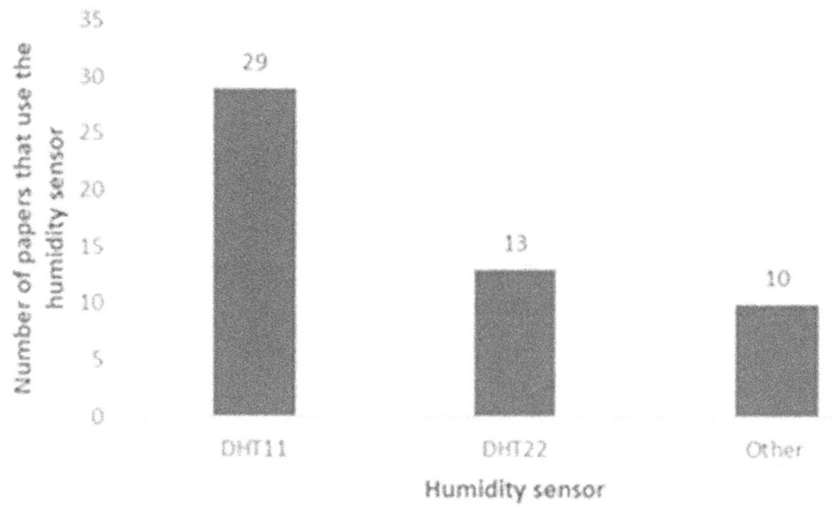

3.2.1.3 Sensors Used to Identify the Health of Weather

Weather sensors are those devices and instruments that measure various physical weather conditions, including temperature, humidity, wind speed, and rainfall. These sensors can be placed on weather stations or individual standalone IoT devices, and they collect data that can be used to make informed decisions about crop selection and weather management. There are various types of weather sensors used to measure and record different weather parameters.

Types of Sensors used to identify Weather Conditions

Here are some common types of sensors that can be used to identify the weather condition for the crop recommendation system as per the requirement of crop:

1. **Thermometers**: Measure temperature, and are commonly used to measure air and water temperature.
2. **Hygrometers**: Measure humidity, or the amount of moisture in the air.
3. **Barometers**: Measure atmospheric pressure, which is the force exerted by the weight of the air.
4. **Anemometers**: Measure wind speed and are used to measure wind speed in units such as miles per hour, kilometers per hour, and meters per second.
5. **Wind vanes**: Measure wind direction, and are typically combined with anemometers to measure both wind speed and direction.
6. **Rain gauges**: Measure rainfall, and are typically used to measure the amount of precipitation in millimeters or inches.
7. **Pyranometers**: Measure solar radiation or the amount of energy from the sun that reaches the Earth's surface.
8. **Weather stations**: Combine multiple sensors to measure and record several weather parameters at once, such as temperature, humidity, air pressure, wind speed and direction, and rainfall.

Weather sensors are typically placed in outdoor locations, where they can gather real-time data about the atmosphere. This information is often used to make weather forecasts, issue warnings about severe weather conditions, and inform decisions about agriculture, transportation, and other industries that are affected by the weather.

3.2.1.3 Development Boards

Development Boards are hardware devices that provide a platform for computation and a development environment for the software tools which can perform calculation and analysis and produces a result based on the input provided. They are the main brain of the IOT system.

These boards consist of crucial components like Processor, Memory, Storage Operating system, and instruction. All the sensors and actuators are connected to these boards. The most utilized development board in smart irrigation system as shown in figure 5 given below.

Figure 5. Different nodes/ development boards used by no. of researchers
(García et Al., 2020)

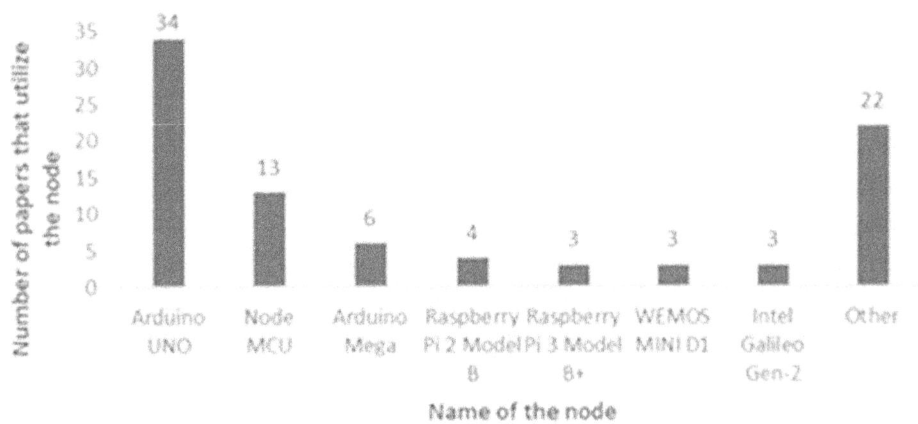

Types of Development Boards

There are various types of development boards that can be categorized based on form factor, size, processor architecture, and features & capabilities.

Below are a few types of development boards that are commonly used for commercial applications:

1. Arduino Boards
2. Raspberry Pi
3. BeagleBone Boards
4. Intel Galileo/Edison Boards
5. STM32 Discovery Boards
6. ESP Boards
7. Teensy Boards

3.3 Integrating Sensors to Identify the Health of Soil and Weather Conditions for Crop Recommendation System

A farmer might use a soil sensor to measure moisture levels in their field, and a weather-based sensor to measure temperature and humidity levels. The data from these sensors can be integrated into a crop recommendation system, which can then provide the farmer with information about which crops will be most suitable for these conditions. Figure 6 shows the flowchart for Integration of IoT in the Crop Recommendation System.

This integration could be described as follows:

1. **Sensor integration**: Combining soil and weather sensors into a single system that can collect data about multiple environmental parameters.
2. **Data acquisition**: Gathering data from the sensors and transmitting it to a central location for analysis.
3. **Cloud connectivity**: Using cloud-based platforms to store, process, and analyze sensor data, and make it accessible to users.
4. **Analytics**: Applying statistical and machine learning algorithms to the sensor data to extract insights and make predictions about the environment.
5. **Visualization**: Creating user-friendly interfaces that display sensor data in real-time, enabling users to make informed decisions based on the information.

Figure 7 shows the data flow of physical information from the hardware device connected in the crop recommendation system.

Integration of soil sensors in the field is very precise work, installation of the small silicon-based sensors and microprocessor chips in the farm and calibration requires technical supervision. Before installation, one should carefully calibrate the sensors to avoid Systematic errors which are hard to deal with and sometimes require the whole system to check up on.

Once these sensors are installed in the field, all of these sensors are connected to the common junction, which is located on the sides of the farm. They are then collectively connected to the server room or the processing channels, which analyze the physical raw data into analyzed data.

Once the data is processed, farmers get information in the user interface and would be the point of interaction of the farmer to the system. Figure 8 illustrates the implementation of the sensors and devices in the field from the ground to the server room.

Well, that was enough for the theory part, let's now jump into the real-life practical implementation of sensors in the field.

- For the monitoring of the water level in the soil or the moisture, moisture detection sensors will be installed in the field.
- For the detection of the heat present in the farm in the soil bed, soil sensors will be installed in the farm, which pass the general level of the temperature of the soil,
- And additionally, other temperature-detecting sensors will be installed to monitor the overall temperature of the place.

Figure 6. Flowchart of integration of IoT in the crop recommendation system

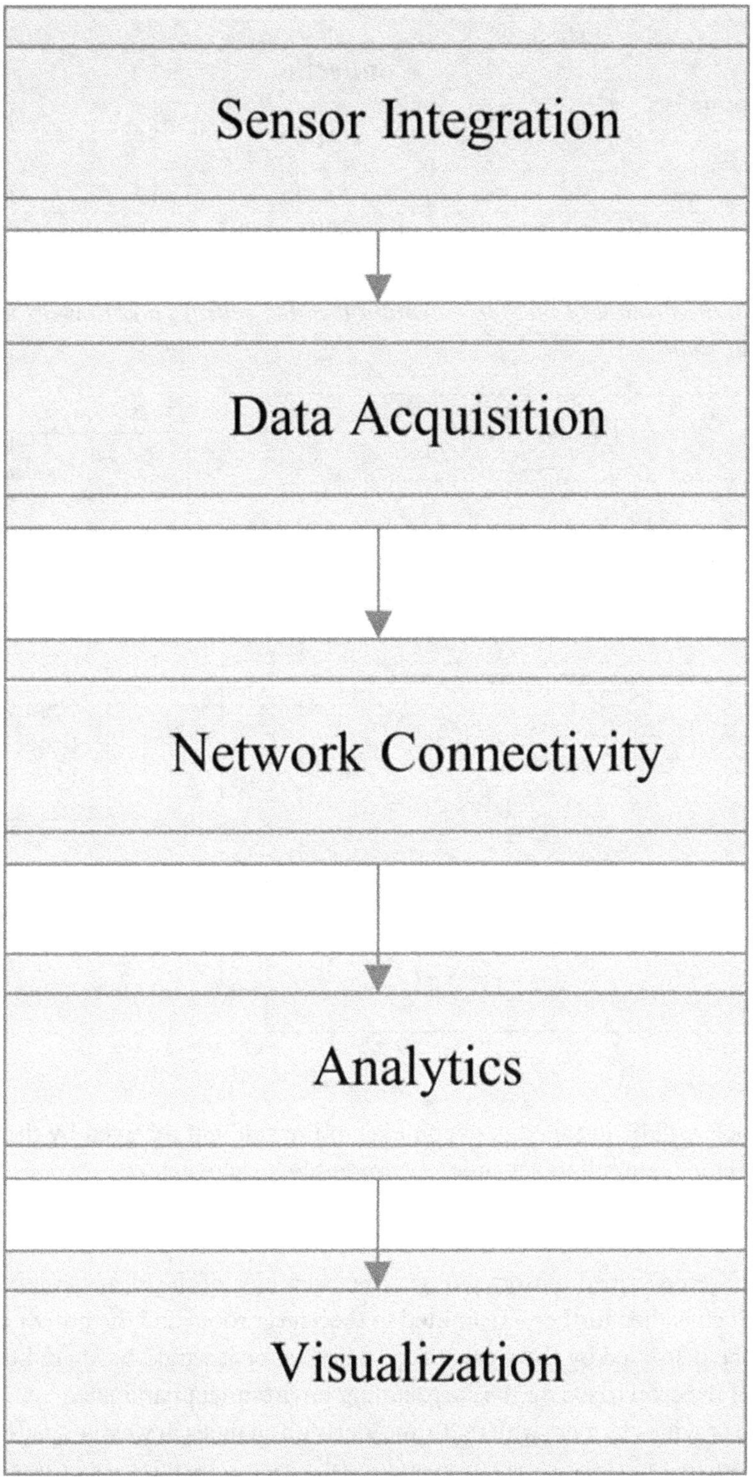

Figure 7. The data flow of physical information

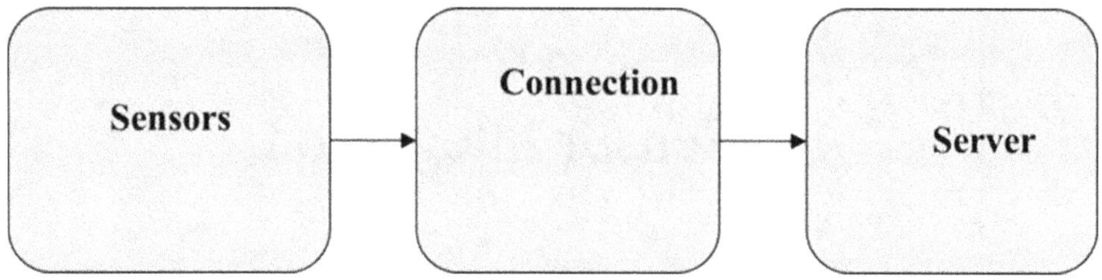

Figure 8. This is an illustration of the implementation of the sensors and devices in the field from the ground to the server room

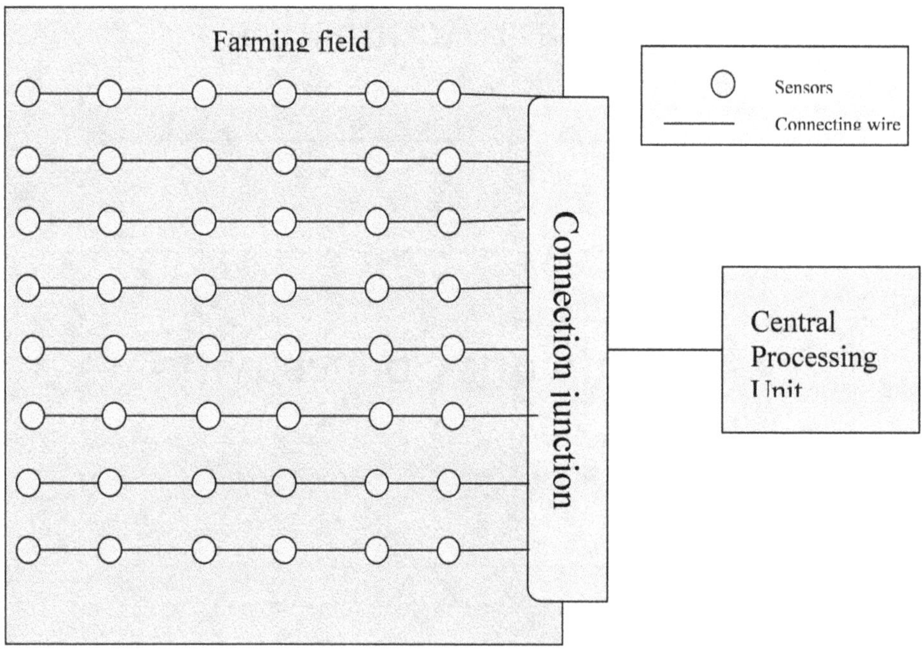

- The solar panels will be installed to get an idea of the sunlight received by the land.
- We can install the windmill to get an idea of monthly air movement.
- The level of rain can be detected by the water sensor.

Using all the above-described sensors will get the exact idea of the climate and environment condition of that land which is then further transmitted to the server room and the proper report will generate which can further be inspected by the Agricultural scientist or it would be shared to trained AI for the recommendation of the crop based on the surrounding environment parameter.

The system also provides farmers with real-time alerts on changes in weather patterns or soil moisture levels, allowing them to adjust their irrigation and fertilization schedules accordingly.

3.4 Implementation Procedure for Integration of Soil and Weather Sensor

Now that we are aware of the use of placement and usage of the sensors. Now let's create a small real-world implementation of IoT systems using actual sensors and microcontrollers.

Now we will create an actual circuit consisting of sensors and controllers

Step 1: Take any microcontroller or single-board computer like Raspberry-Pi, ESP, or Arduino board as per availability and requirements.
Step 2: Arrange some resistors, jumper wires, and breadboard for establishing the connection.
Step 3: Note the physical environment reading, which one has to obtain from the surroundings like moisture, temperature, humidity, sunlight, ph level, etc.
Step 4: Now connect the sensors to the motherboard through the breadboard via jumper wires.
Step 5: Once the connection is established, using programming languages like Python or Java, or any which is supported by the main board, program the IoT system to enable its functioning, and add some logics that empowers the Iot system with automation like dispatching the water when the moisture content is less in the soil, etc.

Properly connect the sensors to the board, and make sure to calibrate and test the sensors before implementing the fields. This is just a small implementation, but this gives the idea of large-scale implementation and more sensors can be connected as per the need of the surroundings and the system.

3.4.1 Circuit Diagram for Moisture and Humid Sensor Connectivity

Below is the circuit diagram of the small-level implementation of small-level sensors. We used a micro-processor through which sensors are connected to a microcontroller via a breadboard and jumper wires.

The diagram contains the following elements as discussed in above section.

- Moisture Detector sensor
- Temperature detector sensor
- Humidity Sensor
- Photocell - to detect sunlight

Hence, making an IoT device that can be used in real-world applications. Figure 9 represents the circuit diagram to connect weather sensor and soil sensor with the development board such as Arduino Uno.

Steps to connect the weather sensor and soil sensor with development board are as follows-

1. Make sure to disconnect the microcontroller board from the power source.
2. Connect the VCC to 5V and GND to GND of the sensor to the microcontroller, respectively.
3. Connect the digital output pin to the D-pins (digital signals pin) and the analogue pin to the A-pin (analogue pin) as shown in the diagram.
4. Double-check the connection and, at last, connect the power supply to the microcontroller board.

Figure 9. Circuit diagram to connect weather sensor and soil sensor with development board

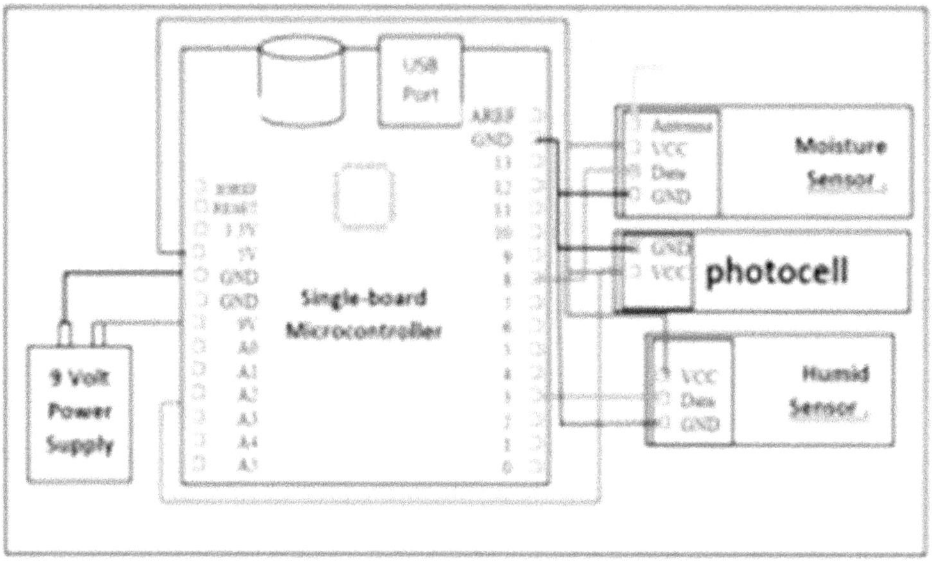

After the completing the connections in all aspects and sensor architecture, the type of output generated will vary. For example, in the above case, the moisture sensor generates digital output, the humid sensor also generates digital output, and the photocell produces analogue signals. According to the type of connection to the microprocessor board.

3.4.2 Analyzing Data From a Combined System

Analyzing data from a proposed connected system, can provide valuable insights and enable data-driven decision-making. It can be possible by using the Artificial Intelligence and Machine Learning technologies. Here, are some key steps for analyzing data from a combined system:

1. **Data Cleaning**: The first step in analyzing data is to clean the data. This involves removing duplicate or irrelevant data and correcting errors or inconsistencies. This ensures that the data is accurate and reliable for analysis.
2. **Data Integration**: Once the data is cleaned, it must be integrated into a single platform. This requires integrating data from multiple sources, such as weather stations, soil sensors, and crop databases. This ensures that the data is complete and consistent for analysis.
3. **Data Analysis**: The next step is to analyze the data. This involves using various analytical techniques, such as statistical analysis, machine learning, and predictive modeling, to extract insights from the data. The analysis should be focused on the specific goals of the system, such as crop selection, irrigation management, and pest control.
4. **Data Visualization**: Once the data is analyzed, it needs to be presented in a way that is easy to understand and interpret. Data visualization techniques, such as charts, graphs, and dashboards, can be used to visualize the data and provide insights into patterns and trends.

5. **Continuous Monitoring**: Analyzing data from a combined system is an ongoing process. The system needs to be monitored regularly to ensure that the data is accurate and up-to-date. Continuous monitoring can also help identify any anomalies or changes in the data that may require further investigation.

4. RESULT ANALYSIS

Result analysis based on prior knowledge and experience of farmers, industries working in agriculture fields, ministries and others gives better process towards the growth of crop. Several studies have been conducted on the use of IoT technology for crop recommendation, and the results have been promising. For instance, climate conditions for rice suggested by Department of Agriculture & Cooperation and Farmers Welfare are shown in following table 2.

Temperature values for the rice plant on three levels that is Low, High and Optimum are recommended as per the study are represented in table 3 given below.

Table 2. Parameter values for the optimum growth of rice (NFSM, 2023)

Sr. No.	Parameter	Requirement for Rice	Optimum Range/ Condition
1.	Humidity	High	40 to 90%
2.	Water	Higher than other crop	Water should stand in the field at a depth of 2-5 CM till the seedlings are well established
			About 5-CM depth of water should be maintained in the field up to the dough stage of the crop
			Before harvesting, water should be **drained out** from the field to allow quick and uniform maturity of grain
3.	Temperature	Average throughout the life period of the crop	Between 21 to 37° C more range according to condition are given in table 3 given below.
4.	Max Temperature	Rice can tolerate	40°C to 42°C

Table 3. Critical temperature for the development of rice plant at different growth stages (NFSM, 2023)

Growth Stage	Critical Temperature (°C)		
	Low	High	Optimum
Germination	16-19	45	18-40
Seedling emergence	12	35	25-30
Rooting	16	35	25-28
Leaf elongation	7-12	45	31
Tillering	9-16	33	25-31
Initiation of panicle primordial	15	-	-
Panicle differentiation	15-20	30	-
Anthesis	22	35-36	30-33
Ripening	12-18	>30	19-20

A study conducted by researchers, found that the use of IoT devices such as soil sensors and weather sensors significantly improved crop yields and reduced water consumption in farming.

The IoT-based crop recommendation was more accurate and reliable than traditional methods of crop recommendation based. Comparison of traditional crop recommendation system and IoT-based crop recommendation system on water usage, growth per tree and overall growth for lemon crop is analyzed and found that the sensor-based system is better as shown in figure 10 given below.

Figure 10. Comparison of traditional crop recommendation system and IoT-based crop recommendation system on Lemon
(Zia, 2021)

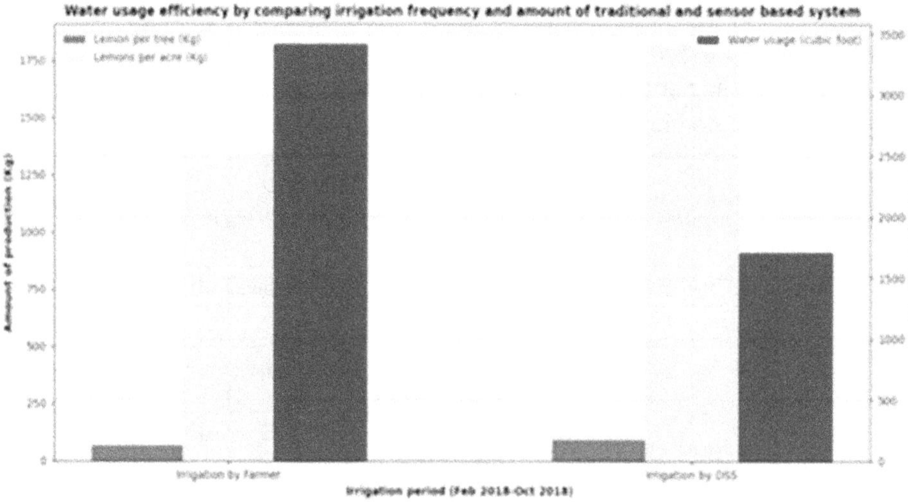

Ministry of Agriculture and Farmers Welfare, provided the Soil Fertility Map that represents the available minerals in the Soil including Iron, Zinc, Sulphur, Ph, Phosphorus, Manganese, etc. These may affect crop production and grade of the crop and may use in the crop recommendation system.

The figure 11 shows the Soil Fertility Maps for Phosphorus for MP state of the India. It represents that soil of Katni, Harda, Sidhi, etc. have phosphorus in high range that is more than 25 kg/ha. While the soil of Betul, Guna, Sheopur, etc. have phosphorus in medium range that is more than 10-25 kg/ha.

The figure 12 shows the Soil Fertility Maps for Manganese for MP state of the India. It represents that soil of in all the state is in sufficient that is >2.0 ppm/ 30867691 ha.

Similarly, the figure 13 shows the Soil Fertility Maps for Nitrogen for MP state of the India. It represents that soil of most areas in MP is low that is <280 kg/ha.

More data related to Social Health Card, National Agriculture Market, Pesticide, Jaivik Kheti, Crops, Scheme, etc. can be found on the website of MoA&FW (Agricoop, 2023 & APS, 2023). According to the various data and report shows by the different ministries and industries crop recommendation system can help farmers by providing the right suggestions according to the analysis of various data/ parameters as discussed in this chapter and others as well.

Figure 11. Soil fertility maps for phosphorus for MP
(SHC, 2023)

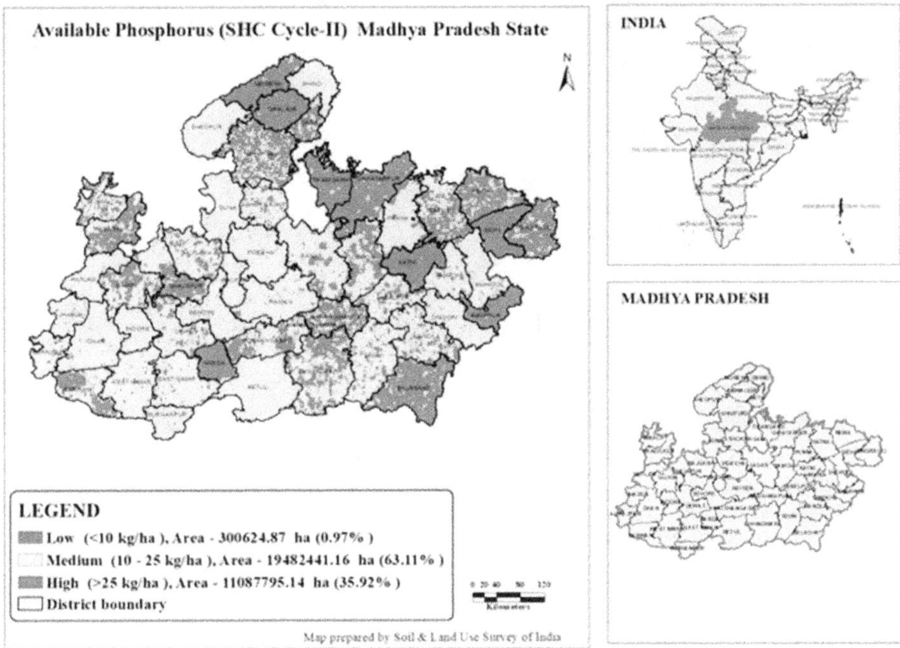

Figure 12. Soil fertility maps for manganese for MP
(SHC, 2023)

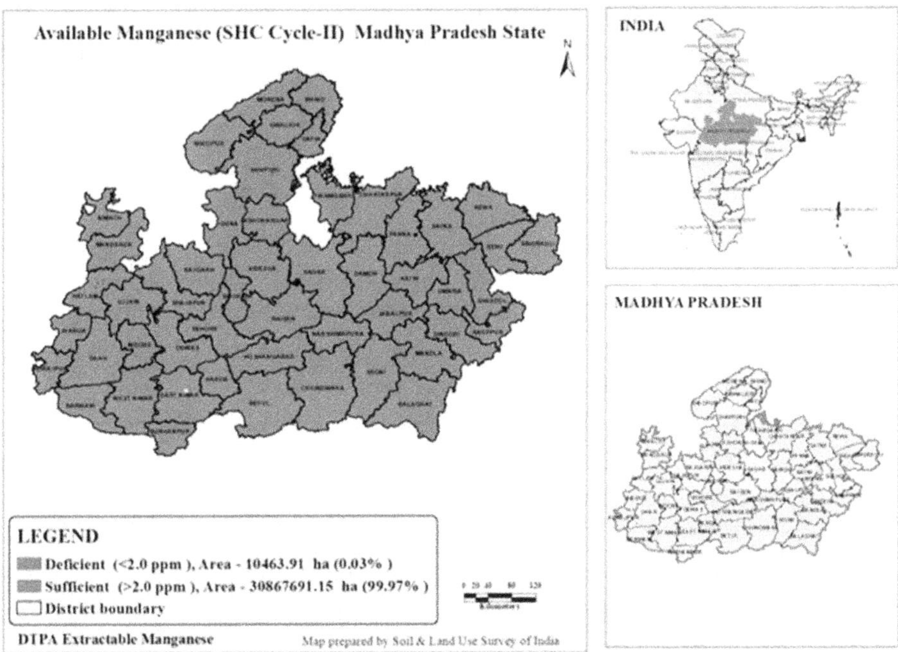

Figure 13. Soil fertility maps for nitrogen for MP
(SHC, 2023)

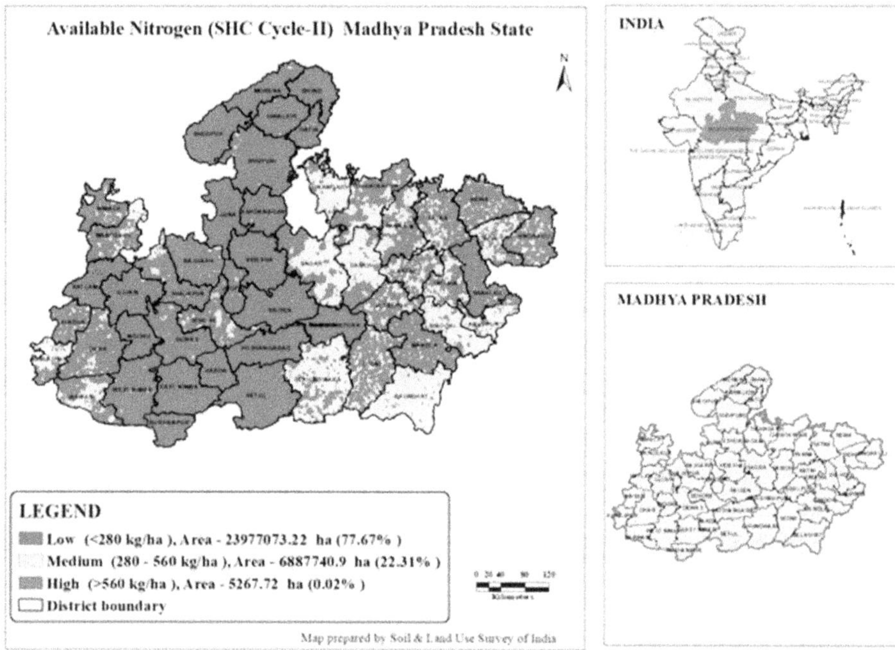

However, it is worth noting that the effectiveness of IoT-based crop recommendation depends on several factors, including the quality and accuracy of the data collected, the availability of internet connectivity, and the technical expertise required to deploy and maintain IoT devices. Therefore, it is essential to carefully consider these factors before adopting IoT technology for crop recommendation. Visualization of these facts and figures is more convenient rather other forms of result representation as discussed in next section.

4.1 Visualization

After data has been collected and analyzed, now the big next step is to visualize the results. Visualization is important because it's the point at which the system interacts with people and presents the analyzed data in a more understandable way. For farmers, this means they can see the data in a way that is meaningful to them and use it to improve their farming practices.

Visualizing the data allows farmers to see different insights and trends that might not have been apparent in the raw data. It also enables them to explore the data in different ways, such as by filtering or zooming in on specific parts of the data. By doing so, they can identify patterns that can help them make better decisions.

Moreover, farmers can adjust the inputs according to their conditions and need and see how it affects the outcomes. This means they can experiment with different approaches to growing crops and see the results in real-time, able to achieve the best possible combination. This is particularly useful when trying to grow different crops, as they can test different methods to see which ones are more profitable.

Figure 14. Visualization of temperature and humidity sensor data

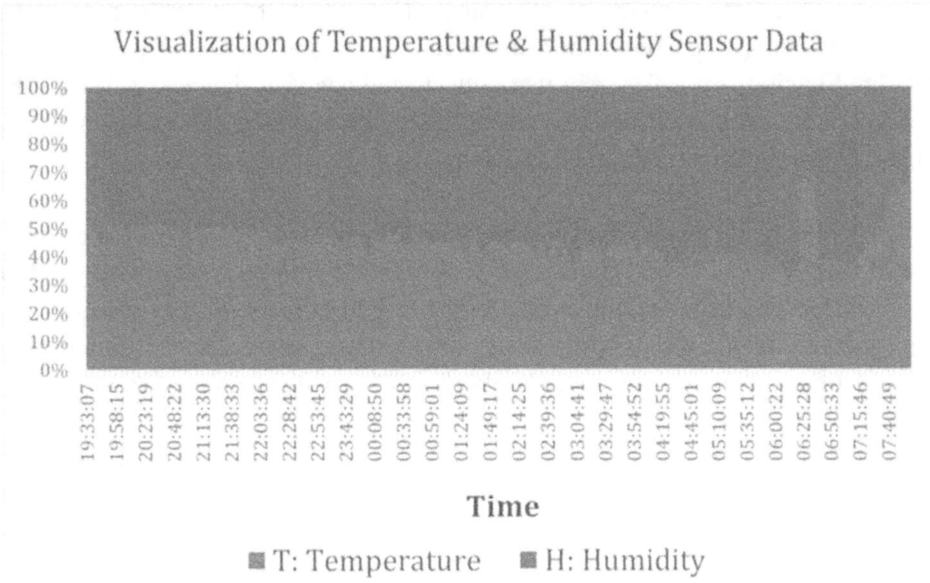

Visualization of temperature and Humidity Sensor Data of crop recommendation system and its comparison is shown in figure 14 as shown below. That can be utilize as per the crop need and stage of growth.

Finally, visualization can help farmers identify crops that might be unprofitable. By looking at the data, they can estimate which crops are not performing well and make adjustments accordingly. This could involve changing the crop type, the soil conditions, or other factors to increase the profitability of the crop.

4.2 Social Impacts

The development and implementation of an IoT system for crop recommendation based on soil and weather patterns can have several positive social impacts, including

1. **Improved Crop Yields**: An IoT system that provides accurate recommendations for crop selection and cultivation can improve crop yields. This can increase farmers' incomes and contribute to food security in the region.
2. **Reduced Environmental Impact**: An IoT system that recommends crops based on soil and weather patterns can reduce the use of fertilizers and pesticides, which can have harmful effects on the environment and human health. This can contribute to sustainable agriculture practices and reduce the environmental impact of agriculture.
3. **Increased Efficiency**: An IoT system that automates data collection and analysis can increase efficiency and reduce the time and effort required for manual data collection and analysis. This can free up time for farmers to focus on other aspects of their farm and contribute to their overall well-being.

4. **Empowering Farmers**: An IoT system that provides accurate and timely information can empower farmers and enable them to make better decisions. This can contribute to their sense of agency and well-being and improve their economic prospects.
5. **Knowledge Sharing**: An IoT system that collects and analyzes data can also enable knowledge sharing among farmers, extension workers, and other stakeholders. This can contribute to a more robust and collaborative agricultural ecosystem.

Overall, the development and implementation of an IoT system for crop recommendation based on soil and weather patterns can have a significant positive social impact, contributing to food security, sustainable agriculture practices, and the overall well-being of farmers and communities.

In the next section, we will discuss the few common challenges faced in the implementation of the crop recommendation systems and what we should remember before implementation and setup.

4.3 Challenges

Despite the benefits of using IoT devices for crop recommendation, there are several challenges associated with this technology. Developing an IoT system for crop recommendation based on soil and weather patterns is a complex task that presents several challenges. Here are some of the challenges that may be faced while implementing such a system:

1. **Data Collection and Management**: Collecting and managing the data required for crop recommendation can be challenging. Soil and weather patterns data need to be collected from multiple sources and then consolidated in a single platform. This requires the integration of different data formats and protocols, which can be a challenge.
2. **Quality of Data**: Data quality is critical in the context of IoT for crop recommendation. The accuracy of the recommendation depends on the quality of the data used. Soil and weather patterns data must be reliable and accurate to provide accurate recommendations.
3. **Interoperability**: Ensuring interoperability between different systems is a challenge in IoT systems. The data must be integrated from various sources, and systems must be interoperable to provide seamless data exchange.
4. **Connectivity**: The IoT system requires an internet connection to collect data and make recommendations. In rural areas where agriculture is prevalent, connectivity can be a significant challenge. Limited or no connectivity can hinder the ability to collect real-time data.
5. **Security**: An IoT system for crop recommendation involves a significant amount of sensitive data. The system must be designed with security in mind to protect the data from unauthorized access.
6. **Cost**: Developing and implementing an IoT system for crop recommendation can be expensive. The system must be designed with cost-effectiveness in mind, to make it viable for farmers to use.
7. **User Adoption**: The success of the system depends on user adoption. The system must be user-friendly, easy to use, and provide value to the farmers. Educating farmers on the benefits of the system is critical to ensure user adoption.

Addressing these challenges requires a holistic approach that takes into account various aspects of the system, including design, data management, connectivity, and security.

4.4 Summary

The IoT system for crop recommendation based on soil and weather patterns can have significant benefits for farmers, communities, and the environment. However, developing such a system presents several challenges, including data collection and management, interoperability, connectivity, security, cost, and user adoption. Although it has some challenges, they can be addressed through a holistic approach that takes into account various aspects of the system, including design, data management, connectivity, and security.

Analyzing, data from a combined system can provide valuable insights and enable data-driven decision-making. It involves data cleaning, integration, analysis, visualization, and continuous monitoring. Data analysis and visualization can help farmers make informed decisions about crop selection, irrigation management, and pest control, leading to increased yields and improved sustainability.

Overall, the development and implementation of an IoT system for crop recommendation based on soil and weather patterns will definitely have a positive social impact, contributing to food security, sustainable agriculture practices, and the overall well-being of farmers and communities. It can also enable knowledge sharing and collaboration among farmers, extension workers, and other stakeholders, leading to a more robust and collaborative agricultural ecosystem.

REFERENCES

Cheruvu, B., Latha, S. B., Nikhil, M., Mahajan, H., & Prashanth, K. (2023). Smart Farming System using NPK Sensor. *9th International Conference on Advanced Computing and Communication Systems (ICACCS)*, Coimbatore, India. 10.1109/ICACCS57279.2023.10112795

Department of Agriculture & Farmers Welfare. (n.d.). *MoA&FW*. Agricoop. https://agridashboard.dac.gov.in/

Department of Agriculture & Farmers Welfare (DA&CFW), Ministry of Agriculture and Farmers Welfare, Government of India. (n.d.). *Farmer's Portal: One stop shop for Farmer. Download Mobile Application (DMA)*. DAFW. https://farmer.gov.in/mobileappsdownload.aspx

Directorate of Economics and Statistics, Ministry of Agriculture and Farmers Welfare. (n.d.). Area and Production Statistics (APS). https://aps.dac.gov.in/Home.aspx

García, L., Parra, L., Jimenez, J. M., Lloret, J., & Lorenz, P. (2020). IoT-Based Smart Irrigation Systems: An Overview on the Recent Trends on Sensors and IoT Systems for Irrigation in Precision Agriculture. *Sensors (Basel)*, *20*(4), 1042. doi:10.339020041042 PMID:32075172

Geetha, M. C. S. (2015). A Survey on Data Mining Techniques in Agriculture. *International Journal of Innovative Research in Computer and Communication Engineering.*, *3*(2), 887–892.

Ghodeswar, U., Balagotra, T., Borale, A., Pawar, K., Parate, T., & Mahajan, S. (2022). Greeniot: A Smart IoT Gateway for Connected Farming. *IEEE 7th International conference for Convergence in Technology (I2CT)*. IEEE. 10.1109/I2CT54291.2022.9824907

Hatcher, W. G., Qian, C., Gao, W., Liang, F., Hua, K., & Yu, W. (2021). Towards Efficient and Intelligent Internet of Things Search Engine. *IEEE Access : Practical Innovations, Open Solutions*, *9*, 15778–15795. doi:10.1109/ACCESS.2021.3052759

Hossain, M. D., Kashem, M. A., & Mustary, S. (2023). IoT Based Smart Soil Fertilizer Monitoring and ML Based Crop Recommendation System. *International Conference on Electrical, Computer and Communication Engineering (ECCE)*, (pp. 1-6). IEEE. 10.1109/ECCE57851.2023.10100744

Ministry of Agriculture and Farmers Welfare. (n.d.). *Soil Health Card (SHC).* Soil Health. https://soilhealth.dac.gov.in/soil-health-map/MADHYA_PRADESH [Accessed: 26/3/2023]

Mishra, D., Khan, A., Tiwari, R., & Upadhay, S. (2018). *Automated Irrigation System-IoT Based Approach. 3rd International Conference on Internet of Things: Smart Innovation and Usages (IoT-SIU)*, Bhimtal, India. 10.1109/IoT-SIU.2018.8519886

National Food Security Mission (NFSM). (n.d.). *Paper on Rice*. Farmer. farmer.gov.in. https://farmer.gov.in/imagedefault/pestanddiseasescrops/rice.pdf (Accessed: 19/2/2023)

Pande, S. M., Ramesh, P. K., Anmol, A., Aishwarya, B. R., Rohilla, K., & Shaurya, K. (2021). *Crop Recommender System Using Machine Learning Approach. 5th International Conference on Computing Methodologies and Communication (ICCMC)*, Erode, India. 10.1109/ICCMC51019.2021.9418351

Population Division of the United Nations Department of Economic and Social Affairs. (2017). *World Population Prospects: The 2017 Revision*. UN. https://www.un.org/en/desa/world-population-projected-reach-98-billion-2050-and-112-billion-2100, (Accessed: 17/2/2021)

Sagarika, G. K., Prasad, S. K., & Kumar, S. M. (2020). Paddy Plant Disease Classification and Prediction Using Convolutional Neural Network. *International Conference on Recent Trends on Electronics, Information, Communication & Technology (RTEICT)*, Bangalore, India. 10.1109/RTEICT49044.2020.9315634

Saleh, A. R. B. M., Venkatasubramanian, S., Paul, N. R. R., Maulana, F. I., Effendy, F., & Sharma, D. K. (2022). Real-time Monitoring System in IoT for Achieving Sustainability in the Agricultural Field. *International Conference on Edge Computing and Applications (ICECAA)*, Tamilnadu, India. 10.1109/ICECAA55415.2022.9936103

Seru, L. K., Seru, & Gandham, S. M. (2022). *Crop recommendation system using machine learning algorithm.* Department of Computer Science and Engineering, School of computing, Sathyabama institute of science and technology.

Small Farmers Agri-Business Consortium (SFAC). (n.d.). Department of Agriculture & Farmers Welfare Ministry of Agriculture & Farmers Welfare. Government of India. Weather Forecast. https://enam.gov.in/web/weather_forecast

Technology Innovation Hub for IoT and IoE by IIT-Bombay. (n.d.). *Agriculture.* TIH. https://www.tih.iitb.ac.in/technology-verticals/agriculture/ [Accessed: 5/1/2023].

Zia, H., Rehman, A., Harris, N. R., Fatima, S., & Khurram, M. (2021). An Experimental Comparison of IoT-Based and Traditional Irrigation Scheduling on a Flood-Irrigated Subtropical Lemon Farm. *Sensors (Basel)*, *21*(12), 4175. doi:10.339021124175 PMID:34204584

Chapter 3
IoT-Based System for Crop Protection From Animals:
Towards Smart Farming

Siddhartha Kumar Arjaria
Rajkiya Engineering College, Banda, India

Aman Verma
Rajkiya Engineering College, Banda, India

Arvind Kumar
Rajkiya Engineering College, Banda, India

Ashwani Kumar Maurya
Rajkiya Engineering College, Banda, India

Praveen Kumar Panth
Rajkiya Engineering College, Banda, India

ABSTRACT

No one can imagine a world without agriculture. Due to overpopulation and deforestation, animal interference in residential areas is increasing day by day affecting human life and property, and causing human-animal conflict. At the present time, farmers in India and other countries are facing a serious threat from wildlife animals which causes less production in crop yielding. It is not possible to stay in the field for 24 hours to guard the crops against animals. In spite of good production, the crops on farms are many times ravaged by animals and it leads to huge losses for the farmers in terms of production, effort, quality, and money. To overcome this problem, an IoT-based crop protection system is being proposed in the present work. The system uses a microprocessor and a set of sensors to detect wild animals approaching near the field. The sensor data is analyzed through a machine learning algorithm and helps the microprocessor to take action. This system helps farmers to protect their crops from animals.

DOI: 10.4018/978-1-6684-8516-3.ch003

INTRODUCTION

More than 65% of people in India as well as many people in the world are living in rural areas which depend on agriculture. Even many urban companies are based on agricultural products. Being dependent on agriculture, animal husbandry is one of the most popular livelihoods in Rural areas. The people keep animals like cows, buffaloes, goats, hens, etc. for earning through various by-products like milk, eggs, wool, etc. from the animals. In rural areas, although these animals help in farming but more often these animals pose a problem for the farmer n crop security. The boundaries of the field are vast and open. The protection of crops is a challenge for the farmers. Great efforts are required from the farmers to make their crops secure. It becomes more difficult in the weather conditions like heavy rain, and peak summer and winter seasons. The efforts needed make the life of the farmer miserable. Despite taking great care, the crops get destroyed due to the frequent entry of animals like cows, nilgai, wild boar, deer, etc. into the fields due to which the farmers do not get even full return of their labor from crop yields.

In order to reduce the effort, labor, and damage of crops and to maximize the income of the farmer, The present work is proposed a technological solution to vigil the crops. With the help of technologies like data analytics & IoT data is collected, pre-processed and analyzed to take the decision. The IoT is being used in agriculture in a variety of ways. Precision farming, monitoring climate conditions, building agriculture drones, and remote sensing smart greenhouses are few examples where IoT is being used for farming. The use of technology reduces the effort of farmers and hence improve the quality of life of farmers and their family.

To vigil the crops 24*7, it is needed to capture the data and inform the farmer on any critical situation in real time. The sensors and camera can be used as data collection in real time and further processed using microprocessor to take effective decision on findings. The Internet of Things play a important role in this. The Internet of Things (IoT) refers to a group of physical objects that are equipped with sensors, software, and other advanced technologies to enable them to communicate and share data with other systems and devices over the Internet. At this time IoT become popular in various industries, agriculture, smart cities etc. Almost every aspect of life where automation is required, are using the capabilities of IoT. Some of the areas where IoT is being used are:

- In industrial automation, IoT is being used to monitor and control industrial equipment to increase efficiency at low cost.
- In the agriculture sector IoT is being used to monitor crop conditions and weather patterns for increased production in low cost and minimum effort.
- IoT is being used in smart homes and smart cities to monitor and control the various equipment and for enhanced security system.

The proposed system comprises cameras and sensors, microprocessor etc. The cameras and PIR sensors are installed in various positions in the field. These camera and sensors are being used to capture the suspicious entry of animals in field. The real time monitoring of the field is done through continuous collection of data and by analyzing the collected data in real time. The data is being used to train the Convolutional Neural Network algorithm to assign the data in one of two classes. The proposed system uses 11 layers of Convolutional Neural Network. If any entry of animals is detected by the system, the proposed system has two ways to secure the crop from the animals.

- To inform the farmers about the entry of animals into crops through alarms and mobile messages.
- To play a pre-recorded roaring sound of dangerous animals like lions etc. to create fear in the animals.

The reminder of the chapter is organized as follows. The section 2 will explore the background work and related work. The section 3 will demonstrate the proposed work. The section 4 includes implementation of algorithm. The section 5 will explore the result and section 6 discussion followed by the conclusion and future scope.

LITERATURE REVIEW

Agriculture is one of the most popular and important work for earning the livelihood all over the world. Agriculture is tedious and laborious task due to the uncertainty involved in the agriculture. The weather, natural disasters like fire and flood make the agriculture very uncertain in terms of production. Apart from it, some other problems like protection of crop from animals, fire and being theft is also posing challenges and make the life of farmer miserable and tough. In the present work a technological solution is being proposed for protection of crops from animals. The presented work proposed a IoT-based system for crop protection from animals. This section explores the related work done by different researchers in this field.

Kumar K.B. P. et al. builds a model to detect the interference using a PIR sensor. This PIR sensor sends the signal to Arduino which gives the signal to the APR9600 module. APR9600 module gives a signal to the speaker and this speaker produces the sound. It is not able to identify whether it is a human, animal, or any non-living object. PIR sensor only detects the motion, if it finds anything is in motion under the sensor area then it sends the signal and starts producing sound.

Karlapudy A. M. et al. has proposed a system that uses a PIR sensor for the detection of motion which send a signal to the raspberry pi. This raspberry pi sends a notification to farmers' phones and provides live streaming. The farmer can see it and provide electricity to the wire fencing. A square wave is generated through a step-up transformer which is connected to a 24-volt battery. Sometimes farmers may be busy and unable to see the notification. If a farmer does not see the notification, then electricity is not provided to the fence. These square waves give a shock through which the animals get away from the field. This shock does not kill the animal. This is very harmful to small animals, and birds. This small current can kill birds and small animals.

Sandeep T. et al. has built a system using Convolutional Neural Network. In this system, the image is taken from the environment and sent to the microprocessor which takes a decision. The trained model is uploaded to the microprocessor. Only test data is provided through the camera to the microprocessor and it gives the decision. According to the decision, it sends a notification to the farmer. It does not have any system that can shoo away the animal from cropland. It only provides information on whether animals are in crop fields or not.

Priyanka R. R. et al. has proposed a system that uses deep learning algorithms to classify and detect animals. The model is trained using Convolutional Neural Network and uploaded into a microcontroller. Based on the classification, it decides whether it is an animal or not. If animals are detected then the Arduino turns on the buzzer which shoos away the animals from the crop field. This model does not provide any information to the farmers about the animals entering their fields.

Srikanth N. et al. has designed a prototype using Arduino Uno, a PIR sensor, a GSM module, and a smoke sensor that can detect the motion of any object and also can detect fire. It provides information about any object entered into the crop field and if any reason crops caught fire, then system detects it. It is also connected to a motor so it can turn on the water supply whenever it detects a fire in a crop field. It sends information to farmers using the GSM module.

Kumar K. B. P. et al. makes a system using the PIR to detect the movement of the animal and send a signal to the Arduino. It diverts the animal by producing sound by using buzzer and the signal is transmitted to GSM which sends an SMS alert to the farmers. This system also used a smoke sensor to detect fire and a DC motor is used to generate the signal receivers of 3V LCD Information. It only detects the presence of any object not sure what type of object is in the field.

Mr. P. Venkateswara Rao et al. has proposed a system in which Arduino is connected to the buzzer and GSM module in the system that captures the object from the camera and then Arduino sends the electric signal to the buzzer to produce sound and simultaneously sends the message to the owner mobile from the GSM module.

N. Penchalaiah et al. has proposed a system in which a DHT11 is attached to Arduino and it senses the temperature of the environment and soil moisture to sense the humidity of the soil. PIR and IR are used to detect the animal if the given threshold value is exceeded then a message is sent to the forest department as well as the farmer immediately and they use LCD to show the readings of the sensor.

P. A. Harsha Vardhini et al. uses raspberry pi for both irrigation systems and detection by using a camera and sensors attached to the microprocessor. If the threshold value of the soil is decreasing then the relay starts the motor in the field that sprinkles the water across the field if the soil threshold value is reached then it automatically switches off the relay for the detection of animals. They use a 360-degree camera which is rotated by a servo motor if the animal is detected by the camera, then the buzzer is ON. All the data of the sensor are stored in a cloud from where they use the mobile application to see the parameters that come from the sensor. In this system, they have to turn ON the device at all times to send the data to the cloud due to this they consume power more and their camera is not enough capable enough to detect the object at night.

Vikhram B. et al. uses an ultrasonic sensor and PIR sensor to detect the presence of any object based on this. If they cross the given threshold APR board is ON and a sound is played to divert the animal. At night time a flashlight is ON to divert the animal. At the same time, the GSM module sends a message to the forest department and makes a call to the farmer, and the LCD is used to display the presence of an animal.

Korche M. et al. uses an LDR to measure the size of the animal. PIR sensors are used to detect the position of the animal. If the animal is detected then from APR board sent a signal is sent to the speaker to produce a sound. At night time, a flashlight is attached to it that is turned ON. The LCD is used to display the presence of the animal a GSM module is used to send a message to the farmer about the entry of the animal.

Navaneetha P. et al. uses the PIR and ultrasonic sensor to detect the presence of an animal in the field if the animal is detected the APR board is turned ON. The system plays a sound to divert the animal for nighttime. A flashlight is attached to it which turns ON when the animal is detected. Then a message to the forest department and a call to the farmer. To power the system, a solar panel is used. The regulated power supply is given to show the presence of animals. They use LCD to display LDR readings.

Sumana P. B. et al. uses a camera attached to Arduino to detect the animal. If the detected object is a domestic animal or bird then a message is only sent to the farmer. But if a wild animal is detected then messages are sent to both the forest officer and the farmer. For sending message system uses noden MCU with a Wi-Fi module. The system also produces the sound using a speaker module when the animal is detected and a laptop or PC, LCD is required to see the detected object from the Arduino.

Abadi M. et al has introduced TensorFlow for large-scale machine learning which is used for building models which can detect and process images in a heterogeneous environment. TensorFlow can also support a variety of applications which is trained on deep neural networks.

David R. et al has demonstrated the TensorFlow lite micro model which is used in embedded system. Embedded systems have a limited amount of hardware in which they perform a specific task. So, they used a unique interpreter-based approach that can work with limited requirements.

Sudheer C. et al. has proposed that smart irrigation systems have proven to be effective in addressing crop damage caused by animals and water resource scarcity in the agricultural sector. The systems are designed using microcontroller such as Arduino, which are connected to sensors like infrared sensors and soil moisture sensors. The sensors help to detect animals in the field and determine when to irrigate the fields. The use of GSM modules allows farmers to monitor the fields from any location at any time, helping to increase efficiency and productivity in the agricultural sector.

Redij J. et al. the use of Arduino Uno and PIR sensors for crop protection and animal detection is a promising approach that has the potential to revolutionize the way farming is done. PIR sensor detects the motion and turns on the camera. This image is processed by a microcontroller. If an animal is detected then it turns on the buzzer and the image is sent to the farmer.

Nanda I. et al. have conducted a comprehensive study on the implementation of a smart irrigation system using Industrial IoT technology for crop protection and irrigation. The system includes DHT11 humidity and temperature sensors, PIR sensors, LDR sensors, and HC-SR04 ultrasonic sensors, all connected to an ARM Cortex-A. These IoT components are useful for agricultural purposes and can be employed for client engagement. The PIR sensors detect movement within a range of 10 meters, and when activated, the primary camera takes a picture and encrypts it using the ARM Cortex-A. The recognized data is then transmitted to the Arduino.

Shashi Kiran V. et al. has developed a device for crop monitoring in agricultural fields, farms, wetlands, and forests using a motion sensor to detect the presence of animals and birds. The PIR sensor immediately signals the Arduino board to take action when an animal or bird enters the farm area. The device comprises of a power supply section, PIR sensor, Arduino, buzzer, and LED, with each unit designed and tested separately. At night, the LED will turn on to detect any animals in the area. This system is using an Embedded PIC Microcontroller with a PIR sensor, PIEZO Buzzer, 12V Adapter, and LED. Whenever an animal attack crops in the agriculture field, the device detects the sound produced by the buzzer and generates a signal to divert the animal. This device is based on a motion-detecting sensor and is designed specifically for crop monitoring.

Munje R. et al. has made a system for intruders to come into the crop field then the PIR sensor is responsible for detecting any motion made by animals or humans that attempt to enter the farm field. Once motion is detected, the sensor sends a signal to the microcontroller, which immediately activates the system. Passive infrared sensors, in particular, use a pair of pyroelectric sensors to sense the heat energy in the environment. When infrared radiation from a human body is detected, the sensor sends a signal to the Arduino microcontroller to activate the alarm system, which includes a speaker. To record the sound of any animal or human, a sound recording module can be used for up to 20 seconds. The microcontroller, which is powered by the power supply module, serves as the main element of the system. A relay unit is used for voltage control. To detect the presence of birds in all directions, an ultrasonic sensor mounted on a servo motor rotates at 360 degrees. Once the ultrasonic sensor detects the presence of birds, it sends a signal to the controller. The signal is then displayed on the display. To notify farmers when movement is detected, a GSM module is used to make calls. The farmer is alerted when animals or birds attempt to enter the farm.

Giordano S. et al. have developed an IOT application for crop protection. To vigil the crop, it uses low pan, wifi, and Zigbee with IoT gateway. An ultrasonic sound device is used to work in partial or total darkness. Using solar panels and a lipo battery to increase the efficiency of the device, system uses a PIR sensor and uses RIOT software.

Iniyaa K. K. et al. have proposed a deep learning-based solution for crop protection. The system protects crops from animals. In their system cameras was used for detecting animals and sound devices are used to divert the animals.

Balakrishna K. et al. proposed a model in which the system has a raspberry pi 4 which takes images through a Pi camera. The detection is performed by RCNN and sends a signal to ESP8266 which is connected to the firebase. Firebase sends a signal to the application. If animals are detected then it turns on the buzzer and LED light simultaneously.

PROPOSED SYSTEM

Agriculture plays a very important role in a country's development. The quality of life of the citizen can be accessed through the quality and quantity of food they consume. In the presented work the issue related to the security of crops from animals is being addressed. The farmers have to be vigilant in their fields during cultivation to protect their crops from the animals. The anna animals pose a big challenge to the farmer. The animals enter the fields and destroy the crop severely. The manual security of crops is a big challenge in terms of effort. The farmer has to be alert throughout the day.

The present section explores the architecture and working of the proposed Internet of Things-based system for crop protection from animals that will help to protect the crop from animals with minimum effort in order to make the farmer's life easy.

Flow Diagram of Proposed System

The flow diagram of the proposed system is demonstrated in Figure 1. The proposed system has Passive Infrared Sensor (PIR) used to detect any object which interferes with the area of the PIR sensor. The PIR sensor detects people within 10m of the sensor. The power supply is typically up to 5V. A Passive Infrared sensor (PIR) is an electronic device that is used to detect the presence of any object by measuring infrared (IR) radiation disturbance of emitted radiation with help of a sensor within its field of view and a corresponding electrical signal is sent to the microprocessor.

If interference occurs in the sensor, then it triggers the camera to capture the image of the object which enters the crop field and this image is processed in the microprocessor to detect which object is entered in the field.

The OV5647 5MP 1080P IR-Cut Camera is connected to the microprocessor. It has an automatic day-night switching mode this camera support night vision. It needs 3.3V power which is provided by the microprocessor. It is easy to attach. This Raspberry Pi IR-Cut camera can capture much higher quality images of day and night with resolutions up to 1080p.

Figure 1. Flow diagram of proposed system

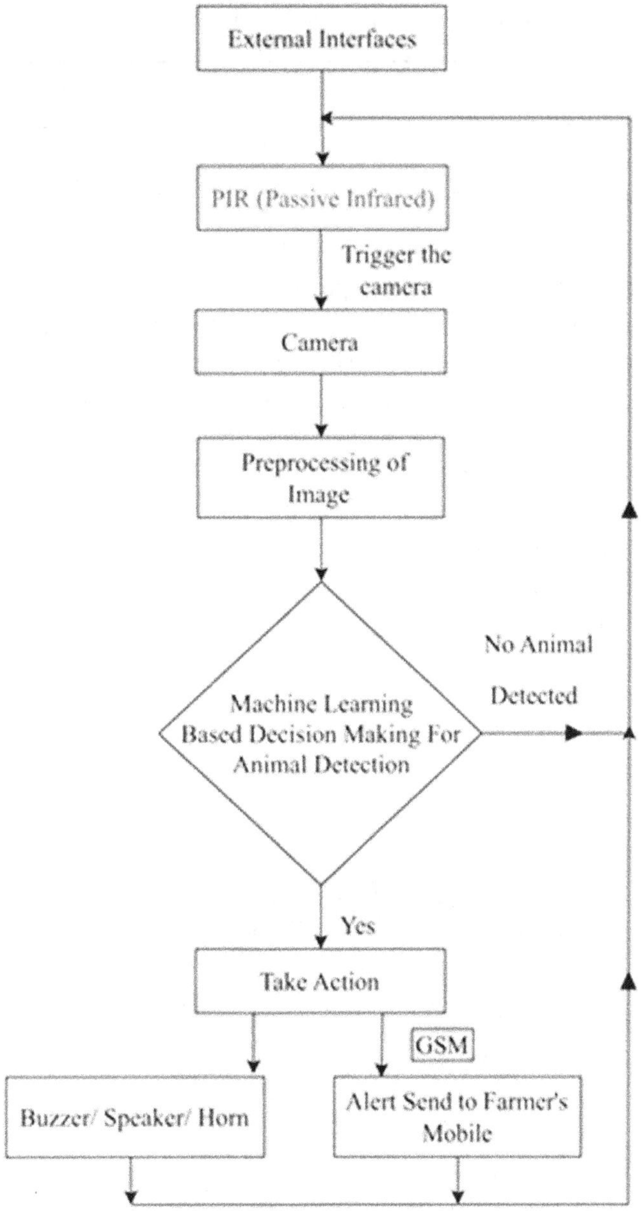

After getting the signal from PIR, it started capturing the image and sends it to the microprocessor for detection. Then the captured image is pre-processed by fixing the dimension of the image. It helps to reduce the complexity of the image and remove unwanted noisy data. This pre-processed image is arranged in a queue and sent to the pre-trained model.

For training and testing the model, the dataset is used which is taken from an open-source platform named Kaggle given in Table 1.

Table 1. Volume of dataset

Image	No. of Images	No. of Training Images	No. of Testing Images	Image Orientation	Purpose
Cow	100	80	20	Front, Back, Left, Right, Group	Detection, Classification
People	100	80	20	Front, Back, Left, Right, Group	Detection, Classification

The pre-processed image sends to the microprocessor (Raspberry Pi). Raspberry Pi is a series of small single-board computers (SBCs). It works on a 5V DC current which is provided by a 5V DC adapter. Due to its low cost, modularity, and open design, it is widely used in many fields. It accepts both HDMI and USB devices. Raspberry Pi is a type of microprocessor that stores a pre-trained model that is working for processing the input data taken as input from sensors, and cameras. This pre-trained model takes a decision on the basis of the class system sending the signal to the buzzer or speaker, and GSM. A palm size fan is used to cool the microprocessor from heating. It works with 5V Direct Current (DC) voltage, and the main purpose of this is to exhaust the hot air. The pre-processed image is further sent to the pre-trained model which is stored in the microprocessor. This model uses Convolutional Neural Network to process the data.

Figure 2. Circuit diagram of system

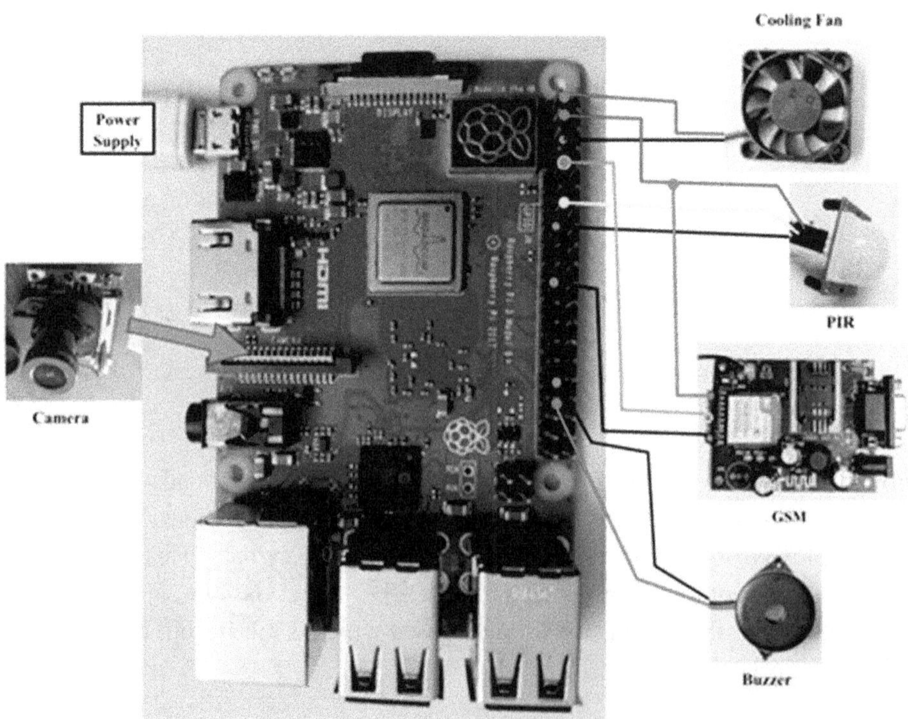

Based on the predictions, the images are given a class. If the provided class is not dangerous for the crop, then it stops the further process of the system and sends a signal to the PIR sensor to run continuously. If the provided class is dangerous for crop fields, then the signal is sent by the Raspberry Pi to the Global System for Mobile Communication.

GSM is used to send notifications to the farmer. The baud rate of the GSM SIM800C modem with antenna module can be set by AT command from 9600 to 115200. GSM GPRS modems have an internal TCP/IP stack and can connect to the Internet via GPRS. It is suitable for SMS, voice, and data transmission applications over M2M interfaces. With simple AT commands, you can make voice calls, compose SMS, read SMS, make incoming calls, join the Internet, etc. This works with a system for sending SMS and calls to the farmer whenever the animals are detected by the model.

The microprocessor also sends the signal to the buzzer or speaker. A loudspeaker is an electro-acoustic transducer, a device that converts an electrical audio signal into an equal sound. It produces an irritating sound to create fear in animals. So that animals will leave the crop field. The connection of all the components with the microprocessor is shown in Figure 2.

IMPLEMENTATION OF ALGORITHM FOR ANIMAL DETECTION

The role of the IoT-based system is to detect the entry of animals into fields and inform the farmer to take the appropriate action on time, to minimize the losses. The system maximizes profit by minimizing efforts and losses. This part demonstrates the implementation of the algorithm in the system. How different modules of the system connect and operates. Some of the steps for operating this system are given below.

Step 1: Any object motion detected in PIR
Step 2: System turns on the camera
Step 3: Image/video collection from the camera
Step 4: The system read the image and then pre-processing image (224*224 pixels)
Step 5: Classification of image in class 0(entry of animal) or class 1
Step 6: If class 0 is provided then system turns on the buzzer and sends a message to the farmer's mobile phone
Step 7: If class 1 then go to step 1.

As shown in Figure 3. the PIR sensor is connected to a microprocessor that is used to detect the presence of an object moving in the field then it triggers the camera. A camera installed on the boundaries of the field is used to take an image or video of field boundaries. The camera is capable to work in day and night both times. The camera is 1080 megapixels that cover approx. 20m distance from the field.

The image is sent to the microprocessor. The microprocessor has an uploaded model which takes an image and pre-processes it into 224 * 224 pixels to send to CNN layers which extracted the feature of an image. This extracted image sends to the Convolutional neural network which makes predictions. Based on prediction it provides the class and label of the image.

If the class 0(animal) and label are given then an animal is present in the field. Now the microprocessor plays a sound through the speaker at the same time a message is sent to the farmer using the GSM module. If class 1(person) then goes to step 1 and there is no alarm or notification sent to the farmer.

Figure 3. Detection of animal

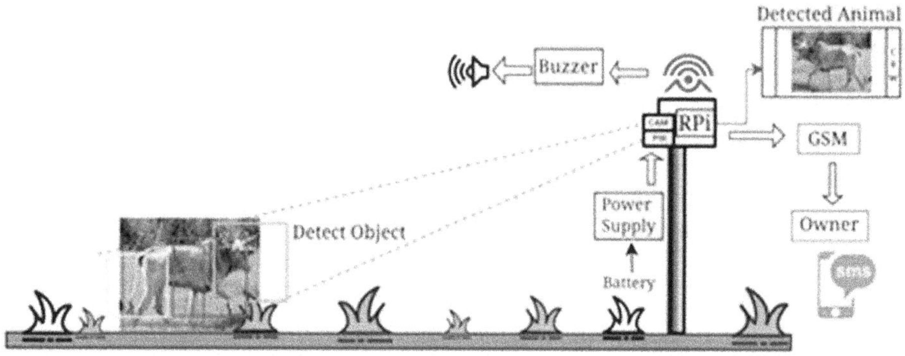

Convolutional Neural Network

The proposed system uses Convolutional Neural Network as classifier. Convolutional Neural Network (CNN) is popular method to capture the inherent properties of input images and classify it correctly. CNNs are mainly suitable for discovery of patterns in images to identify objects. They can also be useful for classifying audio, time-series, and signal data. CNNs can perform both the feature extraction and the classification, resulting in a highly organized and feature-dependent model output. The proposed work implements the CNN model as summarizes in Figure 4. The figure 4 describes the number and type of layers used in Convolutional Neural Network.

Significance of The CNN and its Respective Layers

- **Convolutional Layers:** These layers perform the most important operation in a CNN, which is convolution. Convolution involves applying a set of filters to the input image to extract features, such as edges, corners, and textures. The output of the convolutional layer is a set of feature maps that capture different aspects of the input image. By stacking multiple convolutional layers, a Convolutional Neural Network can learn increasingly complex and abstract features. The given formula is used to calculate the learnable parameters.

$$((s * k * d) + 1) * p$$

Where s is width of filter, k is height of filter, d is number of filters in previous layer, and p is number of filters in the current layer.

- **Pooling Layers:** These layers reduce the size of the feature maps by down-sampling them. This helps to reduce the computational cost of the network and prevent overfitting. The most common type of pooling is max pooling, which selects the maximum value from each sub-region of the feature map.
- **Flatten Layers:** This layer flattens the multi-dimensional output from previous layer which is calculated by the equation:

Figure 4. Summary of CNN model

```
Model: "sequential"
_________________________________________________________________
 Layer (type)                Output Shape              Param #
=================================================================
 conv2d (Conv2D)             (None, 222, 222, 32)      896

 max_pooling2d (MaxPooling2D  (None, 111, 111, 32)     0
 )

 conv2d_1 (Conv2D)           (None, 109, 109, 64)      18496

 max_pooling2d_1 (MaxPooling  (None, 54, 54, 64)       0
 2D)

 conv2d_2 (Conv2D)           (None, 52, 52, 128)       73856

 max_pooling2d_2 (MaxPooling  (None, 26, 26, 128)      0
 2D)

 conv2d_3 (Conv2D)           (None, 24, 24, 128)       147584

 max_pooling2d_3 (MaxPooling  (None, 12, 12, 128)      0
 2D)

 flatten (Flatten)           (None, 18432)             0

 dense (Dense)               (None, 512)               9437696

 dense_1 (Dense)             (None, 3)                 1539

=================================================================
Total params: 9,680,067
Trainable params: 9,680,067
Non-trainable params: 0
_________________________________________________________________
```

Number of Neurons = w * h * f

Where w is width, h is height and f are filter size of previous layer.

- **Fully Connected Layers:** These layers are used at the end of the network to perform classification or regression. The output of the last convolutional layer is flattened into a vector and fed into one or more fully connected layers, which perform a linear transformation and apply a non-linear activation function. The output of the final fully connected layer is the prediction of the network. The number of learnable is calculated by the equation:

No. of learnable parameters in dense layer = Output size * Input size + 1

RESULT

The presented work provides a technological solution for the security of crops using IoT. The present system achieves an average accuracy of 87%. It helps farmers to protect their crops with minimal vigil in crop fields 24 *7. This system reduces labor and also does not harm the animals. The Table 2 presents a comparison of proposed system with some other algorithms.

Table 2. Algorithm Performance Comparison with State-Of-The-Art

Author Name	Paper Name	Accuracy (%)
Proposed model	Proposed work	87.91
K. Balakrishna et al.	Application of IoT and machine learning in crop protection against animal intrusion	85.22
S. Giordano et al.	IoT detection against wild animal's attacks	84.51
Iniyaa K. K. et al.	Crop protection from animals using deep learning	77.45

CONCLUSION AND FUTURE SCOPE

The present work proposes an IoT-based system for crop protection. The main objective of the work is to minimize the losses incurred during agriculture and maximizing the production and profit. It also helps to improve the quality of life of farmers and their family members. The sensors and cameras are being used for detecting and capturing the image. A deep learning-based classification technique is being used in the microprocessor memory to classify the captured image. If the system detects the animal entry, it sends an alert signal to the farmer's mobile and also starts to sound in order to prevent the entry of animals into the field. This IoT-based system achieves an average accuracy of 87%. The present system can further be enhanced by training the model with more different animal samples. The functionality of proposed system can further be enhanced by adding more sensors like smoke sensors, temperature sensors, soil moisture sensors, and ph. sensors, etc. to monitor the quality of soil and water needed.

REFERENCES

Abadi, M., Barham, P., Chen, J., Chen, Z., Davis, A., Dean, J., Devin, M., Ghemawat, S., Irving, G., Isard, M., Kudlur, M., Levenberg, J., Monga, R., Moore, S., Murray, D. G., Steiner, B., Tucker, P., Vasudevan, V., Warden, P., & Zheng, X. (2016). TensorFlow: A system for large-scale machine learning. In *Proceedings of the 12th USENIX Symposium on Operating Systems Design and Implementation (OSDI 16) Savannah*, GA, USA. 10.48550/arXiv.1605.08695

Arjaria, S., Sahu, R., Agrawal, S., Khare, S., Agarwal, Y., & Chaubey, G. (2021). Hand Gesture Identification System Using Convolutional Neural Networks. In *Proceedings of the International Conference on Artificial Intelligence and Machine Vision (AIMV)*, Gandhinagar, India. 10.1109/AIMV53313.2021.9670906

Balakrishna, K., Mohammed, F., Ullas, C. R., Hema, C. M., & Sonakshi, S. K. (2021). Application of IoT and machine learning in crop protection against animal intrusion, (pp. 221–238). Machine Learning Techniques for Sustainable Agriculture. doi:10.4018/978-1-7998-7488-6.ch010

David, R., Duke, J., Jain, A., Reddi, V. J., Jeffries, N., Li, J., Kreeger, N., Nappier, I., Natraj, M., Regev, S., Rhodes, R., Wang, T., & Warden, P. (n.d.). Tensorflow lite micro: Embedded machine learning on TinyML systems. https://doi.org//arXiv.2010.08678 doi:10.48550

Giordano, S., Seitanidis, I., Ojo, M., Adami, D., & Vignoli, F. (2018). IoT solutions for crop protection against wild animal attacks. In *2018 IEEE International Conference on Environmental Engineering (EE)* (pp. 1-5). IEEE. 10.1109/EE1.2018.8385275

Harsha Vardhini, P. A., Koteswaramma, N., & Murali Chandra Babu, K. (2019). IoT-based Raspberry Pi Crop Vandalism Prevention system. [IJITEE]. *International Journal of Innovative Technology and Exploring Engineering*, *9*(1), 3188–3192. doi:10.35940/ijitee.A9156.119119

Iniyaa, K. K., Divya, J. K., Devdharshini, S., & Sangeethapriya, R. (2021). Crop protection from animals using deep learning. *International Journal of Progressive Research in Science and Engineering*, *2*(3), 41–44. https://journals.grdpublications.com/index.php/ijprse/article/view/240

Karlapudy, A. M., Kommasani, R. R., Malla, K. B., & Vasam, S. P. (2019). IoT-based agricultural fields protection from animals. [IJRASET]. *International Journal for Research in Applied Science and Engineering Technology*, *7*(5), 3086–3090. doi:10.22214/ijraset.2019.5509

Korche, M., Tokse, S., Shirbhate, S., Thakre, V., & Jolhe, S. P. (2021). Smart Crop Protection System. [IJLES]. *International Journal of Latest Engineering Science*, *4*(4), 101–106. https://www.ijlesjournal.org/2021/volume-4%20issue-4/ijles-v4i4p101.pdf

Kumar, K. B. P., Bhavitha, T., Karishma, S., Pavithra, M., & Kumar, M. P. (2019). IoT-based crop monitoring from animals. *Journal of Emerging Technologies and Innovative Research*, *6*(4), 173–177. https://www.jetir.org/view?paper=JETIRAK06049. doi:10.24247/jetir.6.4.30

Munje, R. K., Warungase, P., Sarode, M., Lohar, V., & Mehare, A. (2022). Developing a Prototype of Smart Crop Protection System from Animals. In A. Pandian & V. Balaji (Eds.), *Intelligent Agriculture: The Role of IoT*. Big Data and AI. doi:10.5281/zenodo.6874633

Nanda, I., Chadalavada, S., Swathi, M., & Khatua, L. (2021). Implementation of IIoT based smart crop protection and irrigation system. In A. Pandian & V. Balaji (Eds.), *Intelligent Agriculture: The Role of IoT* (pp. 1–13). Big Data and AI. doi:10.4018/IJIA.2021010101

Navaneetha, P., Ramiya Devi, R., Vennila, S., & Manikandan, P. (2020). IoT-Based Crop Protection System against Birds and Wild Animal Attacks. *International Journal of innovative research in Technology (IJIRT).* http://www.ijirt.org/master/publishedpaper/IJIRT149166_PAPER.pdf

Penchalaiah, N., Pavithra, D., Bhargavi, P., & Madhuri, D. P. (2020). Smart irrigation and crop protection from wild animals. *Journal of Engineering Sciences (JES).* https://jespublication.com/upload/2020-1104126.pdf

Priyanka, R. R., Mahesh, M., Pallavi, S. S., Jayapala, G., & Pooja, M. R. (2018). Crop protection by an alert-based system using deep learning concept. International Journal of Scientific Research in Computer Science. *Engineering and Information Technology*, *6*(6), 33–38. doi:10.26438/ijsrcse/v6i6.4749

Raspberry Pi Foundation. (n.d.). Raspberry Pi. https://www.raspberrypi.org/

Reddy, M. C. Mohan, Raju, K., Kodi, K., Anapalli, B., & Pulla, M. (2020). Smart Crop Protection System from Living Objects and Fire Using Arduino. Science. *Technology and Development Journal*, *23*(2), 287–292. doi:10.32508tdj.v23i2.1081

Redij, J., Shitap, P., Singh, S., Zagade, D., & Chougule, S. (2022). Smart Crop Protection System from Animals. In A. Pandian & V. Balaji (Eds.), *Intelligent Agriculture: The Role of IoT, Big Data and AI.* IJCRT. https://www.ijcrt.org/papers/IJCRTO020033.pdf

Sandeep, T., Manushree, B., Rahul, S., & Bharath, T. (2022). Crop protection from animals using CNN. *International Research Journal of Modernization in Engineering Technology and Science, 1*(1), 12-18. https://www.irjmets.com/uploadedfiles/paper/issue_3_march_2022/19639/final/fin_irjmets1646900628.pdf

Shashi, V., Manoj, N., Hemanth, M., Namith, M. N., & Surekha, M. (2022). Smart Crop Protection System from Birds and Animals. In A. Pandian & V. Balaji (Eds.), *Intelligent Agriculture: The Role of IoT, Big Data and AI.* IJRTC. https://ijcrt.org/papers/IJCRT2207476.pdf

Srikanth, N., Aishwarya, K., Kavita, H. M., Rashmi Reddy, K., & Soumya, D. B. (2019). Smart crop protection system from animals and fire using Arduino. *International Journal of Electrical, Electronics, and Computer Engineering.* https://www.technoarete.org/common_abstract/pdf/IJERECE/v6/i4/Ext_90718.pdf

Sudheer, C. L., Kaveri, K., Kavya, M., Nath, G. K., Javeed, P. M., & Kumar, B. N. (2023). Crop Protection and Monitoring from Animal Attacks by Using IoT Solutions. In A. Pandian & V. Balaji (Eds.), *Intelligent Agriculture: The Role of IoT, Big Data and AI.* IJCRT. https://www.ijcrt.org/papers/IJCRT2303875.pdf

Sumana, P. B., Sanjana, R., Sharanya, M., & Harish, N. J. (2021). Smart crop protection system from wild animals and birds using IoT. *International Journal of Advanced Research, Ideas and Innovations in Technology (IJARIIT), 7*(2), 2232-2237. https://www.ijariit.com/manuscripts/v7i4/V7I4-1724.pdf

Vikhram, B., Revathi, B., Shanmugapriya, R., Sowmiya, S., & Pragadeeswaran, G. (2017). Animal detection system in farm areas. [IJARCCE]. *International Journal of Advanced Research in Computer and Communication Engineering*, *6*(1), 1–5. https://ijarcce.com/upload/2017/march-17/IJARCCE%20137.pdf

World Bank. (n.d.). [Data set]. World Bank. https://data.worldbank.org/

Chapter 4
Deep Learning-Based Agriculture Monitoring and Forecasting

Ishaan Mehta
https://orcid.org/0009-0002-8215-550X
Pandit Deendayal Energy University, India

Santosh Kumar Bharti
https://orcid.org/0000-0002-0627-6433
Pandit Deendayal Energy University, India

Honey Mehta
Pandit Deendayal Energy University, India

Rajeev Kumar Gupta
https://orcid.org/0000-0002-5317-9919
Pandit Deendayal Energy University, India

ABSTRACT

Smart agriculture has become crucial in meeting the increasing dietary needs of a growing population, particularly in countries where agriculture has significant economic impact. Deep learning techniques, such as convolutional neural networks and recurrent neural networks, have been extensively researched and applied in agriculture in recent years. In this study, recent research articles on deep learning in agriculture over the past five years are analyzed to identify key contributions and challenges. The study has also explored agriculture parameters monitored by the internet of things and used them to train the deep learning algorithms for analysis. The study compares various factors across different studies, including the agriculture area of focus, dataset used, deep learning model and framework, data preprocessing and augmentation methods, and accuracy of results.

DOI: 10.4018/978-1-6684-8516-3.ch004

INTRODUCTION

Deep learning-based agriculture monitoring and forecasting is a field of study that uses deep learning (DL) techniques to analyze and predict agricultural patterns and outcomes. This technology has potential to revolutionize the way we approach farming by providing farmers with real-time data and insights into crop health, yield, and soil moisture levels (Muhammad et al., 2021). The DL algorithms are using large data to identify patterns and make predictions. In agriculture, this can include data such as satellite imagery, weather patterns, soil moisture levels, and crop health metrics. By analyzing these data, deep learning models can provide the insights that can help farmers to make informed decisions about planting, harvesting, and crop management (Garg and Rani, 2020).

The deep learning applications in agriculture includes namely, crop yield prediction, disease detection, and irrigation management, etc. By predicting crop yields, farmers can adjust planting and harvesting schedules to maximize their output. Disease detection can help farmers to identify and treat crop diseases before they become widespread, saving time and resources. And by monitoring soil moisture levels as well as weather patterns, farmers can optimize the irrigation systems to conserve water and increase crop yields (Muhammad et al. 2021). Overall, the deep learning-based agriculture monitoring and forecasting has potential to revolutionize the agriculture industry by providing real-time data and insights that can help them to make decisions with improve crop yields (Sladojevic et al., 2020).

Figure 1. Challenges of smart agriculture
https://www.researchgate.net/figure/Selected-Challenges-of-Smart-Agriculture_fig1_346358538

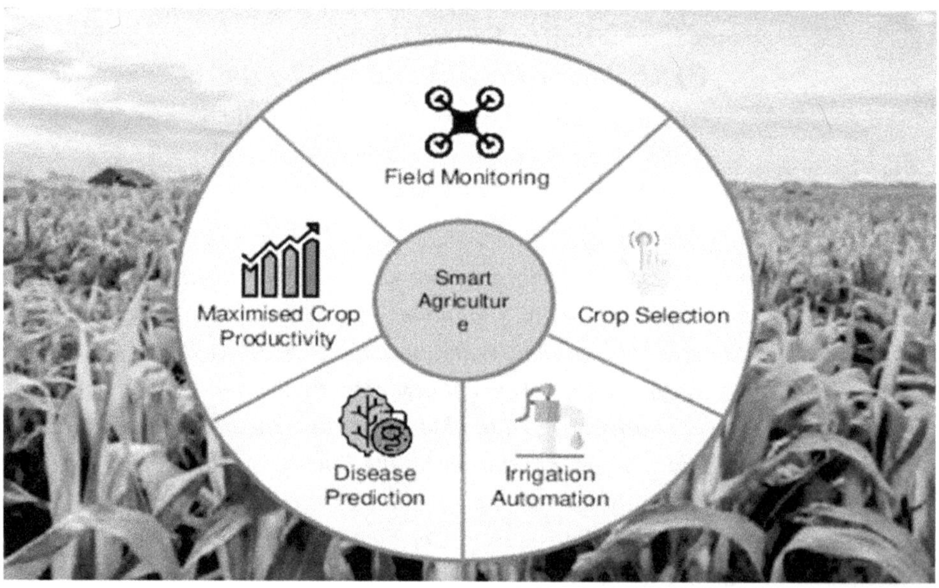

Smart agriculture having several challenges as shown Figure 1. Thus, there is a need of smart agriculture monitoring and forecasting system that can help farmers in following ways:

Increasing food production: The global population is expected to reach 9.7 billion by 2050, so food production needs to increase by 70% to meet the necessity. Agriculture monitoring and forecasting can help farmers optimize their yields and increase production, thereby contributing to food security (Sladojevic et al., 2020) Resource optimization: Agriculture monitoring and forecasting can help farmers optimize use of resources like water, fertilizers, and pesticides. By providing real-time data on soil moisture levels and weather patterns, farmers can adjust their irrigation systems as well as reduce water usage. Similarly, by using predictive analytics to identify potential crop diseases, farmers can reduce the amount of pesticides they use, which can have a positive impact on the environment (Fountas et al., 2020).

Cost reduction: Agriculture monitoring and forecasting can help farmers reduce costs by optimizing their use of resources and improving their crop yields. By using real-time data to make informed decisions, farmers can reduce waste and increase their profits (Muhammad et al., 2021).

Risk mitigation: Agriculture is a high-risk industry, with weather patterns, pests, and diseases all posing a threat to crops. Agriculture monitoring and forecasting can help farmers mitigate these risks by providing early warning systems for weather events and identifying potential crop diseases before they become widespread (Boulent, et al. 2019).

Sustainable agriculture: Agriculture monitoring and forecasting can help farmers adopt sustainable practices by reducing their reliance on chemical inputs and optimizing the use of resources. By adopting sustainable practices, farmers can reduce their environmental impact and ensure the long-term viability of their operations (Fountas et al., 2020; Boulent et al. 2019).

Agriculture monitoring and forecasting is crucial for adapting to climate change. As weather patterns become more unpredictable, farmers need to have access to real-time data to make informed decisions about planting, irrigation, harvesting (Wang et al., 2019). By using predictive analytics to identify potential risks, farmers can take proactive measures to protect their crops and reduce losses (Mohanty, 2019).

Agriculture monitoring and forecasting plays significant role in improving livelihoods of farmers, particularly in developing countries. By providing farmers with access to real-time data and insights, they can make better-informed decisions about their crops and improve their yields. This can help to lift them out of poverty and improve their overall quality of life (Mohanty, 2019).

Finally, agriculture monitoring and forecasting is essential for ensuring global food security. With the world's population continuing to grow, it is essential that we find ways to produce more food with limited resources. Agriculture monitoring and forecasting can help farmers to achieve higher yields and increase food production, which will be critical in meeting the growing demand for food in the future.

Motivation

Agriculture is the important field for the development and prosperity of country. As every passing day, the need of crops are increasing and on the other hand land area is decreasing due to increase in population. For the accurate forecasting, it is the necessity to keep all the parameters with us (Fountas et al., 2020). This might be done by humans but, it is not as accurate as deep learning techniques. It is very hard for humans to process so much data at a time with high accuracy (Ma et al., 2018).

Therefore, there is a need of deep learning based Agriculture Monitoring and Forecasting system as shown in Figure 2.

Figure 2. Deep learning for agriculture
https://pyimagesearch.com/2022/08/10/computer-vision-and-deep-learning-for-agriculture/

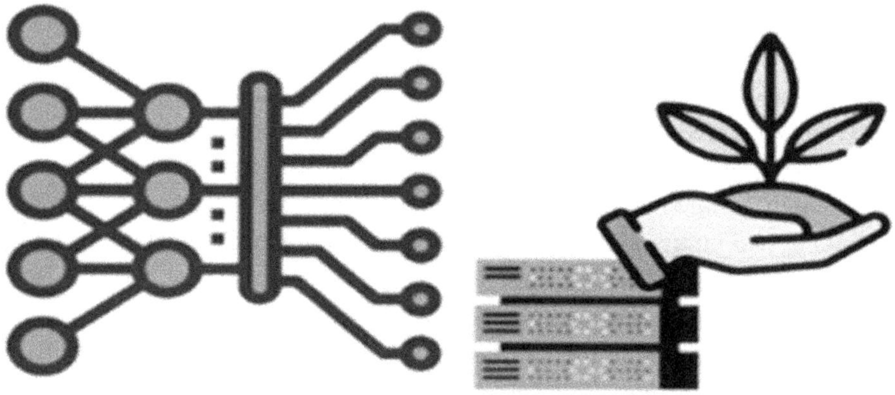

Deep learning models can process large amounts of data quickly and efficiently, providing real-time insights into crop health, soil moisture, weather patterns, and other factors that impact agriculture. Humans would not be able to analyze such a large amount of data as quickly and accurately as deep learning models can do. Deep learning models can provide accurate and consistent analyses of data, reducing the risk of human error and bias (Mohanty, 2019). Human-based approaches can be prone to errors and inconsistent, particularly when dealing with large datasets. Deep learning models can analyze vast amounts of data from various sources, such as satellite imagery, weather data, and soil sensors, providing a comprehensive view of agriculture conditions. Humans would not be able to cover such a large area and analyze data from multiple sources simultaneously (Liu, et al., 2021). Deep learning models can adapt to changing environmental conditions and incorporate new data sources as they become available, providing up-to-date information that can help farmers make informed decisions. Human-based approaches may not be as adaptable to changing conditions. Deep learning-based approaches can be more cost-effective than human-based approaches, particularly when dealing with large-scale agriculture operations. By reducing the need for manual labor and improving efficiency, deep learning can help farmers reduce their costs and increase profits (Muhammad et al., 2021).

It can operate around the clock without fatigue, making them suitable for continuous monitoring of crop conditions. Human-based approaches are limited by the availability and endurance of workers, which can lead to gaps in monitoring and delays in decision-making. Furthermore, deep learning-based agriculture monitoring and forecasting can also identify patterns and relationships in data that may not be apparent to humans.(Mohanty, Hughes, Salathé 2019) The models can recognize complex patterns and relationships between different variables, such as soil moisture, temperature, and humidity, which may not be easily discernible by humans. This can lead to more accurate predictions and insights, ultimately resulting in better decision-making for farmers (Sladojevic et al. 2020). While human-based approaches to agriculture monitoring and forecasting are still valuable. However, deep learning-based approaches can provide significant advantages in terms of speed, accuracy, consistency, adaptability, and cost-effectiveness, ultimately leading to more efficient and sustainable agriculture practices (Liu et al., 2021).

There have been several studies that have compared the effectiveness of deep learning-based agriculture monitoring and forecasting with human-based approaches, and some of these studies have reported positive results for deep learning models (Garg and Rani, 2020). A study published in the journal Computers and Electronics in Agriculture compared the performance of deep learning models and human experts in identifying plant diseases from images (Wang, et al. 2019). The results showed that the deep learning models achieved higher accuracy than the human experts in identifying and classifying plant diseases. Santos (2020), compared the effectiveness of deep learning models and human experts in predicting soybean yields. The results showed that the deep learning models outperformed the human experts in predicting soybean yields, particularly in years with extreme weather conditions (Santos, et al. 2020). Mohanty (2019), compared the effectiveness of deep learning models and human experts in predicting crop yields in China. The results showed that the deep learning models achieved higher accuracy than the human experts in predicting crop yields, particularly in regions with complex terrain and weather patterns. These studies suggest that deep learning-based agriculture monitoring and forecasting can provide significant advantages over human-based approaches, particularly in terms of accuracy and efficiency (Mohanty, 2019; 2018).

LITERATURE SURVEY

In this section, we have discussed the past study of the domain of agriculture monitoring and forecasting in various aspect namely, economic importance, crop classification, different detection methods, etc. At the end, we have presented a table 1 for State-of-the-art studies on deep learning-based agriculture monitoring.

Economic importance of Agriculture Monitoring and Forecasting

Food Security: Agriculture monitoring and forecasting are essential for food security, especially in developing countries where agriculture is the primary source of income and livelihood for a significant portion of the population (Mohanty, 2018). Timely and accurate monitoring and forecasting of crop production help policymakers and stakeholders to make informed decisions on food import and export, ensure food availability, and prevent food price spikes (Ma, et al. 2018).

Risk Management: Monitoring and forecasting help farmers, governments, and agribusinesses manage risks associated with weather, pests, and diseases, which can significantly impact crop production and yield (Muhammad, et al. 2021). Timely and accurate forecasting allows for the implementation of appropriate risk management strategies, such as the use of drought-resistant crop varieties, integrated pest management, and crop insurance (Sladojevic et al. 2020).

Market Analysis: Agriculture monitoring and forecasting provide insights into the supply and demand dynamics of the agricultural market (Boulent, et al. 2019). This information helps farmers and agribusinesses make informed decisions on crop pricing, marketing, and distribution. Additionally, monitoring and forecasting can help policymakers identify and mitigate market inefficiencies, such as monopolies, that negatively impact the agricultural sector's overall economic performance (Ma, et al. 2018).

Policy Formulation: Agriculture monitoring and forecasting play a crucial role in policy formulation, especially regarding agricultural production and land use. Policy decisions based on timely and accurate monitoring and forecasting of crop yield and production can help optimize resource allocation, promote sustainable agriculture, and improve overall economic growth (Santos, et al. 2020).

Investment and Financing: Agriculture monitoring and forecasting can also attract investment and financing to the agricultural sector. Timely and accurate information on crop yield and production can help investors and financiers assess the potential profitability and risk associated with investing in agriculture (Wang, et al. 2019). This can encourage investment and financing for new agricultural projects, such as irrigation systems, seed production, and agro-processing (Ma, et al. 2018).

Climate Change Mitigation: Agriculture monitoring and forecasting can help mitigate impact of climate change on agriculture. With increasing frequency along with severity of extreme weather events, such as droughts and floods, monitoring and forecasting can help farmers adapt to changing climatic conditions. For example, monitoring and forecasting can help farmers adopt climate-smart agriculture practices, such as the use of drought-tolerant crops and water conservation measures, to minimize the impact of climate change on farming (Santos, et al. 2020).

Trade and Export: Agriculture monitoring and forecasting are essential for trade and export of agricultural products. Timely and accurate information on crop yield and production can help exporters meet the demand for agricultural products in international markets (Altalak, et al. 2022) Additionally, monitoring and forecasting can help policymakers negotiate trade agreements and tariffs that benefit the agricultural sector (Santos, et al. 2020).

The timely and accurate monitoring and forecasting of crop yield and production enable farmers, policymakers, and stakeholders to make informed decisions that optimize resource allocation, minimize risks, and promote sustainable agricultural practices, ultimately contributing to economic development (Thai-Nghe, et al. 2022).

Table 1. State-of-the-art studies on deep learning-based agriculture monitoring

Title	Year	Deep Learning Technique	Data Used	Application
"Crop yield prediction using deep learning: A review"	2021	CNN, LSTM	Satellite Imagery, Weather Data	Crop Yield Prediction
"Agricultural Monitoring and Management using Deep Learning: A Review"	2020	CNN, LSTM, GAN	Satellite Imagery, Sensor Data, Weather Data	Crop Monitoring, Disease Detection, Yield Estimation
"Forecasting Crop Yield Using Deep Learning: A Systematic Review"	2020	LSTM, CNN, GRU	Satellite Imagery, Weather Data	Crop Yield Prediction
"Deep Learning Techniques for Plant Disease Detection and Diagnosis: A Review"	2020	CNN, R-CNN, YOLO, Faster R-CNN	Plant Images, Disease Datasets	Disease Detection, Diagnosis
"Deep Learning Techniques for Crop Classification: A Review"	2019	CNN, R-CNN, VGG, InceptionV3	Satellite Imagery	Crop Classification
"Deep Learning Approaches for Crop Yield Prediction and Nitrogen Status Estimation in Precision Agriculture"	2019	CNN, LSTM	Satellite Imagery, Sensor Data	Crop Yield Prediction, Nitrogen Status Estimation
"Deep Learning for Plant Disease Detection: A Comprehensive Review"	2019	CNN, VGG, ResNet	Plant Images, Disease Datasets	Disease Detection, Diagnosis
"Deep learning-based crop recognition and disease detection"	2018	CNN, SVM	RGB Images, NIR Images	Crop Recognition, Disease Detection
"A review of deep learning models for crop recognition"	2018	CNN, VGG, ResNet, GoogleNet	RGB Images	Crop Recognition

Applications of Deep Learning Model for Agriculture Monitoring and Forecasting

In this, we have discussed several applications of deep learning model for agriculture domain as shown in Figure 3.

Figure 3. Deep learning applications in agriculture [14]

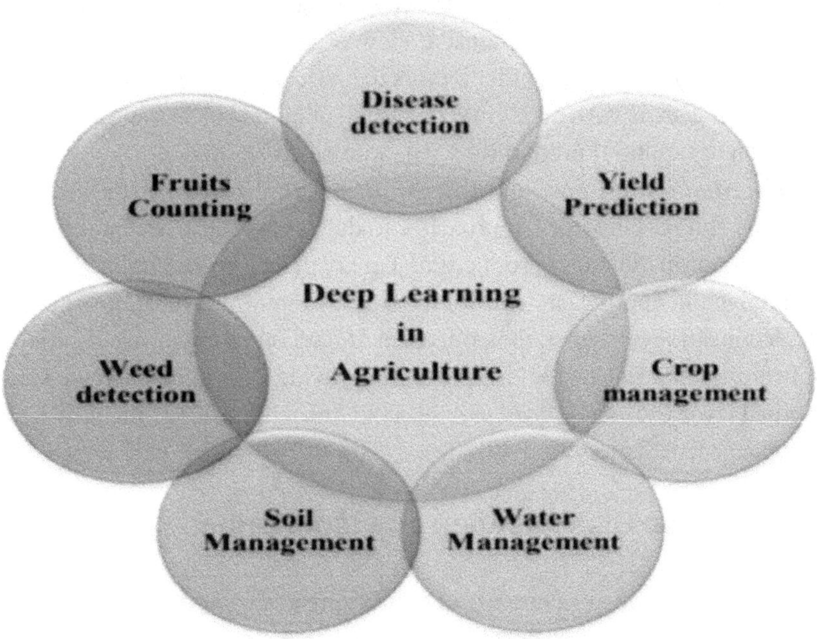

Remote Sensing: Remote sensing involves use of satellites, airplanes, or drones for collecting data on land use, crop type, vegetation health, soil moisture, and other variables that affect crop yield and production. The data collected is analyzed to monitor crop growth and health, predict yield, and detect anomalies such as pests, diseases, and weather-related damage (Boulent, et al. 2019)

Ground-based Sensors: Ground-based sensors such as weather stations, soil moisture probes, and leaf wetness sensors provide localized data on weather conditions, soil moisture levels, plant growth that are important for monitoring and forecasting crop yield and production (Liu, et al. 2021).

Crop Modeling: Crop modeling involves using mathematical models to simulate growth including development of crops under different environmental conditions. Crop models incorporate data on climate, soil, crop type, and management practices to predict yield and production, as well as to identify optimal planting and harvesting times (Mohanty, 2018).

Machine Learning: Machine learning algorithms can be trained on large datasets of historical weather and yield data to predict future crop yield and production. Machine learning can also be used for analyzing remote sensing data to detect anomalies and predict yield (Mohanty, 2019).

Participatory Approaches: Participatory approaches involve engaging farmers and other stakeholders in the monitoring and forecasting process. Participatory approaches can involve the use of local knowledge and traditional forecasting methods, such as the observation of natural phenomena, to supplement data from remote sensing and ground-based sensors (Santos, et al. 2020).

Weather Forecasting: Weather forecasting is an important method used in agriculture monitoring and forecasting. Accurate weather forecasts can help farmers plan and adjust crop management practices, such as planting and harvesting times, irrigation, and pest control, to optimize yield and minimize risks associated with adverse weather conditions (Ma, et al. 2018).

Geographic Information Systems (GIS): GIS is computer-based system that combines spatial data with other data sources to provide comprehensive view of agricultural landscapes. GIS can be used to map crop types, land use, soil types, and other factors that affect crop yield and production. Iinformation can be used for identifying areas with high yield potential, plan irrigation systems, and assess the impact of land use changes on agricultural productivity (Sladojevic, et al. 2020).

Pest and Disease Monitoring: Pest and disease monitoring involves the use of remote sensing, ground-based sensors, and participatory approaches to detect and track pest and disease outbreaks that can significantly impact crop yield and production. Timely detection and tracking of pests and diseases can help farmers implement appropriate control measures to minimize crop losses (Mohanty, 2018).

Yield Surveys: Yield surveys involve the collection of data on crop yield and production from a representative sample of farms or fields. Yield surveys can provide comprehensive view of crop production at national or regional level and can be used to assess the impact of policy changes or climate variability on agricultural productivity (Santos, et al. 2020).

Deep Learning for Crop Classification and Monitoring

Crop classification and monitoring is a process of identifying and categorizing different crops and monitoring their growth and development over time. It involves using various techniques and tools to gather data on crop types, their health, and their yield potential (Thai-Nghe, et al. 2022).

Crop classification involves identifying different types of crops grown in a particular area. This can be done using remote sensing techniques like satellite imagery or aerial photography, as well as ground-based observations. The classification process typically involves using algorithms to analyze the data and identify different crop types based on their spectral characteristics. Crop monitoring involves tracking the growth and development of crops over time. This can be done using a variety of techniques, including remote sensing, ground-based sensors, and visual observation. Monitoring can help identify potential issues such as pests, disease, or drought, allowing farmers to take corrective action before significant crop damage occurs. (Santos, et al. 2020)

Crop classification and monitoring are essential for improving agricultural production and ensuring food security. By providing farmers with accurate information about crop types and their health, farmers can make informed decisions about crop management practices along with maximize their yields. Additionally, crop monitoring can help detect early signs of environmental stress and enable farmers to adapt their practices to mitigate effects of climate change (Muhammad, et al. 2021). Crop classification with monitoring can also aid in land use planning and resource management. By identifying the types and distribution of crops in a particular region, policymakers can make informed decisions about land use and resource allocation. For example, if a region is predominantly growing water-intensive crops, policymakers may prioritize water conservation measures to ensure sustainable resource use (Mohanty, 2018).

Advances in technology have greatly improved the accuracy and efficiency of crop classification and monitoring. Remote sensing platforms such as satellites and drones can quickly capture large amounts of data, while machine learning algorithms can analyze the data to identify crop types, health status. Ground-based sensors, such as weather stations and soil moisture sensors, can provide real-time data on environmental conditions that impact crop growth. (Wang, et al. 2019).

ROLE OF DEEP LEARNING IN CROP CLASSIFICATION AND MONITORING

Deep learning has a significant role in crop classification and monitoring. The technique of deep learning involves use of neural networks with several layers to learn along with recognize patterns in data. In case of crop classification and monitoring, deep learning algorithms can analyze remote sensing images and other data sources to identify and classify different crops, as well as to monitor their growth and health over time. One application of deep learning in crop classification is to use satellite imagery and other remote sensing data to identify and classify different crops. This can be useful for monitoring changes in land use, predicting yields, and understanding the impact of environmental factors on crop growth (Fountas, 2020)

Another application of deep learning in crop monitoring is to use sensor data, such as data from drones, to monitor crop growth and health. Deep learning algorithms can analyze this data to identify areas of the field that require additional attention, such as areas that may be experiencing water stress or nutrient deficiencies (Santos, et al. 2020) Deep learning algorithms can be used to develop predictive models for crop yield estimation. By analyzing historical data on weather patterns, soil conditions, crop growth, these models can provide accurate predictions of crop yields, which can help farmers to make informed decisions regarding planting, harvesting (Ma, et al. 2018). Furthermore, deep learning can be used to develop precision agriculture techniques.

By analyzing data on crop growth and soil conditions, deep learning algorithms can provide farmers with recommendations on optimal amount of fertilizer, water, and other inputs required for each individual crop. This can help to reduce waste and improve crop yields (Boulent, et al. 2019). Deep learning has significant potential in crop classification and monitoring, yield estimation, precision agriculture. With increasing availability of remote sensing data and sensor technologies, deep learning algorithms will become increasingly important tools for farmers and researchers in the agricultural sector (Liu, et al. 2021).

TYPES OF DEEP LEARNING ALGORITHMS FOR CROP CLASSIFICATION AND MONITORING

Convolutional Neural Network (CNN)

CNN is a DL network consisting of many convolutional techniques, layers, convolutional techniques. It is multi-layered neural network based on animal visual cortex as shown in Figure 4. CNN is mainly used for image processing, tagging. In many computer vision studies, CNNs have been used for tasks such as image classification, object identification, image classification, speech recognition, text, video recognition, examining medical images. CNN architecture consists of convolutional layers, pooling layers (Ma, et al. 2018).

Figure 4. CNN architecture
(Altalak, et al. 2022)

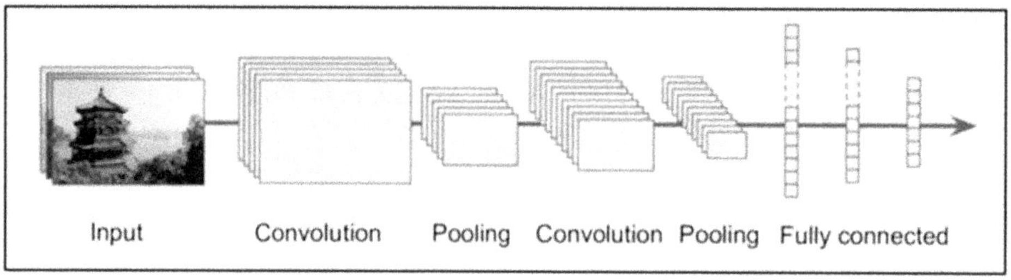

Convolutional Layer

Convolutional layers are the most basic and important layers in CNNs.

The resulting matrix of pixels for the provided image or object is transformed or multiplied.

Activation map for this image. The main advantage of activation maps is that they preserve all features of given image while limiting amount of data that must be processed simultaneously.(Wang, Liu, Chen, et al. 2019) Data is combined into a feature detector matrix and different image variants are generated using different levels of feature detectors. Convolutional models are trained with backpropagation to achieve lowest possible error at individual layer. Depth and padding are defined as the set of errors with the minimum number of errors. Convolutional layers are responsible for extracting visual features (Mohanty, Hughes, and Salathé 2018).

Pooling Layer

This is an important step in trying to reduce the size of the workbook while preserving key features and reducing security. This reduces number of learnable features in model and thus helps solve overfitting problem. Pooling allows CNN to combine different parts of images to recognize given object, even if it is distorted or at different angle. There are several methods of pooling, including maximum pooling, average pooling, random pooling, and spatial pyramid. Most common method is maximum pooling (Muhammad, Amjad, Khan et al. 2021), (Garg A and Rani A 2020).

Fully Connected Layer

This is last layer used by the neural network. Most matrices are flattened before being sent to neurons. In this case it is difficult to keep track of the data content because there are many hidden processes with different weights for the output of each neuron. All causes and calculations appear here (Wang, Liu, Chen, et al. 2019), (Liu, Yan, Tian, et al. 2021)

Convolutional Neural Networks, are a type of deep learning architecture used for computer vision tasks. They have several advantages, such as being robust to input variation, using a hierarchical representation, achieving state-of-the-art performance, and parameter sharing. However, there are also some disadvantages, including the need for large amounts of data, being computationally expensive, having low interpretability, and being prone to overfitting.

Recurrent Neural Network (RNN)

RNNs are neural array models that perform particularly well in important tasks like language modelling, speech recognition, machine translation. RNNs differ from normal neural networks in their use of information from connected networks; this tool is important in many applications where structure is private with file system contains important information. For example, in order to understand the words in a sentence, one must first understand their meaning. Therefore, RNN can be viewed as short circuit consisting of input layer x, hidden (state) layer s, output layer y. Long Short-Term Memory (LSTM) is an RNN architecture designed to model time and their long-term relationships with better performance than RNNs.(Wang, Liu, Chen, et al. 2019),(Altalak et al. 2022). Figure 5 shows the generic architecture of RNN Model.

Figure 5. The generic structure of RNN
(Altalak, Ammad uddin, Alajmi, et al. 2022)

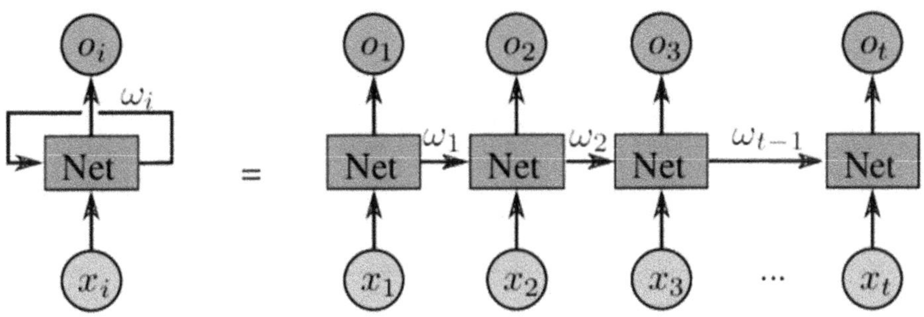

The Core Idea behind LSTMs:

Forget Gate: What part of memory to "forget" – (zero means forget this bit).

$$f_t = \sigma\left(W_f \cdot \left[h_{t-1,} x_t\right] + b_f\right)$$

Input Gate: What bits to insert into the next states

$$i_t = \sigma\left(W_i \cdot \left[h_{t-1,} x_t\right] + b_i\right)$$

What content to store into the next state

$$\tilde{C}_t = \tanh\left(W_C \cdot \left[h_{t-1,} x_t\right] + b_c\right)$$

Output Gate: What part of cell to output?

$$o_t = \sigma\left(W_o\left[h_{t-1,}x_t\right] + b_o\right)$$

Figure 6. The architecture of LSTM
(Altalak et al. 2022)

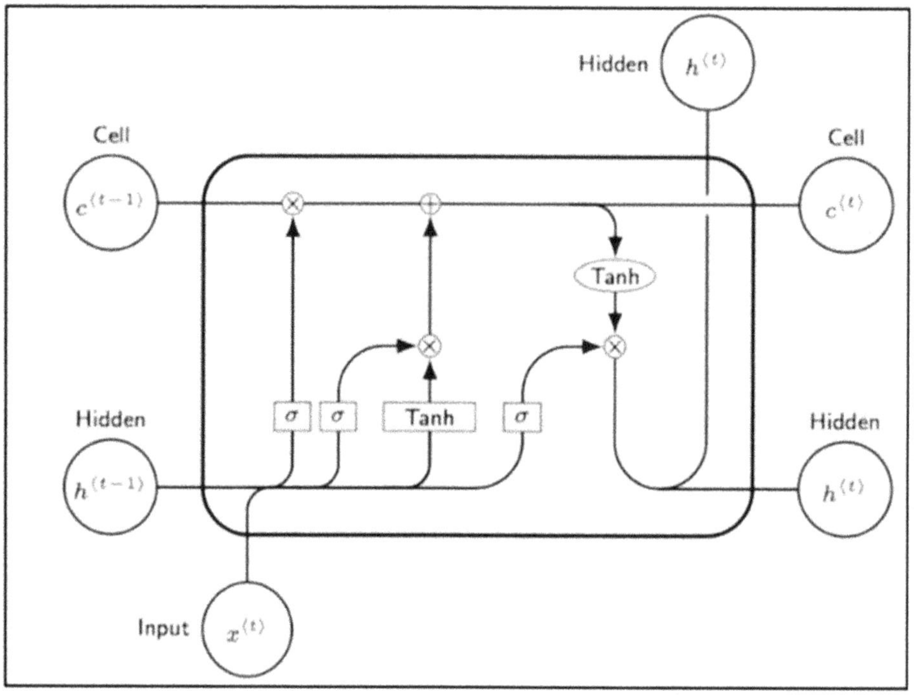

Recurrent Neural Networks, are deep learning models used for tasks involving sequential data. They have several advantages, including the ability to model sequences, flexibility, statefulness, and strong performance. However, there are also some disadvantages, such as being computationally expensive, suffering from gradient vanishing and exploding, lacking parallelism, and struggling with long-term dependencies. Overall, RNNs are a powerful tool for modeling sequential data, but they require careful design and training to overcome their challenges.

CNNs and LSTMs are two types of neural networks used in deep learning that have distinct strengths and weaknesses. CNNs are used for image and video recognition tasks due to their ability to extract spatial features from grid-like structures such as images. LSTMs, on the other hand, are used for tasks involving sequential data due to their ability to capture long-term dependencies and context in the data. CNNs are limited to grid-like structures and lack the ability to handle variable-length input sequences, while LSTMs are computationally expensive and can suffer from vanishing and exploding gradients when trained on long sequences. When choosing between CNNs and LSTMs, it is important to consider the specific task and nature of the data being analyzed. Figure 6 shows the architecture of LSTM model.

Deep Learning for Crop Yield Prediction

Overview of Crop Yield Prediction

Crop yield prediction is the process of forecasting the amount of crop production that will be obtained from a particular agricultural field or region in a given growing season. Crop yield prediction is a vital task for farmers, agronomists, and policymakers, as it enables them to plan for food security, market prices, and resource allocation(Garg A and Rani A 2020) The traditional approach to crop yield prediction involves manual observation and data collection, which is a time-consuming and labor-intensive process. However, with the advent of modern technology, crop yield prediction has become more efficient and accurate (Boulent, et al. 2019)

There are various methods used for crop yield prediction, including statistical models, machine learning algorithms, and remote sensing techniques. Statistical models use historical data and other relevant information like weather data, soil type, and fertilizer usage to predict crop yields. ML algorithms can learn from data with improve prediction accuracy over time. Remote sensing techniques, for example, satellite imagery, aerial photography, provide valuable information on crop health, biomass, and vegetation indices that can be used to predict yields.(Thai-Nghe, Tri, Hoa, 2022)

Some of the benefits of crop yield prediction include:

I. Better resource allocation: Accurate yield predictions allow farmers to plan for optimal resource allocation such as water, fertilizer, and pesticides.
II. Improved food security: Accurate yield predictions can help governments and policymakers plan for food security by ensuring that there is enough food to meet demand.
III. Better risk management: Yield predictions can help farmers and investors manage risk by providing early warning of potential yield losses due to weather events or other factors.

Crop yield forecast using deep learning

- DL is subfield of ML that uses ANN to model and solve complex problems. In recent years, DL has emerged as a powerful tool for crop yield prediction, offering high level of accuracy, efficiency compared to traditional statistical models.(Boulent, et al. 2019)
- Main advantage of DL in crop yield prediction is its ability to learn complex relationships between variables and to make accurate predictions based on large datasets. By processing vast amounts of data, deep learning models can identify patterns, relationships that might be difficult for detecting using traditional statistical methods.(Mohanty, et al. 2019)
- Deep learning techniques can be used for different types of data, including remote sensing data, meteorological data, soil data, and crop management data. For example, Convolutional Neural Networks (CNNs) can be used to process satellite imagery and detect changes in vegetation patterns and health, which can be used to predict crop yields. Recurrent Neural Networks (RNNs) can be used to model temporal data, such as weather patterns or crop development over time, to predict crop yields accurately (Fountas, et al. 2020).
- The use of deep learning in crop yield prediction has several benefits, including:

- Improved accuracy: Deep learning models can achieve higher accuracy than traditional statistical models, particularly for complex and non-linear relationships between variables (Wang et al., 2019).
- Efficient data processing: Deep learning models can process vast amounts of data quickly, making it possible to handle large datasets efficiently.
- Real-time predictions: Deep learning models can be trained and deployed in real-time, allowing for quick updates and adjustments based on changing conditions (Santos et al. 2020).

Types of deep learning algorithms for crop yield prediction:

1. CNN: CNNs are majorly used for image recognition tasks and can be used for crop yield prediction by analyzing remotely sensed data. CNNs can be trained to detect features related to crop health and growth patterns, which can be used to predict yield.
2. RNN: RNNs are generally used for sequence prediction tasks, such as language modeling and time series analysis. RNNs can be used for crop yield prediction by analyzing time-series data, such as weather data, soil moisture, and crop development stages, to predict yield.
3. LSTM networks: LSTMs are type of RNN that can capture long-term dependencies in time-series data. LSTMs can be used for crop yield prediction by analyzing time-series data and predicting yield based on weather, soil, and crop development.
4. Generative Adversarial Networks (GANs): GANs are commonly used for image generation tasks and can be used for crop yield prediction by generating synthetic images that represent different yield scenarios. These images can be used to train CNNs to predict yield based on remotely sensed data.
5. Autoencoders: Autoencoders are commonly used for dimensionality reduction tasks and can be used for crop yield prediction by compressing high-dimensional data, such as satellite imagery or weather data, into a lower-dimensional representation that can be used to predict yield.

Deep Learning for Pest and Disease Detection and Monitoring

Overview of Pest, Disease Detection Along With Monitoring

Pest, disease detection, monitoring are essential aspect of plant health management in agriculture. Timely detection and monitoring of pests and diseases can prevent crop damage and reduce yield loss, resulting in improved productivity and economic outcomes for farmers. There are several methods used for pest and disease detection and monitoring in agriculture (Boulent, et al. 2019).

Visual inspection: This is a traditional method of pest and disease detection, where farmers or trained personnel visually inspect the plants for signs of damage or disease symptoms. This method is often labor-intensive and may not detect pests or diseases in the early stages (Mohanty, et al. 2018).

Traps and baits: These are devices that are used to attract and capture pests or insects. Traps and baits are often used for pest monitoring and can be useful for detecting the presence of pests in the early stages (Sladojevic, et al. 2020).

Remote sensing: Remote sensing techniques, for instance satellite imagery, aerial photography, or drones, can be used to detect and monitor changes in vegetation patterns or plant health. These techniques can provide early warning of potential pest or disease outbreaks and allow for targeted intervention (Garg A and Rani A 2020).

DNA-based methods: DNA-based methods, for example polymerase chain reaction or loop-mediated isothermal amplification, can be used to detect pests or diseases at the molecular level. These methods are highly specific and sensitive and can detect pathogens even in low concentrations (Ma, et al. 2018)

Machine learning: ML algorithms can be trained on large datasets of images, data, or sensor data to detect and classify pests and diseases accurately. These algorithms can learn to detect subtle changes in plant health and detect pests or diseases that may not be visible to the naked eye (Thai-Nghe, Tri, Hoa, 2022).

Role of deep learning in pest and disease detection and monitoring

Deep learning is playing increasingly important role in pest and disease detection and monitoring in agriculture. DL algorithms can analyze vast amounts of data, identify patterns and features that may not be visible to the naked eye. It makes them well-suited for early detection and monitoring of pests and diseases, which is essential for effective plant health management (Boulent, et al. 2019).

Deep learning can be used for pest and disease detection and monitoring in several ways, including:

Image analysis: Deep learning algorithms can analyze images of plants and detect subtle changes in plant health that may indicate presence of pests or diseases. For example, Convolutional Neural Networks (CNNs) can analyze satellite imagery or high-resolution images of plants to detect changes in vegetation patterns, leaf color, or shape (Altalak, et al. 2022).

Sensor data analysis: Deep learning algorithms can analyze data from sensors, such as temperature, humidity, or soil moisture, to detect changes that may indicate the presence of pests or diseases. For example, Recurrent Neural Networks (RNNs) can analyze time-series data from sensors to detect changes in environmental conditions that may favor the development of pests or diseases (Thai-Nghe, Tri, Hoa, 2022).

Disease classification: Deep learning algorithms can be trained to classify images of plants based on the presence or absence of specific diseases. For example, a CNN can be trained on images of plants with and withoutt particular disease to classify new images accurately (Wang, et al. 2019).

Decision support: Deep learning algorithms can provide decision support for farmers and agronomists by predicting the likelihood of pest or disease outbreaks based on environmental conditions or historical data (Mohanty, et al. 2018)

Deep Learning for Weather Forecasting and Climate Change Adaptation

Overview of Weather Forecasting and Climate Change Adaptation

Weather forecasting is the process of predicting atmospheric conditions, such as temperature, precipitation, wind speed, and humidity, for a specific location and time period. Weather forecasting plays a crucial role in agriculture, as it allows farmers to plan their planting, harvesting, and other farm operations more effectively. Climate change adaptation refers to process of preparing for and adapting to impacts of climate change, like more frequent, severe weather events, rising temperatures, and changing precipitation patterns. Climate change adaptation is essential for ensuring resilience, sustainability of agriculture, as well as for maintaining food security in the face of changing climate conditions (Garg A and Rani A 2020)

Weather forecasting and climate change adaptation are closely related, as accurate weather forecasts are essential for effective climate change adaptation. Farmers and other stakeholders need timely and accurate weather information to make decisions about crop selection, planting times, irrigation schedules, and other farm operations.

Weather forecasting and climate change adaptation rely on a range of data sources, including satellite imagery, ground-based weather stations, and climate models. Advanced technologies such as machine learning and artificial intelligence are increasingly being used to analyze these data sources and improve the accuracy of weather forecasts and climate change projections (Fountas, et al. 2020).

DL has potential to revolutionize weather forecasting along with climate change adaptation in several ways. Here are some roles of deep learning in weather forecasting and climate change adaptation.

Role of Deep Learning in Weather Forecasting and Climate Change Adaptation

Enhancing weather forecasting accuracy: DL can improve the accuracy of weather forecasting models by analyzing vast data with identifying patterns along with relationships that may not be apparent to human analysts. Deep learning algorithms can analyze data from multiple sources, including satellite imagery, ground-based weather stations, and weather models, to make more accurate and reliable weather forecasts (Mohanty, et al. 2019)

Improving climate change projections: Deep learning can analyze climate data from multiple sources to create more accurate climate change projections. By identifying patterns and relationships between various climate variables, deep learning algorithms can improve the accuracy of climate models and help predict future climate trends with greater precision (Garg A and Rani A 2020).

Developing early warning systems: Deep learning can be used to develop early warning systems for extreme weather events, like hurricanes, floods, and droughts. By analyzing historical weather data, identifying patterns that precede these events, deep learning algorithms can help predict and mitigate their impacts (Boulent, et al. 2019)

Precision agriculture: Deep learning can help farmers make better decisions by presuming them with more accurate weather forecasts and climate projections. By analyzing weather data and soil moisture data, deep learning algorithms can help farmers optimize irrigation schedules and make more informed decisions about planting and harvesting.

Case Study

Time Series Prediction: LSTM has been used extensively for time series prediction. For instance, in a study titled "Long Short-Term Memory Networks for Traffic Flow Forecasting: A Comparative Study," LSTM models were trained on historical traffic data to predict future traffic flow.

Natural Language Processing: LSTM has been used in various NLP applications, such as text classification, sentiment analysis, and machine translation. In a study titled "Long Short-Term Memory Networks for Machine Reading," LSTM was used to improve machine reading comprehension.

Speech Recognition: LSTM has been used in speech recognition systems to improve accuracy. In a study titled "A Comparison of LSTM and GRU Networks in Speech Recognition," LSTM was compared with GRU (Gated Recurrent Unit) networks, and LSTM was found to perform better.

Image Captioning: LSTM has been used in image captioning systems to generate descriptive captions for images. In a study titled "Show and Tell: A Neural Image Caption Generator," an LSTM network was trained to generate captions for images.

Stock Price Prediction: LSTM has also been used for stock price prediction. In a study titled "Stock Price Prediction Using LSTM, RNN, and GRU Neural Networks," LSTM was found to outperform other types of neural networks in predicting stock prices. The result of the case study is shown in Figure. 7.

Experimental Results

In this chapter, we present the results and analysis of our study on forecasting crop production using Long Short-Term Memory (LSTM), a powerful deep learning technique. The aim of our research was to leverage historical data on crop production and related factors to develop accurate predictions for future agricultural yields. This section highlights the key findings and insights gained from our LSTM-based forecasting model.

Using the trained LSTM model, we successfully generated predictions for future crop production. By incorporating relevant input features, such as weather data, soil characteristics, and agricultural practices, we obtained forecasts for specific time periods. These predictions showcased a seasonal trend, aligning closely with the actual crop production values observed in the testing dataset. The visual representation of the predicted values alongside the actual values offered a compelling comparison, highlighting the accuracy and precision of our LSTM-based forecasting approach.

X-axis shows the no. of instances that are predicted. Y-axis shows the amount of crop over the period of time, it is in the 10^7 units at y-axis.

Figure 7. Forecasting of crop production

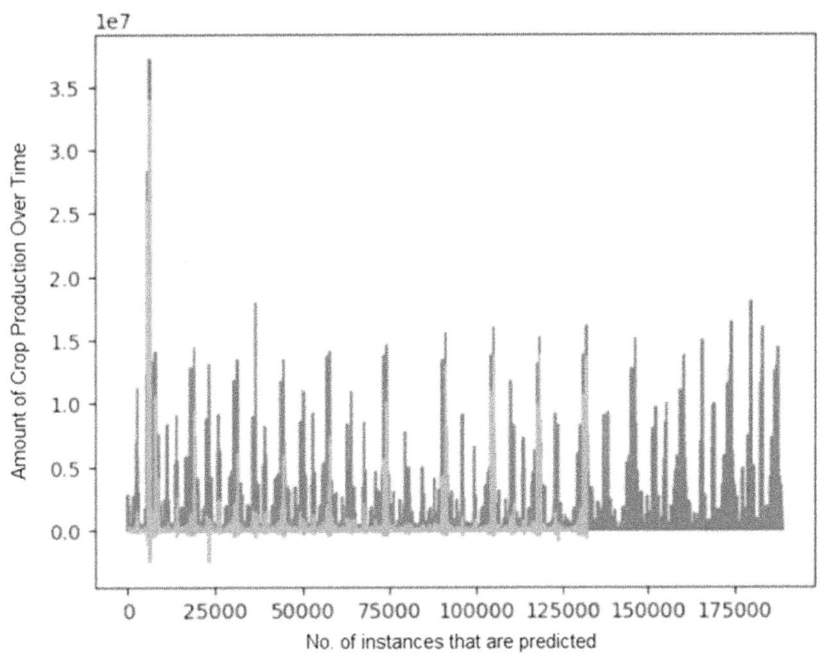

To gain deeper insights, we conducted an in-depth analysis of the forecasting results. We examined the significance of individual input features by investigating feature importance or weights assigned by the LSTM model. This analysis provided valuable insights into the factors that had the most influence on crop production. By identifying key drivers, such as temperature, rainfall, or fertilizer usage, we contributed to a better understanding of the dynamics affecting agricultural yields. Furthermore, we explored any observed discrepancies between the predicted and actual values, uncovering potential areas for further investigation and improvement in future research.

Our research demonstrates the efficacy of LSTM in forecasting crop production as shown in Figure 7. The results obtained from our LSTM-based model exhibit promising accuracy, offering valuable insights into future agricultural yields. By incorporating advanced deep learning techniques into the realm of crop production forecasting, our study contributes to the growing field of data-driven agriculture. This chapter provides a comprehensive analysis of the results, laying the foundation for future research and practical applications in optimizing crop production and ensuring food security.

Challenges and Limitations of Deep Learning-Based Agriculture Monitoring and Forecasting

While deep learning has great potential to improve weather forecasting and climate change adaptation, there are several technical challenges and limitations that must be considered. Here are some of them:

Data quality and quantity: Deep learning algorithms require large high-quality data to learn patterns, relationships accurately. On the contrary, data quality can vary widely, especially in developing countries with limited weather station networks. Also, some regions may not have sufficient data, making it difficult to train accurate models (Boulent, et al. 2019)

Model complexity: Deep learning models are highly complex, with many parameters that need to be fine-tuned. Training process can be computationally intensive and require large amounts of computing power, memory. This can be challenge for organizations with limited resources (Ma, et al. 2018)

Interpretability: DL models can be challenging to interpret, making it difficult to understand why they make specific predictions. This can be significant barrier to adoption, as stakeholders may be hesitant to trust a model they don't understand fully (Liu et al. 2021).

Generalization: Deep learning models are highly sensitive to the data they are trained on. They may perform well on the training dataset but may not generalize well to new data or different regions (Garg A and Rani A 2020).

Limited historical data: Historical weather and climate data are essential for developing accurate models, but in some regions, availability of historical data may be limited or of poor quality (Muhammad et al. 2021)

Uncertainty: Weather forecasting with climate change projections are inherently uncertain due to complexity and unpredictability of Earth's climate system. Deep learning models can help improve accuracy, but they cannot eliminate uncertainty entirely (Wang, et al. 2019).

As deep learning continues to advance, its potential impact on weather forecasting and climate change adaptation is significant. Here are some future directions in which deep learning can be used to further improve weather forecasting and climate change adaptation:

Better data collection and management: To improve the accuracy of deep learning models, more high-quality data is needed. Efforts to expand weather station networks and improve data collection and management will be critical (Mohanty, 2019)

Improved interpretability: To gain trust and widespread adoption, deep learning models must be more transparent and interpretable. Researchers are developing techniques to explain the reasoning behind deep learning predictions, which will help stakeholders understand and trust the models (Santos, et al. 2020).

Integration with other techniques: While deep learning is a powerful tool for analyzing large datasets, it is just one of many approaches that can be used to improve weather forecasting and climate change adaptation. Integrating deep learning with other techniques, such as traditional statistical models and expert knowledge, can help mitigate some of the limitations of deep learning models (Altalak, et al. 2022).

Applications beyond weather forecasting: Deep learning has applications beyond weather forecasting, such as land use mapping and crop yield prediction. These applications can also benefit from the development of more advanced deep learning techniques (Thai-Nghe, et al. 2022)

CONCLUSION

To recapitulate, DL has the potential to revolutionize weather forecasting and climate change adaptation by improving the accuracy and reliability of weather and climate models, developing early warning systems for extreme weather events, and optimizing agricultural practices. While there are technical challenges and limitations that must be addressed, the application of deep learning in these areas has shown promising results. As we continue to collect more data and improve the interpretability of deep learning models, we can leverage this technology to prepare for impacts of climate change with ensure sustainability of agriculture. However, it is important to recognize that deep learning is just one tool in a larger toolbox of approaches for improving weather forecasting and climate change adaptation. By combining deep learning with other techniques and approaches, we can achieve even greater success in mitigating the impacts of climate change and ensuring a sustainable future.

REFERENCES

Altalak, M., Ammad uddin, M., Alajmi, A., & Rizg, A. (2022). Smart Agriculture Applications Using Deep Learning Technologies: A Survey. *Applied Sciences (Basel, Switzerland), 12*(12), 5919. doi:10.3390/app12125919

Boulent, J., Ge, Y., & Zhu, H. (2019). Deep Learning Techniques for Crop Classification: A Review. *Computers and Electronics in Agriculture, 162*, 219–236. doi:10.1016/j.compag.2019.04.014

Fountas, S., Anastasiou, K., & Katsoulas, N. (2020). Forecasting Crop Yield Using Deep Learning: A Systematic Review. *Agronomy (Basel), 10*(4), 571. doi:10.3390/agronomy10040571

Garg, A., & Rani, A. (2020). Agricultural Monitoring and Management using Deep Learning: A Review. *Computers and Electronics in Agriculture, 173*, 105374. doi:10.1016/j.compag.2020.105374

Liu, Q., Yan, Q., Tian, J., & Yuan, K. (2021). Key technologies and applications in intelligent agriculture. *Journal of Physics: Conference Series, 1757*(1), 012059. doi:10.1088/1742-6596/1757/1/012059

Ma, J., Zhuang, D., & Gao, Y. (2018). A review of deep learning models for crop recognition. *Precision Agriculture, 19*(4), 693–723. doi:10.100711119-017-9513-9

Mohanty, S. P., Hughes, D. P., & Salathé, M. (2018). Using Deep Learning for Image-Based Plant Disease Detection. *Frontiers in Plant Science*, *9*, 1420. doi:10.3389/fpls.2018.01420 PMID:27713752

Mohanty, S. P., Hughes, D. P., & Salathé, M. (2019). Deep Learning for Plant Disease Detection: A Comprehensive Review. *Computers and Electronics in Agriculture*, *162*, 139–153. doi:10.1016/j.compag.2019.03.016

Muhammad, A., Amjad, A., & Khan, I. (2021). Crop yield prediction using deep learning: A review. *Computers and Electronics in Agriculture*, *184*, 106024. doi:10.1016/j.compag.2020.106024

Santos, L., Santos, F., Mendes, J., Costa, P., Lima, J., Reis, R., & Shinde, P. (2020). Path planning aware of robot's center of mass for steep slope vineyards. *Robotica*, *38*(4), 684–698. doi:10.1017/S0263574719000961

Sladojevic, S., Arsenovic, M., & Anderla, A. (2020). Deep Learning Techniques for Plant Disease Detection and Diagnosis: A Review. *Computers and Electronics in Agriculture*, *166*, 105073. doi:10.1016/j.compag.2019.105073

Thai-Nghe, N., Tri, N. T., & Hoa, N. H. (2022). Deep Learning for Rice Leaf Disease Detection in Smart Agriculture. In *Artifificial Intelligence in Data and Big Data Processing* (pp. 659–670). Springer International Publishing. doi:10.1007/978-3-030-97610-1_52

Wang, Y., Liu, H., & Chen, Y. (2019). Deep Learning Approaches for Crop Yield Prediction and Nitrogen Status Estimation in Precision Agriculture. *IEEE Access : Practical Innovations, Open Solutions*, *7*, 92025–92036. doi:10.1109/ACCESS.2019.2922124

Chapter 5
Computer Vision and Machine Learning for Smart Farming and Agriculture Practices

Kassim Kalinaki
https://orcid.org/0000-0001-8630-9110
Islamic University in Uganda, Uganda

Wasswa Shafik
https://orcid.org/0000-0002-9320-3186
Ndejje University, Uganda

Tar J. L. Gutu
Soroti University, Uganda

Owais Ahmed Malik
Universiti Brunei Darussalam, Brunei

ABSTRACT

The advent of cutting-edge techniques such as Computer Vision (CV) and Artificial Intelligence (AI) have sparked a revolution in the agricultural industry, with applications ranging from crop and livestock monitoring to yield optimization, crop grading and sorting, pest and disease identification, and pesticide spraying among others. By leveraging these innovative techniques, sustainable farming practices are being adopted to ensure future food security. With the help of CV, AI, and related methods, such as Machine Learning (ML) together with Deep Learning (DL), key stakeholders can gain invaluable insights into the performance of agricultural and farm initiatives, enabling them to make data-driven decisions without the need for direct interaction. This chapter presents a comprehensive overview of the requirements, techniques, applications, and future directions for smart farming and agriculture. Different vital stakeholders, researchers, and students who have a keen interest in this field would find the discussions in this chapter insightful.

DOI: 10.4018/978-1-6684-8516-3.ch005

INTRODUCTION

Agriculture, aside from technology and oil, remains a leading contributor to many nations' GDPs and is the world's most established and vital industry. It has consistently been the primary provider of food and livestock for billions of people (A. Kumar et al., 2019; Thilakarathne et al., 2022). As the global populace continues to soar, projections estimate that by 2050, it will have reached a staggering 9.8 billion individuals (United Nations, 2017). This consequently means that arable lands are expected to decrease due to increased urbanization which will raise concerns as to whether the anticipated increase in the demand for food and agricultural production will be met.

To meet such massive demand due to the increased population, a 70% increase in the current production of agricultural foods will have to be realized by the year 2050 (Kakani et al., 2020). Achieving that status will, however, prove to be a tough challenge due to reasons such as decreased cultivable lands, rising food demands, labor requirements, and limited financial capital (Thilakarathne et al., 2022). Hence, multi-stakeholder initiatives from the industry, academia, along with other research and development entities are needed to fill up this gap by developing and implementing innovative solutions. These solutions are aimed at increasing crop yields with minimal labor and limited cultivable lands, thereby providing much-needed food security for the increased population.

Currently, the realm of information and communication technology, along with its related technologies like fifth generation and beyond, CV, internet of things (IoT), Big Data, AI, Edge/Fog/cloud Computing, and image processing (Kakani et al., 2020; Shafik et al., 2021; Yang et al., 2021), is transforming the agricultural sector, fostering the development of innovative solutions to boost crop yields and satisfy the needs of a burgeoning population (Uddin and Bansal, 2021). The integration of these technologies has brought forth the emergence of "smart farming" and "precision agriculture," novel concepts revolutionizing the way we cultivate and harvest, which involve the use of advanced technologies to optimize decision-making in farm management and enhance the efficiency and effectiveness of agricultural tasks (Pathan et al., 2020). So, by harnessing the capabilities of these technologies, farmers can implement sustainable practices that increase crop productivity and combat the effects of food insecurity caused by poor planning, uneven harvesting, inadequate irrigation, low crop yields, and unpredictable weather events such as droughts.

Recently, the field of CV has been garnering substantial interest in the realm of agriculture, owing to its capacity to lower the expenses associated with food production through adaptable and intelligent automation mechanisms, thereby helping farmers and other key stakeholders increase crop productivity. By allowing machines to perceive and understand the environment in a way similar to humans, CV, combined with image acquisition through remotely configured camera sensors, holds enormous potential for enhancing the overall performance of the farming and agricultural sector through contactless and scalable solutions (Uddin and Bansal, 2021).

As a critical technology, AI and its related technologies, including ML and DL, are gaining traction in the realm of smart farming and agriculture. Their adaptability, speed, automation, precision, and predictive capabilities enable them to emulate human thinking attributes, allowing farmers and other key stakeholders to make precise and timely decisions that lead to increased crop yields in both quality and quantity (Ayoub Shaikh et al., 2022) In contrast to traditional methods, these technologies offer several key advantages, including decreased equipment expenses, amplified computational capabilities, and the ability to collect and process large amounts of agricultural and farm-related data, thus enabling efficient monitoring and timely decision-making (Raval et al., 2022).

Furthermore, the fusion of the above techniques together with other support technologies such as wireless sensor networks (WSNs), robotics, big data, and IoT, among others, have been widely applied in areas such as crop monitoring, pest and diseases detection and prevention, plant phenotyping, yield estimation and maximization, weed control, livestock monitoring, farm security monitoring among others (Patrício and Rieder, 2018; Raval et al., 2022). By leveraging the power of these technologies, food security is guaranteed, meeting the demands of an ever-increasing population while also bolstering the agricultural sector's contribution to the GDP of many countries. This chapter mainly explores CV, ML, and DL application overview in smart farming and agriculture to aid different stakeholders in instituting precise and timely decisions that enhance the agricultural sector. Our other contributions include the following.

- An in-depth analysis of the specific requirements and considerations for CV and ML/DL in the context of smart farming and agriculture, highlighting the data acquisition, data pre-processing, hardware, and software infrastructure, ML/DL processing, and the user interface.
- A thorough examination of the various ML/DL techniques and their utilization in the realm of intelligent agriculture and farming, including supervised and unsupervised ML as well as DL techniques.
- A detailed discussion of the different applications of CV in smart farming and agriculture, including drone crop monitoring, crop yield analysis, crop sorting and grading, pesticide autonomous spraying, plant phenotyping, vegetation information forecasts, irrigation management, livestock farming, and soil analysis.
- A discussion of existing challenges of CV, ML/DL in smart farming and agriculture, including plant database unavailability and limited size, limited technical expertise and generalizability, security and privacy concerns, and minimal large-scale deployments in robust environments.
- A discerning examination of the potential future developments and opportunities for the implementation of these technologies in the realm of smart farming and agriculture.

The remainder of this chapter is organized as follows into seven sections. Section 2 delves into the necessary considerations of CV and ML/DL in the realm of smart farming and agriculture, which includes data acquisition, data pre-processing, hardware and software infrastructure, ML/DL processing and the user interface. Section 3 presents an in-depth analysis of the various ML/DL techniques and their applications in smart farming and agriculture, which are categorized into supervised and unsupervised ML as well as DL techniques. The utilization of CV in smart farming and agriculture is examined in section 4, detailing applications such as drone crop monitoring, crop yield analysis, crop sorting and grading, pesticide autonomous spraying, plant phenotyping, vegetation information forecasts, irrigation management, livestock farming, and soil analysis. The challenges faced by CV and ML/DL in smart farming and agriculture are portrayed in section 5, including plant database unavailability and limited size, limited technical expertise and generalizability, security and privacy concerns, and minimal large-scale deployments in robust environments. Future research prospects are highlighted in section 6, and the final thoughts are presented in section 7.

CV, ML/DL, and Their Requirements in Smart Farming and Agriculture

The field of CV allows computers/machines to perceive and understand captured images (video and still) in the same way humans do. Over the years, the scope of CV has become prominent through the introduction of ML/DL to encompass undertakings like object detection, tracking, and facial recognition, among others (Kakani et al., 2020). ML is a subfield of AI that involves teaching computer programs to improve their accuracy in predicting different outcomes. There are two main categories of ML: supervised learning and unsupervised learning. In supervised learning, the computer program is given a set of labeled training data, and the goal is to predict the correct labels for new, unseen data. Unsupervised learning, on the other hand, involves discovering hidden patterns and relationships in the data without the use of labeled data (Jagtap et al., 2022).

DL, on the other hand, is a type of ML that is based on artificial neural networks (ANNs). DL algorithms are capable of automatically extracting and hierarchically learning different features from raw datasets, allowing them to make decisions on their own by identifying patterns and relationships (Nasiri et al., 2022). Compared to traditional ML methods, DL techniques have a more nonlinear nature, making them ideal for mastering complex scenes and achieving superior computational performance (Jin et al., 2022). In the agriculture and farming sector, the application of CV, ML and DL techniques has the potential to play a crucial role in the development of smart agriculture systems. These technologies can be used to automate various tasks and make more data-driven decisions about farming and crop management. However, specific requirements must be met to make this possible; these requirements in this section as description below.

Data Acquisition

The data used to train DL and ML models in agriculture can come from various sources, such as sensors, drones, and satellite imagery. The sensors can include soil moisture sensors, weather sensors, and crop sensors, which can measure variables such as temperature, humidity, precipitation, and the presence of pests and diseases. Drones and satellites can be used to capture aerial images of crops and fields, which can be used to monitor crop health, detect changes in vegetation, and estimate yields. The data acquired from these sources should be accurate, reliable, and relevant to the problem being investigated.

Data Preprocessing

Once the data is acquired, it must be pre-processed to make it suitable for DL and ML models. This involves various steps, such as data cleaning, normalization, and data annotation. Data annotation involves labeling the data with relevant tags and categories, which can be a time-consuming and specialized task. Pre-processing ensures that the data is in a format that can be used by the DL and ML models. The pre-processed datasets are then split into training, testing, and validation sets to train, evaluate and fine-tune the model.

Hardware and Software Infrastructure

Smart agriculture systems require various hardware and infrastructure, including sensors, drones, robots, and other equipment. These systems must communicate with each other and with the DL/ML models used to make predictions and decisions. The sensors and other equipment used should be accurate, reliable, and relevant to the problem being investigated. The hardware infrastructure used for DL/ML models should be capable of handling large amounts of data, and the computation power required to train and test the models can be significant. Cloud-based solutions can be used to provide the necessary hardware and infrastructure.

ML and DL Processing

There are many different ML and DL algorithms (supervised and non-supervised) that can be used in smart agriculture, allowing for processing and further analysis of the captured datasets to differentiate unwanted regions from the regions of interest through techniques such as classification and segmentation (Hornberg, 2017). Algorithmic approaches such as Neural Networks (NN), Random Forests (RF), Decision Trees (DT), and Support Vector Machines (SVM) can be utilized. However, it is essential to choose a suitable algorithm for the task at hand and to have a good understanding of how it works. The models are tested and validated, and different parameters are then fine-tuned to get the best computational performance. Using various metrics such as F-1 score, recall, kappa, precision, and intersection over union (IoU), among others, the performance of the model is evaluated, after which the model is deployed.

User Interface

Smart agriculture systems should have an easy-to-use user interface that allows farmers and agriculture professionals to access and analyze data, make predictions, and take action based on that information. Overall, the utilization of CV and ML and DL in the realm of smart agriculture requires a combination of data collection and labeling, hardware and infrastructure, different ML and DL algorithms, and a user-friendly interface for farmers and other key stakeholders.

ML and DL Techniques and Their Applications in Smart Agriculture

Supervised ML Techniques

These are ML techniques where the model is trained on annotated/labeled datasets which include input data along with the corresponding correct output. These techniques are mainly used to solve classification and regression problems (Uddin & Bansal, 2022). For example, a supervised ML could be trained on images of crops and their corresponding labels ("healthy" or "unhealthy"). The model can then be used to predict the label for new and unlabeled image datasets. Examples of supervised ML techniques used in agriculture include SVM, K-Nearest Neighbor (K-NN), RF, Naïve Bayes (NB), and DT. Additionally, Artificial Neural Networks (ANN) are also mentioned in this section, as depicted in Figure 1, even when they can be unsupervised, as will be discussed later. Detailed explanations of the above methods and their applications in smart agriculture are given below.

Support Vector Machines

These supervised ML algorithms can be deployed for regression or classification problems. To understand how SVMs work, it's helpful to think of them as drawing a line (in two dimensions) or a plane (in three dimensions) to separate different classes of data points. This line or plane is called a "decision boundary" and is chosen so that it maximally separates the other categories of data points (Kok et al., 2021). So, for example, if we are trying to classify different types of fruit based on their size and color, an SVM could be used to draw a decision boundary between apples and oranges based on their size and color features.

In agriculture, SVMs can be utilized to analyze a plethora of data, such as meteorological data, soil data, and data from sensors and other types of monitoring equipment. For example, Somasundaram et al. (2022) predicted crop yield based on temperature, rainfall, and soil nutrients using SVM, Multivariate Regression, and RF ML techniques. In their findings, both the SVM and RF provided similar which were better than the Multivariate Regression. Chakraborty et al. (2021) have shown the effectiveness of using SVMs to diagnose diseases and pests in crops by analyzing images of the affected plants and as well as classify the type of disease or pest based on visual characteristics. Their multiclass SVM was able to recognize the disease type from the original leaf image among 500 images with 96% accuracy.

He et al. (2022) have shown promising results when SVMs are deployed to optimize irrigation schedules by predicting when and how much water is needed based on weather and soil conditions. In their novel hybrid SVM-whale optimization algorithm, soil moisture was estimated at varying depths. Results showed an improvement at a depth of 60cm, which provides farmers with better guidance for precision irrigation and smart farming. Figure 1 depicts the different ML/DL techniques in smart farming and agriculture and has been detailed in this section.

Figure 1. ML/DL techniques in smart farming and agriculture

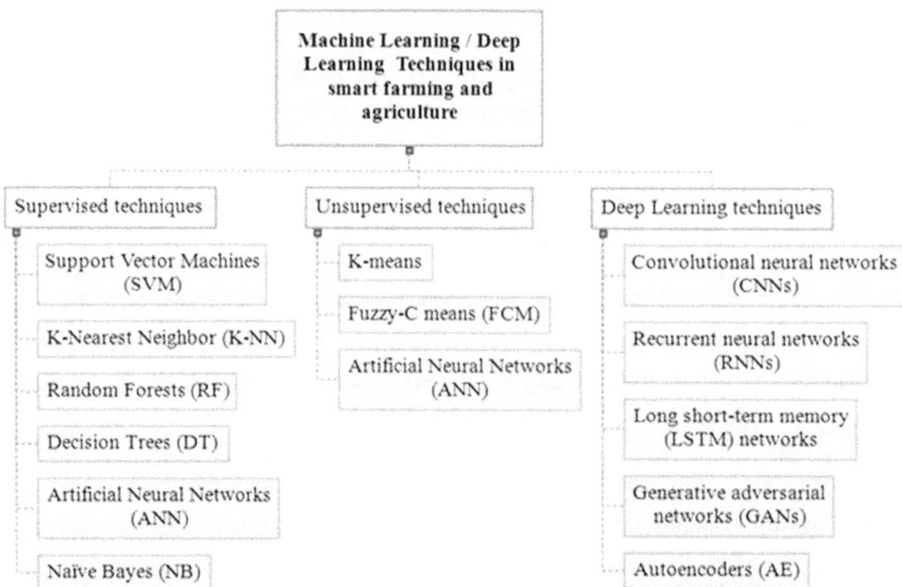

K-NN

This classification and regression technique in ML works by identifying the K closest data points within a training dataset to a given test point and then assigning the class based on the majority class among the K nearest neighbors (Jagtap et al., 2022). In agriculture, K-NN can be used for various tasks, such as predicting crop yields, detecting pests and diseases, and identifying the optimal conditions for crop growth. For instance, a farmer with a crop yield prediction problem of a particular crop might use K-NN based on features such as soil type, weather conditions, and previous yields (Narayan Mishra et al., 2022). The deployed algorithm finds the K nearest data points based on these features. Then, it uses the majority class of these K nearest neighbors to predict the yield for the current year results showed an accuracy of 97.05%.

K-NN has been deployed to detect pests and diseases in crops (Goel and Nagpal, 2022). The K-NN was used to K-NN to classify images of plants as either healthy or infected based on features such as leaf shape, color, and texture. Based on these features, the deployed algorithm finds the K nearest data points (previous images of healthy and infected plants). Then, it classifies the current image based on the majority class of its K nearest neighbors. Results from surveyed studies showed an average accuracy of 68%. Moreover, K-NN has also been used in crop classification, such as the classification of wheat grains (Agarwal et al., 2023). For example, when classifying the texture, the K-NN achieved an accuracy of 55% which was compared with other classifiers such as NB, SVM, and Multi-Layer Perceptron.

RF

Used for classification and regression tasks, this ML algorithm creates many decision trees (hence the name "forest"), each of which is trained on a random subset of the data. The predictions made by the individual decision trees are then combined to make a final prediction (Jagtap et al., 2022). This algorithm has found fame in various agricultural applications, such as crop yield predictions using weather parameters with an accuracy of up to 90% compared to SVM (Aruna Devi et al., 2022). Moreover, RF can predict pests and diseases such as Brown spots, Bacterial leaf blight, and Leaf blasts in their proposed pipeline. Their proposed model achieved an accuracy of 91.47% in rice disease classification at an early stage (Saha and Ahsan 2021). Furthermore, Geetha et al. (2020) have applied RF in identifying the optimal conditions for plant growth, leading to improved crop yield. Their RF model achieved an accuracy of approximately 96% compared with the NB model under different weather patterns. Finally, RF has shown tremendous success when used to map crop types, cropping patterns, and croplands (Tariq et al., 2022). Images taken by different orbital platforms (Landsat 8 and sentinel-2) were used; results show that the RF and DT models perform very well in mapping different types of crops, crop patterns, and croplands. Overall. RF is beneficial for dealing with large datasets and for handling complex relationships between features being examined.

DT

DTs are an ML algorithm that creates a tree-like model of decisions and their possible consequences to predict the best action to take in each situation (Jagtap et al., 2022). They can be used in agriculture and farming to make data-driven decisions like predicting crop yields (Aditya Shastry et al., 2022), indicating the success of different farming techniques, and identifying the best times to plant and harvest crops

(Iniyan et al., 2023). DT has been used to identify pests and diseases, recommend the most effective treatment methods, cropping patterns, and crop types (Tariq et al., 2022; Varshney et al., 2022). Decision trees are beneficial in situations where many variables can affect the outcome of a decision, as they can help to identify the most important factors and make predictions based on that information. They are also useful because they are easy to understand and interpret, making them useful for farmers without a data science background.

NB

This ML algorithm is used for classification tasks. It is based on applying a statistical theorem known as Bayes' theorem to describe the probability of an event occurring based on certain conditions (Ayoub Shaikh et al., 2022). In agriculture, the NB algorithm has been used to classify crops based on specific characteristics and redact the yield based on past data with promising results (Aditya Shastry et al., 2022). Additionally, the algorithm has been used to classify a crop as prone to a particular disease based on factors such as the type of soil it is grown in and the local climate (I. Ahmed and Yadav, 2022; Sagana et al., 2022). Results from the NB, SVM, K-NN, RF, and ANN ensemble showed better accuracy than individual models. Moreover, the NB algorithm has shown great potential when used in the analysis of satellite imagery to identify and classify different types of crops, which can be helpful in precision agriculture, which involves applying data and technology to optimize crop production and increase efficiency (Iniyan et al., 2023); a summarized table 1 illustrates supervised ML methods and their applications in smart agriculture.

Table 1. Summary of Supersized ML Techniques and their applications in smart agriculture

Supervised ML	Application in smart agriculture	References
SVM	Crop yield prediction, pest and disease detection, soil moisture and irrigation schedule optimization, analysis of meteorological data, and soil data.	(Chakraborty et al., 2021; He et al., 2022; Somasundaram et al., 2022)
K-NN	Predicting crop yields, detecting pests and diseases, crop classification	(Agarwal et al., 2023; Goel & Nagpal, 2022; Pavani & Augusta Sophy Beulet, 2022)
RF	Crop yield prediction, pest and disease detection and plant growth optimal conditions, cropping patterns, crop types, and cropland mapping	(Aruna Devi et al., 2022; V. B. Devi et al., 2022; Geetha et al., 2020; Tariq et al., 2022)
DT	Crop yield prediction, pest and disease detection, identifying the best times to plant and harvest crops, cropping patterns, crop types, and cropland mapping	(Aditya Shastry et al., 2022; Iniyan et al., 2023; Tariq et al., 2022; Varshney et al., 2022)
NB	Crop classification, pest, and disease detection, and Crop production optimization.	(I. Ahmed & Yadav, 2022; Iniyan et al., 2023; Sagana et al., 2022; Somasundaram et al., 2022)
ANN	Crop yield prediction, pest and disease detection, identifying optimal irrigation and fertilization practices	(Fu et al., 2018; Hara et al., 2021; V. Kumar & Laxmi, 2022)

Overall, supervised ML techniques discussed and summarized above can be valuable tools for farmers and agriculture professionals to make informed, data-driven decisions about their operations to improve crop management practices.

Unsupervised ML techniques

Unsupervised ML is a type of ML where the data used for training is not labeled, and the algorithm tries to find patterns or relationships within the data (Jagtap et al., 2022). These techniques are used to discover hidden structures or patterns in data without guidance or prior knowledge. Some of the unsupervised ML techniques used in agriculture are discussed below.

K-Means

Based on their similarity, this clustering algorithm can be used in agriculture to group data points (such as crops or soil samples). For example, let's say we have a dataset of soil samples from different locations on a farm, and we want to group the samples based on their nutrient content. We can use k-means to partition the soil samples into k clusters, where each cluster represents a group of soil samples with similar nutrient profiles. To use k-means, we first need to select the number of clusters (k) we want to create. We would then randomly initialize k "centroids" (representative points) for the clusters. Next, we would assign each soil sample to the cluster with the closest centroid based on the distance between the sample and the centroid. It would update the centroids by calculating the mean of all the data points in each cluster and repeating the process until the centroids converge (i.e., they stop changing). In agriculture, K-means clustering can help farmers identify different soil types on their farms and understand the nutrient requirements of different crops (Gupta et al., 2022). It can also help identify farm areas that may require fertilization or other treatments to improve crop yield and in pest and disease detection/diagnosis (Javidan et al., 2023; Selvaraj et al., 2019; Somasundaram et al., 2022). K-means is helpful in identifying soil heavy metal pollution with good classification and evaluation accuracy (Huang et al., 2023).

Fuzzy-C Means (FCM)

FCM, a variant of the widely-utilized k-means clustering algorithm, stands apart by enabling each data point to be assigned to multiple clusters, as opposed to the singular cluster limitation of k-means (Ayoub Shaikh et al., 2022). In FCM, each data point is assigned a membership value for each cluster, indicating the extent to which a point aligns with a specific cluster and the degree of belonging. These association coefficients are often represented as fuzzy variables, hence the name "fuzzy" C-means. FCM is often used in agriculture to aid farmers in grouping fields with similar soil types, as this could help identify fields that might be more suitable for certain crops (Srinivasan et al., 2022).

ANNs

These mimic the complexity of the human brain to create a powerful class of ML algorithms. Comprised of interconnected "neurons" that enable information processing and transmission, ANNs can be trained to recognize patterns and make informed decisions based on input data, making them a powerful tool for solving a wide range of problems (Ayoub Shaikh et al., 2022). For example, iculture, ANNs can be utilized in avenues such as crop yield prediction (Hara et al., 2021) and detecting pests and diseases, which could be useful for farmers who are trying to prevent or mitigate the impact of pests and diseases on their crops (V. Kumar & Laxmi, 2022), and identifying optimal irrigation and fertilization practices

(Fu et al., 2018). Additionally, with deep neural networks, ANNs have been used to classify plant species Wakchaure et al., 2023). Table 2 summarizes the unsupervised ML methods utilized in smart agriculture and farming.

Table 2. Summary of unsupervised ML methods used in smart agriculture and farming

Unsupervised ML	Applications	References
K-means	Soil type identification, crop yield improvement, pest, and disease identification and diagnosis, soil heavy metal classification	(Gupta et al., 2022; Huang et al., 2023; Javidan et al., 2023; Selvaraj et al., 2019; Somasundaram et al., 2022)
Fuzzy-C means	Soil type identification, plant disease detection	(Kumar Sahu & Pandey, 2023; Srinivasan et al., 2022)
ANN	Crop yield prediction, pest and disease detection, identifying optimal irrigation and fertilization practices	(Fu et al., 2018; Hara et al., 2021; V. Kumar & Laxmi, 2022)

DL

Several DL models have been developed specifically for use in agriculture, whose choice largely hinges on the present specific task and the available data type. Some types of DL models and their applications in smart agriculture have been summarized in Table 3.

Table 3. Summary of DL network types and their applications in smart farming and agriculture

DL network type	Applications in smart farming and agriculture	References
Convolutional neural networks (CNNs)	They are used for image classification tasks, such as identifying pests and diseases in crops, grading fruits, and plant species identification.	(Ismail & Malik, 2022; Jiao et al., 2022; Malik et al., 2021)
Recurrent neural networks (RNNs)	Well-suited for analyzing time-series data, such as weather patterns or crop growth over time	(Boechel et al., 2022)
Long short-term memory (LSTM) networks	They are particularly good at modeling long-term dependencies in data, making them well-suited for tasks such as crop yield prediction	(Saini & Nagpal, 2022)
Generative adversarial networks (GANs)	They are used to generate realistic images of crops, which can be helpful for tasks such as data augmentation or simulating different crop growth scenarios.	(Lu et al., 2022)
Autoencoders (AE)	Used for dimensionality reduction, which can be helpful for tasks such as feature selection in large datasets	(Kulkarni et al., 2022)

APPLICATIONS OF CV IN SMART AGRICULTURE AND FARMING

This section presents some of the CV applications in agriculture since it is the core of several economies across the globe. The farmers are being pushed to utilize the existing novel artificial intelligence models to protect and boost crop production as the world's population and food demand grow. Several

CV applications have been tested and accepted in several ways, including plant health monitoring, plant health detection and harvesting, weeding, planting, and unconventional weather scrutiny, which are demonstrated below.

Drone Crop Monitoring

Drones are aircraft that can semi or autonomously operate without a human. This technology was first developed for military purposes but has expanded to almost the entire human lifestyle. Using a pre-installed camera, they have massive data capture abilities since it can fly and cover significant distances (Hafeez et al., 2022). Exploiting geo-sensing competencies, a camera with CV competencies can be trained to detect adverse situations; for example, using satellites and DL, accurate crop health information can be obtained, aerial views of entire agricultural fields, and soil conditions (Mathivanan and Jayagopal 2022). Semantic segmentation and picture annotation are techniques for detecting and defining a specific detail that help farmers make early decision about their farms, thus, crop yield prediction (Safonova et al., 2021).

Crop Yield Analysis

Crop yield investigation uses a variety of ML, such as DL technologies, to create cutting-edge computer approaches, such as weed control threshold utilizing red, green, and blue images from drones to facilitate proper crop yield detection (Ercolini et al., 2022). The drone data, satellite images, nitrogen levels, meteorological conditions, and soil elements are some of the inputs into smart CV algorithms trained to handle vast amounts of complex data. The outcomes of such analyses enable the farmer to make critical predictions ahead of time (Go et al., 2022). Zhou et al. (2022) proposed a technique that excellently checks for novel rice yield indices and plans for unanticipated natural disasters related to food shortages. Khalfi et al. (2020) unveiled a novel method to estimate kiwifruit yield throughout a designated orchard; a wholly automated and non-intrusive computer vision system. The system, which features an optical sensor affixed to a minitractor, slowly and meticulously surveys the orchard. This innovative solution has been successfully deployed in commercial operations for nearly two years.

Crop Sorting and Grading

Farming has a significant demand for CV machines driven by artificial intelligence. Agriculture has benefited from the ability to automate tedious and time-consuming processes like separating the right from the wrong crops and identifying which ones are best for shipping (Shevchenko and Aliiev, 2022). As CV can determine the longevity of the crops and the prevalence of viruses in the crops, resulting in less agricultural loss, it is evolving into a crucial component. In their study, a deep CNN, presenting an approach and assessment of a smart sorting system for bell pepper entailing crops and grading phases was availed (Mohi-Alden et al. 2022). The rating of the fruits and vegetables is evaluated to ascertain which batches of the product should be transported first and which ones may sustain for a more extended period and can be shipped to distant locations. A real-time system for visual inspection and grading fruits was developed by (Ismail and Malik 2022), which allowed efficient segmentation of several instances of fruits from images and accurate grading of the fruits.

Pesticide Autonomous Spraying

Spraying insecticides and pesticides to save crops and animals are traditional agricultural practices. However, CV-equipped drones are becoming increasingly popular in the international market. This is evident when (K. G. Devi et al., 2022), reviewed several current approaches availed to the farmers to improve the agricultural grade. Some of these techniques include, (Tahmasebi et al., 2022) suggesting using a drone with a color-based vision structure to capture and detect. K et al. (2022) designed an automatic pesticide spraying drone embedded control system, and autonomous triangulation smart spraying drone (Wang et al. 2022). Meng et al. (2022) devised an autonomous droplet distribution of drone-centered sprayer. These machines can detect and monitor diseased crops and spray the appropriate pesticide on animals. As a result of such contributions, the workforce is healthy and not exposed to pesticides and pesticides through inhalation.

Plant Phenotyping

Phenotyping is widely used to detect agricultural features for precision agriculture. It is an efficient way to manage sophisticated artificial algorithms (Lube et al., 2022). These algorithms are combined with image processing characteristics to remove extraneous crop data from photos while retaining only the relevant details on precise measurements (Uryasheva et al., 2022). Techniques like depth estimation, color enhancement, and identifying and segmenting the region of interest are options for producing accurate results for further investigation with the gathered data (Tripodi et al., 2022). These practices have significantly improved crop breeds compared to the traditional approach through digitalizing and AI applications in agriculture toward next-generation plant phenotypes.

Vegetation Information Forecasts

Drones with CV capabilities collect data on trees spread across several acres of land. A few examples of potential detection-based activities include tree classification, stem analysis data, health, yield estimation, identification, and drainage information, among others (Bojanowski et al., 2022). Even with occlusion produced by leaves or branches and object size, modern DL approaches achieve good accuracy (Stepanov et al., 2022).

Irrigation management

CV can be used to improve irrigation management by analyzing images of crops and soil to detect patterns and anomalies in crop growth. The technology can detect areas of the field that are not receiving enough water and identify areas where the irrigation system is not functioning correctly. Bini et al. (2020) have practically shown that one way to use CV for irrigation management is through drones outfitted with cameras. Field images can be acquired, then analyzed by an algorithm to identify areas needing irrigation. The algorithm can also optimize irrigation schedules by determining the most efficient times to water based on weather conditions and crop growth patterns. Smart irrigation schemes can be optimized using unmanned aerial vehicles and ground vehicles, as depicted in Figure 2.

Figure 2. Autonomous vehicles, (A) unmanned aerial vehicle, (B) unmanned ground vehicle used in irrigation
(Bini et al., 2020)

A B

CV Systems in Livestock Farming

CV systems can be used in livestock farming to improve the efficiency and productivity of operations. Some examples of how a CV can be used in livestock farming include monitoring animal health (such as lameness, signs of injury or disease, and odd behaviors), automated feeding, livestock identification, heat detection, and livestock welfare (Li et al., 2021). As a result, CV gives measurable data regarding animal welfare that can be utilized to ensure on-farm animal welfare compliance (Divakar et al., 2022). In addition, such systems provide farmers with valuable information about the health and well-being of their livestock monitoring as demonstrated in Figure 3.

Figure 3. CV System for Monitoring Animals
(Li et al., 2021).

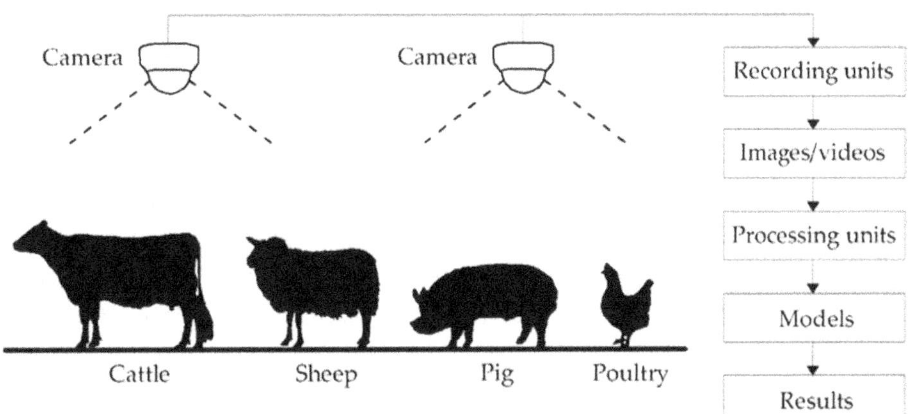

Soil Analysis

By analyzing soil images, relevant information can be extracted using CV, including identifying soil type, determining soil moisture content, and detecting any abnormalities such as rocks, debris, or compaction. This information can aid farmers in optimizing irrigation and fertilization practices and making more informed decisions regarding planting and crop management. One method of utilizing CV for soil analysis involves using drones equipped with cameras to capture images of fields, which are subsequently analyzed by an algorithm to determine the soil type and moisture content (Sharma et al., 2021).

CHALLENGES OF CV, ML, AND DL IN SMART FARMING AND AGRICULTURE

CV applications are used in almost all experimental phases of smart agriculture in plant and animal industry domains. This section presents challenges that are still prevalent in CV.

Plant Database Unavailability and Limited Size

Acquiring leaf images of specific plant infections poses a significant challenge, resulting in the limited availability of plant data sets. The effectiveness of ML and DL algorithms depends on how many standard datasets can be used to train them. With only a handful of studies reporting thousands of images intended for research activities, such a small dataset size presents a major obstacle that limits the generalizability of the ML and DL models. To overcome this limitation, most ML and DL methods use a significant portion of the data for training. This data unavailability issue requires further investigation to improve the efficacy of ML/DL methods in pests and disease detection (Giua et al., 2022). Unsupervised grading algorithms are the most effective when working with unstructured image libraries since they can assist classification tasks without training samples.

Limited Technical Expertise and Generalizability

CV technology, an interdisciplinary realm encompassing computer science, pattern recognition, AI, and other related fields, has made remarkable strides. However, the area still grapples with limitations in generalizability and specialized knowledge. Moreover, with the increasing adoption of agricultural automation, there is a rising need for skilled professionals and experts. Thus, there is an imperative need to foster multidisciplinary talents, drive the development of new technologies, and encourage cross-disciplinary research, all to enhance the intricate landscape of agricultural production (Tian et al., 2020). Also, to fully leverage the potential of CV technology in automating agricultural activities, it will be necessary to address and overcome the various challenges and opportunities that lie ahead. This applies to scientific research, education and training, and the promotion of real-world applications, all requiring specialized skills and expertise.

Security and Privacy Concerns

Integrating CV technology in smart farming and agriculture has created new opportunities for precision farming and crop management. However, it poses security and privacy risks (Alli et al., 2021; Kalinaki et al., 2023). These include the potential for unauthorized monitoring of farm operations and sensitive crop data and the safety and security risks of autonomous vehicles and drones (Abbasi et al., 2022). Moreover, the vulnerability of "smart farms" to cyber-attacks could result in crop failures and financial losses. Therefore, measures must be implemented to protect the privacy and security of farm operations and data. The increasing use of CV in agriculture highlights the need for robust security protocols.

Minimal Large-Scale Deployments in Robust Environments

Most existing CV, ML and DL techniques and approaches are trialed in a lab or experimental setup. When these results are translated to real-world scenarios, differences can arise due to environmental factors, leading to discrepancies between actual and experimental data. The use of visual technology in agricultural automation requires interdisciplinary collaboration and must contend with various perturbations that can greatly affect practical outcomes (Tian et al., 2020). Ensuring these technologies' reliable and precise performance in real time remains a challenge.

Future Directions of CV, ML, and DL in Smart Farming and Agriculture

Creating robust systems becomes crucial as the CV field advances at an accelerated rate within smart farming and agriculture. These systems must be built on a foundation of high-quality data generation, processing, and transfer and robust security measures to thwart potential attacks and security breaches. In addition, integrating various digital technologies, including the IoT, cloud/edge/fog computing, and WSN, presents many security concerns, such as data integrity and privacy. Therefore, these systems must be equipped with security features for each technology, ensuring data and device integrity, availability, and accuracy.

From land preparation to harvest, various stakeholders leverage innovative methods to extract valuable insights from images and objects through detection and tracking to optimize production. As CV, ML/DL models continue to evolve and become more sophisticated, they can tackle sophisticated agricultural computations with precision and exactness, thus ensuring optimal performance. Therefore, creating standard datasets for agricultural picture analysis and building unsupervised grading algorithms are critical tasks that must be completed in the future (Puri et al., 2022). Ultimately, the efficacy of these solutions will hinge on the robustness of their business cases and scalability.

Finally, transfer learning- which entails leveraging knowledge gained from one domain for another-should be delved into more deeply to surmount the challenges presented by the intricate and ever-changing data demands for enhancing precision in areas such as pest and disease prediction. New techniques, such as optical flow frames, GANs, LSTMs, 3D convolution, and temporal pooling, may be examined to harness their capability in automating plant pest and disease detection. However, it is essential to devise more solid and comprehensive methodologies to make headway in this field. Strategies like data augmentation - artificially expanding the training dataset - could be considered a potential solution.

CONCLUSION

In conclusion, the application of CV, ML and DL in smart farming and agriculture has the potential to revolutionize the way we grow and harvest crops. This chapter provided a compact review of the different techniques, applications, challenges, and future directions of deploying CV, ML and DL techniques in smart agriculture. As discussed, these technologies can help farmers and other key stakeholders optimize their operations, increase yields, and reduce costs by automating tasks such as crop monitoring, pest and disease detection, crop sorting and grading, vegetation forecast, phenotyping, soil analysis, and irrigation management. In addition, these technologies can also help farmers make more informed decisions by providing real-time data and insights about their crops and fields. However, it is essential to note that these technologies are still in the nascent stages of development, and further research is needed to unlock their potential fully. Nonetheless, the future of smart farming and agriculture is optimistic as advancements in these technologies continue to be made. By leveraging CV and ML and DL in smart farming and agriculture, there's renewed hope in meeting the food demands of an increasing population coupled with the reduction in cultivable lands due to urbanization. The discussions detailed in this chapter will be helpful in a mix of different audiences, including researchers, students, and other key stakeholders with vested interests in this area.

REFERENCES

Abbasi, R., Martinez, P., & Ahmad, R. (2022). The digitization of agricultural industry – a systematic literature review on agriculture 4.0. *Smart Agricultural Technology*, *2*, 100042. doi:10.1016/j.atech.2022.100042

Aditya Shastry, K., Sanjay, H. A., & Sajini, M. C. (2022). Decision Tree Based Crop Yield Prediction Using Agro-climatic Parameters. *Lecture Notes in Electrical Engineering*, *789*, 87–94. doi:10.1007/978-981-16-1338-8_8

Agarwal, D., Sweta, & Bachan, P. (2023). Machine learning approach for the classification of wheat grains. *Smart Agricultural Technology*, *3*, 100136. doi:10.1016/j.atech.2022.100136

Ahmed, I., & Yadav, P. K. (2022). Plant disease detection using machine learning approaches. *Expert Systems: International Journal of Knowledge Engineering and Neural Networks*, 13136.

Alli, A. A., Kassim, K., Mutwalibi, N., Hamid, H., & Ibrahim, L. (2021). Secure Fog-Cloud of Things: Architectures, Opportunities and Challenges. In M. Ahmed & P. Haskell-Dowland (Eds.), *Secure Edge Computing* (1st ed., pp. 3–20). CRC Press. doi:10.1201/9781003028635-2

Aruna Devi, M., Suresh, D., Jeyakumar, D., Swamydoss, D., & Lilly Florence, M. (2022). Agriculture Crop Selection and Yield Prediction using Machine Learning Algorithms. *Proceedings of the 2nd International Conference on Artificial Intelligence and Smart Energy, ICAIS 2022*, (pp. 510–517). IEEE. 10.1109/ICAIS53314.2022.9742846

Ayoub Shaikh, T., Rasool, T., & Rasheed Lone, F. (2022). Towards leveraging the role of machine learning and artificial intelligence in precision agriculture and smart farming. *Computers and Electronics in Agriculture*, *198*, 107119. doi:10.1016/j.compag.2022.107119

Bini, D., Pamela, D., & Prince, S. (2020). Machine Vision and Machine Learning for Intelligent Agrobots: A review. *ICDCS 2020 - 2020 5th International Conference on Devices, Circuits and Systems*, (pp. 12–16). IEEE.

Boechel, T., Policarpo, L. M., Ramos, G. de O., da Rosa Righi, R., & Singh, D. (2022). Prediction of Harvest Time of Apple Trees: An RNN-Based Approach. *Algorithms 2022, 15*(3), 95.

Bojanowski, J. S., Sikora, S., Musiał, J. P., Woźniak, E., Dąbrowska-Zielińska, K., Slesiński, P., Milewski, T., & Łączyński, A. (2022). Integration of Sentinel-3 and MODIS Vegetation Indices with ERA-5 Agro-Meteorological Indicators for Operational Crop Yield Forecasting. *Remote Sensing, 14*(5), 1238.

Chakraborty, S., Paul, S., & Rahat-Uz-Zaman, M. (2021). Prediction of apple leaf diseases using multi-class support vector machine. *International Conference on Robotics, Electrical and Signal Processing Techniques*, (pp. 147–151). IEEE. 10.1109/ICREST51555.2021.9331132

Devi, K. G., Kumar, C. S., & Kishore, B. (2022). A Survey on the Design of Autonomous and Semi Autonomous Pesticide Sprayer Robot. *El-Cezeri, 9*(1), 371–381.

Devi, V. B., Prabavathi, R., Subha, P., & Meenaloshini, M. (2022). An Efficient and Robust Random Forest Algorithm for Crop Disease Detection. *2022 International Conference on Communication, Computing and Internet of Things, IC3IoT 2022 - Proceedings*. IEEE. 10.1109/IC3IOT53935.2022.9767937

Divakar, M. S., Elayidom, M. S., & Rajesh, R. (2022). Forecasting crop yield with deep learning based ensemble model. *Materials Today: Proceedings, 58*, 256–259. doi:10.1016/j.matpr.2022.02.109

Ercolini, L., Grossi, N., & Silvestri, N. (2022). A Simple Method to Estimate Weed Control Threshold by Using RGB Images from Drones. *Applied Sciences 2022, Vol. 12, Page 11935, 12*(23), 11935.

Fu, J., Li, P., & Wang, Y., Phu Do, X., Bok Choi -, S., Sun, F., Ma, W., Li, H., & Wang, S. (2018). Research on water-fertilizer integrated technology based on neural network prediction and fuzzy control. *IOP Conference Series. Earth and Environmental Science, 170*(3), 032168.

Geetha, V., Punitha, A., Abarna, M., Akshaya, M., Illakiya, S., & Janani, A. P. (2020). An Effective Crop Prediction Using Random Forest Algorithm. *2020 International Conference on System, Computation, Automation and Networking, ICSCAN 2020*. IEEE. 10.1109/ICSCAN49426.2020.9262311

Giua, C., Materia, V. C., & Camanzi, L. (2022). Smart farming technologies adoption: Which factors play a role in the digital transition? *Technology in Society, 68*, 101869. doi:10.1016/j.techsoc.2022.101869

Go, S. H., Lee, D. H., Na, S. il, & Park, J. H. (2022). Analysis of Growth Characteristics of Kimchi Cabbage Using Drone-Based Cabbage Surface Model Image. *Agriculture, 12*(2), 216.

Goel, L., & Nagpal, J. (2022). A Systematic Review of Recent Machine Learning Techniques for Plant Disease Identification and Classification. *IETE Technical Review*.

Gupta, H. P., Chopade, S., & Dutta, T. (2022). Computational Intelligence in Agriculture. *Emerging Computing Paradigms*, 125–142.

Hafeez, A., Husain, M. A., Singh, S. P., Chauhan, A., Khan, M. T., Kumar, N., Chauhan, A., & Soni, S. K. (2022). Implementation of drone technology for farm monitoring & pesticide spraying: A review. *Information Processing in Agriculture*.

Hara, P., Piekutowska, M., & Niedbała, G. (2021). Selection of Independent Variables for Crop Yield Prediction Using Artificial Neural Network Models with Remote Sensing Data. *Land 2021, 10*(6), 609.

He, B., Jia, B., Zhao, Y., Wang, X., Wei, M., & Dietzel, R. (2022). Estimate soil moisture of maize by combining support vector machine and chaotic whale optimization algorithm. *Agricultural Water Management, 267*, 107618. doi:10.1016/j.agwat.2022.107618

Hornberg, A. (2017). Front Matter. Handbook of Machine and Computer Vision.

Huang, H., Huang, S., & Du, Q. (2023). Evaluation of soil heavy metal pollution based on K-means and SVM. *International Journal of Environmental Science and Technology, 2023*, 1–10. doi:10.100713762-022-04727-2

Iniyan, S., Akhil Varma, V., & Teja Naidu, C. (2023). Crop yield prediction using machine learning techniques. *Advances in Engineering Software, 175*, 103326. doi:10.1016/j.advengsoft.2022.103326

Ismail, N., & Malik, O. A. (2022). Real-time visual inspection system for grading fruits using computer vision and deep learning techniques. *Information Processing in Agriculture, 9*(1), 24–37. doi:10.1016/j.inpa.2021.01.005

Jagtap, S. T., Phasinam, K., Kassanuk, T., Jha, S. S., Ghosh, T., & Thakar, C. M. (2022). Towards application of various machine learning techniques in agriculture. *Materials Today: Proceedings, 51*, 793–797. doi:10.1016/j.matpr.2021.06.236

Javidan, S. M., Banakar, A., Vakilian, K. A., & Ampatzidis, Y. (2023). Diagnosis of grape leaf diseases using automatic K-means clustering and machine learning. *Smart Agricultural Technology, 3*, 100081. doi:10.1016/j.atech.2022.100081

Jiao, L., Xie, C., Chen, P., Du, J., Li, R., & Zhang, J. (2022). Adaptive feature fusion pyramid network for multi-classes agricultural pest detection. *Computers and Electronics in Agriculture, 195*, 106827. doi:10.1016/j.compag.2022.106827

Jin, X., Zhang, J., Kong, J., Su, T., & Bai, Y. (2022). A Reversible Automatic Selection Normalization (RASN) Deep Network for Predicting in the Smart Agriculture System. *Agronomy 2022, 12*(3), 591.

Kakani, V., Nguyen, V. H., Kumar, B. P., Kim, H., & Pasupuleti, V. R. (2020). A critical review on computer vision and artificial intelligence in food industry. *Journal of Agriculture and Food Research, 2*, 100033. doi:10.1016/j.jafr.2020.100033

Kalinaki, K., Thilakarathne, N. N., Mubarak, H. R., Malik, O. A., & Abdullatif, M. (2023). Cybersafe Capabilities and Utilities for Smart Cities. In *Cybersecurity for Smart Cities* (pp. 71–86). Springer. doi:10.1007/978-3-031-24946-4_6

Kok, Z. H., Mohamed Shariff, A. R., Alfatni, M. S. M., & Khairunniza-Bejo, S. (2021). Support Vector Machine in Precision Agriculture: A review. *Computers and Electronics in Agriculture, 191*, 106546. doi:10.1016/j.compag.2021.106546

Kulkarni, P., Wyawahare, M., Karwande, A., Kolhe, T., Kamble, S., & Joshi, A. (2022). Agricultural Field Analysis Using Satellite Hyperspectral Data and Autoencoder. *Communications in Computer and Information Science, 1576 CCIS*, (pp. 363–375). IEEE.

Kumar, A., Sarkar, S., & Pradhan, C. (2019). Recommendation system for crop identification and pest control technique in agriculture. *Proceedings of the 2019 IEEE International Conference on Communication and Signal Processing, ICCSP 2019*, (pp. 185–189). IEEE. 10.1109/ICCSP.2019.8698099

Kumar, V., & Laxmi, V. (2022). Pests Detection Using Artificial Neural Network and Image Processing: A Review. *International Conference on Sustainable Computing and Data Communication Systems, ICSCDS 2022 - Proceedings*, (pp. 462–467). 10.1109/ICSCDS53736.2022.9760976

Kumar Sahu, S., & Pandey, M. (2023). An optimal hybrid multiclass SVM for plant leaf disease detection using spatial Fuzzy C-Means model. *Expert Systems with Applications*, *214*, 118989. doi:10.1016/j.eswa.2022.118989

Li, G., Huang, Y., Chen, Z., Chesser, G. D., Purswell, J. L., Linhoss, J., & Zhao, Y. (2021). Practices and Applications of Convolutional Neural Network-Based Computer Vision Systems in Animal Farming: A Review. *Sensors, 21*(4), 1492.

Lu, Y., Chen, D., Olaniyi, E., & Huang, Y. (2022). Generative adversarial networks (GANs) for image augmentation in agriculture: A systematic review. *Computers and Electronics in Agriculture*, *200*, 107208. doi:10.1016/j.compag.2022.107208

Lube, V., Noyan, M. A., Przybysz, A., Salama, K., & Blilou, I. (2022). MultipleXLab: A high-throughput portable live-imaging root phenotyping platform using deep learning and computer vision. *Plant Methods*, *18*(1), 1–22. doi:10.118613007-022-00864-4 PMID:35346267

Malik, O. A., Faisal, M., & Hussein, B. R. (2021). Ensemble Deep Learning Models for Fine-grained Plant Species Identification. *2021 IEEE Asia-Pacific Conference on Computer Science and Data Engineering, CSDE 2021*. IEEE. 10.1109/CSDE53843.2021.9718387

Mathivanan, S. K., & Jayagopal, P. (2022). Utilizing satellite and UAV data for crop yield prediction and monitoring through deep learning. *Acta Geophysica*, *70*(6), 2991–3004. doi:10.100711600-022-00911-7

Mekhalfi, M. L., Nicolò, C., Ianniello, I., Calamita, F., Goller, R., Barazzuol, M., & Melgani, F. (2020). Vision System for Automatic On-Tree Kiwifruit Counting and Yield Estimation. *Sensors 2020, 20*(15), 4214.

Meng, Y., Zhong, W., Liu, Y., Wang, M., & Lan, Y. (2022). Droplet Distribution of an Autonomous UAV-based Sprayer in Citrus Tree Canopy. *Journal of Physics: Conference Series*, *2203*(1), 012022. doi:10.1088/1742-6596/2203/1/012022

Mohi-Alden, K., Omid, M., Soltani Firouz, M., & Nasiri, A. (2022). Design and evaluation of an intelligent sorting system for bell pepper using deep convolutional neural networks. *Journal of Food Science*, *87*(1), 289–301. doi:10.1111/1750-3841.15995 PMID:34940977

Mustak, A., Ma, H., Song, L., & Jin, Y. (2022). *Embedded Design of Automatic Pesticide Spraying Robot Control System.*

Narayan Mishra, K., Kumar, S., Patel, N. R., Shripathi Rao, M., Singh, A., Subba Reddy, N., & Acharya, D. U. (2022). Crop prediction using machine learning. *Journal of Physics: Conference Series*, *2161*(1), 012033. doi:10.1088/1742-6596/2161/1/012033

Nasiri, A., Omid, M., Taheri-Garavand, A., & Jafari, A. (2022). Deep learning-based precision agriculture through weed recognition in sugar beet fields. *Sustainable Computing: Informatics and Systems, 35*, 100759. doi:10.1016/j.suscom.2022.100759

Pathan, M., Patel, N., Yagnik, H., & Shah, M. (2020). Artificial cognition for applications in smart agriculture: A comprehensive review. *Artificial Intelligence in Agriculture, 4*, 81–95. doi:10.1016/j.aiia.2020.06.001

Patrício, D. I., & Rieder, R. (2018). Computer vision and artificial intelligence in precision agriculture for grain crops: A systematic review. *Computers and Electronics in Agriculture, 153*, 69–81. doi:10.1016/j.compag.2018.08.001

Pavani, S., & Augusta Sophy Beulet, P. (2022). Prediction of Jowar Crop Yield Using K-Nearest Neighbor and Support Vector Machine Algorithms. *Lecture Notes in Electrical Engineering, 792*, 497–503. doi:10.1007/978-981-16-4625-6_49

Puri, D., Kumar, A., Virmani, J., & Kriti. (2022). Classification of leaves of medicinal plants using laws' texture features. *International Journal of Information Technology : an Official Journal of Bharati Vidyapeeth's Institute of Computer Applications and Management, 14*(2), 931–942. doi:10.100741870-019-00353-3

Raval, M. S., Chaudhary, S., & Adinarayana, J. (2022). Computer Vision and Machine *Learning in Agriculture*, 97–126.

Safonova, A., Guirado, E., Maglinets, Y., Alcaraz-Segura, D., & Tabik, S. (2021). Olive Tree Biovolume from UAV Multi-Resolution Image Segmentation with Mask R-CNN. *Sensors 2021, 21*(5).

Sagana, C., Keerthika, P., Thangatamilan, M., Kamali, R., Nanthini, K., & Maghathani, S. (2022). Identification of Suitable Crop Based on Weather Condition. *2022 International Conference on Computer Communication and Informatics, ICCCI 2022*. IEEE. 10.1109/ICCCI54379.2022.9740905

Saha, S., & Ahsan, S. M. M. (2021). Rice Disease Detection using Intensity Moments and Random Forest. *2021 International Conference on Information and Communication Technology for Sustainable Development, ICICT4SD 2021 - Proceedings*, (pp. 166–170). IEEE. 10.1109/ICICT4SD50815.2021.9396986

Saini, P., & Nagpal, B. (2022). *Deep-LSTM Model for Wheat Crop Yield Prediction in India.*

Selvaraj, M. G., Vergara, A., Ruiz, H., Safari, N., Elayabalan, S., Ocimati, W., & Blomme, G. (2019). AI-powered banana diseases and pest detection. *Plant Methods, 15*(1), 1–11. doi:10.118613007-019-0475-z

Shafik, W., Mojtaba Matinkhah, S., Shokoor, F., & Nur Sanda, M. (2021). Internet of Things-Based Energy Efficiency Optimization Model in Fog Smart Cities. *JOIV : International Journal on Informatics Visualization, 5*(2), 105–112. doi:10.30630/joiv.5.2.373

Sharma, A., Jain, A., Gupta, P., & Chowdary, V. (2021). Machine Learning Applications for Precision Agriculture: A Comprehensive Review. *IEEE Access : Practical Innovations, Open Solutions, 9*, 4843–4873. doi:10.1109/ACCESS.2020.3048415

Shevchenko, I., & Aliiev, E. (2022). Precise grading and sorting of sunflower plant materials in industrial facilities. *Journal of Central European Agriculture, 23*(2), 327–341. doi:10.5513/JCEA01/23.2.3345

Somasundaram, R. S., Nagamani, K., Florence, M. L., & Swamydoss, D. (2022). *Estimation and prediction of crop yielding rate using machine learning techniques.* 2022 International Conference on Computer Communication and Informatics, ICCCI, Coimbatore, India.

Srinivasan, R., Shashikumar, B. N., & Singh, S. K. (2022). Mapping of soil nutrient variability and delineating site-specific management zones using fuzzy clustering analysis in eastern coastal region, India. *Photonirvachak (Dehra Dun), 50*(3), 533–547. doi:10.100712524-021-01473-9

Stepanov, A., Dubrovin, K., & Sorokin, A. (2022). Function fitting for modeling seasonal normalized difference vegetation index time series and early forecasting of soybean yield. *The Crop Journal, 10*(5), 1452–1459. doi:10.1016/j.cj.2021.12.013

Tahmasebi, M., Gohari, M., & Emami, A. (2022). An autonomous pesticide sprayer robot with a color-based vision system. *International Journal of Robotics and Control Systems, 2*(1), 115–123. doi:10.31763/ijrcs.v2i1.480

Tariq, A., Yan, J., Gagnon, A. S., Riaz Khan, M., & Mumtaz, F. (2022). Mapping of cropland, cropping patterns and crop types by combining optical remote sensing images with decision tree classifier and random forest. *T and F Online.* Http://Www.Tandfonline.Com/Action/JournalInformation?Show=aimsScope&journalCode=tgsi20#.VsXpLiCLRhE/

Thilakarathne, N. N., Bakar, M. S. A., Abas, P. E., & Yassin, H. (2022). A cloud enabled crop recommendation platform for machine learning-driven precision farming. *Sensors, 22*(16), 6299.

Tian, H., Wang, T., Liu, Y., Qiao, X., & Li, Y. (2020). Computer vision technology in agricultural automation—A review. *Information Processing in Agriculture, 7*(1), 1–19. doi:10.1016/j.inpa.2019.09.006

Tripodi, P., Nicastro, N., Pane, C., Cammarano, D., Tripodi, P., Nicastro, N., Pane, C., & Cammarano, D. (2022). Digital applications and artificial intelligence in agriculture toward next-generation plant phenotyping. *Crop & Pasture Science, 74*(6), 597–614. doi:10.1071/CP21387

Uddin, M. S., & Bansal, J. C. (2021). *Introduction to computer vision and machine learning applications in agriculture*, 1–8.

Uddin, M. S., & Bansal, J. C. (Eds.). (2022). *Computer Vision and Machine Learning in Agriculture, 2.* doi:10.1007/978-981-16-9991-7

United Nations. (2017). *World population projected to reach 9.8 billion in 2050, and 11.2 billion in 2100.* United Nations. https://www.un.org/en/desa/world-population-projected-reach-98-billion-2050-and-112-billion-2100/

Uryasheva, A., Kalashnikova, A., Shadrin, D., Evteeva, K., Moskovtsev, E., & Rodichenko, N. (2022). Computer vision-based platform for apple leaves segmentation in field conditions to support digital phenotyping. *Computers and Electronics in Agriculture, 201*, 107269. doi:10.1016/j.compag.2022.107269

Varshney, D., Babukhanwala, B., Khan, J., Saxena, D., & Singh, A. K. (2022). Plant Disease Detection Using Machine Learning Techniques. *2022 3rd International Conference for Emerging Technology, INCET 2022*. IEEE.

Wakchaure, M., Patle, B. K., & Mahindrakar, A. K. (2023). Application of AI techniques and robotics in agriculture: A review. *Artificial Intelligence in the Life Sciences*, *3*, 100057. doi:10.1016/j.ailsci.2023.100057

Wang, X., Wang, S., Huang, J., Lv, S., & Li, X. (2022). Design of an autonomous navigation intelligent spraying robot. *IEEE 6th Information Technology and Mechatronics Engineering Conference, ITOEC 2022*, (pp. 496–499). IEEE.

Yang, Z., Jianjun, L., Faqiri, H., Shafik, W., Talal Abdulrahman, A., Yusuf, M., & Sharawy, A. M. (2021). Green internet of things and big data application in smart cities development. *Complexity*, *2021*, 2021. doi:10.1155/2021/4922697

Zhou, J., Lu, X., Yang, R., Chen, H., Wang, Y., Zhang, Y., Huang, J., & Liu, F. (2022). Developing novel rice yield index using UAV remote sensing imagery fusion technology. *Drones, 6*(6), 151.

Chapter 6
Deep Learning-Based Air Pollution Forecasting System Using Multivariate LSTM

K. Lavanya
Vellore Institute of Technology, India

Narayan R. Prathik
Vellore Institute of Technology, India

ABSTRACT

Air quality forecasting is an important way to safeguard public health since it provides ample notice of harmful air pollutants in a given location. Exhaust emissions from gas-guzzling vehicles, industrial pollution, and wildfires obfuscate some of the country's most dramatic vistas and can pose a significant hazard. If forecasts are trustworthy and accurate enough, they can be a crucial component of an air quality control system that complements more conventional emissions-based strategies. This study uses multivariate LSTM Time series forecasting techniques to understand and apply multiple variables together to contribute more accuracy to forecasting due to its proven track record of success with time-series data. The factors that are found to have a major effect on the air quality are dew point, wind speed, temperature, snowfall and rain conditions. In particular, this model focuses on predicting and forecasting the concentrations of PM2.5 (Atmospheric Particulate Matter with a diameter of less than 2.5mm) as this pollutant has a major effect on human health worldwide.

INTRODUCTION

In most major cities across the world, air pollution is found to have a substantial effect on living conditions.. In the existence of air pollution control technology, accurate air pollution estimation is a crucial initial step that promotes the economic and social advancement of emerging nations. Pollution is generally classified into two categories: (1) pollution caused by natural disasters like forest fires and volcanic eruptions, as a result of which oxides and sulphates of carbon and nitrogen are emitted as air pollutants.; and

DOI: 10.4018/978-1-6684-8516-3.ch006

(2) artificial pollution generated through some human activities, such as the burning of oils, discharges from industrial production processes, and traffic emissions, has gained a lot of attention because of its negative impact on human health, other types of species, and the environment.

Particulate matter is one of the air pollutants that harms human health. Serious health problems are caused by heavy metals and compounds that are carcinogenic, such as mercury, lead, and cadmium. When particulates such as benzo(a)pyrene are inhaled for an extended period, they cause cancer. This paper mainly focuses on the pollution caused by Particulate Matter 2.5 or $PM_{2.5}$. $PM_{2.5}$ is a term used to describe airborne particles which are found to have a diameter less than 2.5 micrometers. There are numerous sources of fine particles. Some are created when gases and particles in the atmosphere interact, while others are directly released into the atmosphere.These Fine particles, due to their tiny dimensions and lightweight, are more likely to remain in the atmosphere longer than larger ones. This increases the likelihood that people and animals will breathe them in and absorb them. Particles less than 2.5 micrometres can escape the nose and throat and reach the lungs profoundly due to their small size. Some of these particles may even enter the circulatory system. Long-term exposure to $PM_{2.5}$ may result in plaque deposits in the arteries, which can eventually induce arterial inflammation and stiffening, which can result in heart attack and stroke, claims a study published in the Journal of the American Medical Association.

The factors chosen for this study are Temperature, Wind Speed, Dew Point, Snowfall and Rainfall. The temperature of the air has an impact on airflow and, consequently, on the movement of air pollution. Air near the surface of the Earth is warmer than air further up in the troposphere because the Earth's surface absorbs solar radiation. Warmer, lighter air rises to the surface, while cooler, heavier air sinks to the upper troposphere. Pollutants are transported from the ground to higher elevations by the phenomenon of convection. In normal weather, dew is more effective than haze or fog at removing airborne particles; On normal, foggy, and hazy days, PM2.5 removal rates were found to be 21.5 percent, 15.2 percent, and 13.7%, respectively. Dew condensation minimise the concentrations of gaseous and particle pollution in the near-surface layer. A raindrop can pick up tens to hundreds of small airborne particulates as it descends through the atmosphere before it reaches the ground. Coagulation, a natural occurrence whereby droplets and aerosols attract one another, can assist in the removal of air pollutants such soot, sulphates, and organic particles.

This study presents an algorithm that uses the concept of multivariate time series forecasting techniques with a Long Short Term Memory of LSTM model. This model is chosen as it requires no prerequisites and can model non-linear functions with neural networks.

LITERATURE SURVEY

As a general observation, The LSTM method is frequently combined with other algorithms to create the best hybrid forecasting model. In their baseline prediction models, (Kaya K & Gündüz Öğüdücü, U (2020)) use LSTM. It has been discovered that this addition will improve the algorithm's performance and give the user the best results. These hybrid models, however, have occasionally been found to be too intricate to comprehend.(Guo Q et al., (2020)) had to deal with the problem that their suggested model was a black box model made up of challenging mathematical calculations. However, the majority of those models were only created for a limited set of circumstances and locales. Other articles have attempted to apply different methods with encouraging results. For instance, (Samal K K et al., (2019)) and (Arsov

M et al., (2021))concentrate on the prophet models and the auto regressive moving average (ARIMA), however it was discovered that these were ineffective when attempting to forecast the levels of air pollution in regions that experience harsh conditions. Some models have peculiar problems that mostly result from the framework of the model itself. Artificial and Wavelet Artificial Neural Networks are the subject of (Guo Q et al., (2020))) ANNs and WANNs.Although the WANN model produced results that were close to ideal and optimal, it was discovered that selecting the model's mother wavelet function without which it cannot make predictions. Finally, in certain instances, the faults are mostly caused by missing data or failing to take specific elements into consideration. It was discovered that the datasets used in (Belavadi S et al., (2020)), (Cordova C H (2021)), (Kothandaraman D et al.,(2022)), and (Zakaira N N et al., (2019)) were inadequate, which reduced the efficacy of their algorithms. This demonstrates the necessity for a forecasting method or algorithm that is more successful virtually always while taking into account frequently ignored aspects. LSTM Multivariate approaches or their variations will be used to address this key point in this study. (Belavadi S et al., (2020)) mainly used Wireless Sensor Networks to record the dataset and LSTM as the Deep Neural Network for forecasting. The forecasting model has an RSME value of 30-40ppm in Bengaluru and an RSME value of 0-5ppm in Amaravati. This is because Bengaluru is an enormous metropolis with high temporal variance. (Cordova C H (2021) incorporated Knowledge Discovery from Databases (KDD) with a multi-level perceptron model and an LSTM model for forecasting. In terms of error rate, the model was found to have an RMSE value as low as 2.192 for the MLP model and 1.732 for LSTM model. (Samal K K et al., (2019)) mainly used SARIMA and Prophet model in their study which resulted in RSME values ranging between 2.55 and 4.13. (Kothandaraman D et al.,(2022)) was more of a comparative study for 6 different algorithms which were Random Forest model, Linear Regression, Ridge and Lasso model, KNN, Adaboost models, and Xgboost model. Out of these models, KNN, XGboost and Random Forests where shown to be better than the other 3 models in the study. (Guo Q et al., (2020)) incorporated Artificial and Wavelet Artificial Neural Networks (ANNs and WANNs). This study concluded that the WANN model performed better than the ANN model. (Kaya K & Gündüz Öğüdücü, U (2020)) used a Deep Flexible Sequential model with the incorporation of Convolutional Neural Networks and LSTM units. The model's efficacy was compared to that of GRU, LSTM, and bi-LSTM, which use sequential data to predict PM10. Based on two performance metrics, MAE and RMSE, DFS Model surpassed the competition. The DFS model prevailed over the Conv-LSTM model while lacking a variable dropout layer structure when it comes to forecasting air pollution (particularly at larger window sizes). (Arsov M et al., (2021)) involves the use of an ARIMA Baseline Model with Recurrent Neural Networks and LSTM to form a hybrid model. The results of this study stand out from the rest in terms of RMSE value, with the lowest error rate going down to 0.0273. (Michaelides S et al., (2017)) involves the use of Synoptic classification algorithms using ANN and linear regression models. The results imply that taking current synoptic circumstances into account is a crucial component of a linear regression-based air pollution forecasting system. (Chen Y et al., (2021)) uses a DLP1- LSTM(Double Layer with Previous One observed data) model and results with R2 score between 0.81-0.91. (Zakaira N N et al.,(2019)) incorporated the use of Markov Chain Models to predict the class labels. (Ejohwomu O A et al., (2022)) studies multiple algorithms for the study and the model that stands out is their hybrid model XGBoost-Random Forest-ARIMA which predicts the PM(2.5) matter adequately well. (Božnar, M. Z., Mlakar, P., et al.,(2022)) mainly compares between Multi-Layer Permultiple and Multiple Regression models. The RMSE value goes as low as 11.25. (Zhalehdoost A & Taleai M (2022).) offers a hybrid model for AQI prediction based on CEEMDAN, FE, and LSTM neural networks. The original, highly erratic, and nonlinear AQI series is first broken down into multiple more stable subseries called

IMF and a residue using CEEMDAN in the suggested model. The proposed model performs better than similar models like LSTM, EEMD-LSTM and ARIMA, due to its lower prediction error and improved fitting impact. Additionally, the proposed model for AQI grade prediction has significantly higher accuracy than current forecasting models, which is crucial in motivating the general public to schedule their outdoor activities and take precautions in advance. In (Wu Z et al., (2022)) Xgboost Regressor (XgbR), Artificial Neural Network (ANN), Random Forest Regressor (RFR), K-Nearest Neighbours Regressor (KNN), Hyperparameter Tuning (HPT), Decision Tree Regressor (DTR), LinearRegression (LR), and Lasso Regression (LaR) were all implemented. The other ML models looked at were outperformed by ML algorithms that combined RFR and HPT and XgbR and HPT, respectively. Using RFR and HPT, a practical air quality estimating system was created. (Isaev E et al., (2022)) to solve air pollution classification, five ML models were built, including one innovative model called SMOTEDNN(Synthetic Minority Oversampling Technique with Deep Neural Network). All five models made use of effective data pre-processing and strict hyperparameter optimisation. Three forecasting models were created to predict air pollution for one step-index based on statistical autoregression. In terms of accuracy, precision, sensitivity, and specificity, each constructed model worked brilliantly. Significantly, the unique model SMOTEDNN outperformed the other models in the current study and earlier research in terms of accuracy (99.90%). Thorough hyperparameter optimisation and rigorous data pre-processing were primarily responsible for this exceptional performance.

BACKGROUND STUDY-LSTM MODEL

Long Short-Term Memory (LSTM) is a type of recurrent neural network that avoids the vanishing gradient issue by the use of backpropagation across time while training the model. Memory blocks in LSTM networks are connected through layers rather than neurons. Three nonlinear gates in the LSTM's building blocks give it greater intelligence than a conventional neuron and a memory for sequences.

These non-linear gate types are as follows:

Input Gate: This gate selects the values from the input to change the state of the memory.
Forget Gate: This gate determines exactly what part of the data should be removed from the block..
Output Gate: This gate determines what should go into the output depending on the input and memory gate.

Each LSTM unit's gates have weights acquired throughout the training process, and each one functions like a miniature state machine that uses a "memory" cell that can hold its state value for a longer period. The model picks up these weights over time. Therefore, when inputs come in, the model gradually learns which inputs are significant and must be remembered, in addition to how each input's cell state should determine the main portion of the output. This makes it possible for the LSTM RNN model to precisely capture long-range relationships in their cell states while also addressing the vanishing gradient problem that typical RNN architectures encounter. The data is organised into supervised learning using an X and Y format during the data pre-processing stage. In layman's terms, it establishes the nature and magnitude of the link (positive/negative impact and the values generated are known as quantification of impact) between a single dependent variable (Y) and several additional independent variables (X).

Figure 1. Basic framework of a multivariate LSTM model

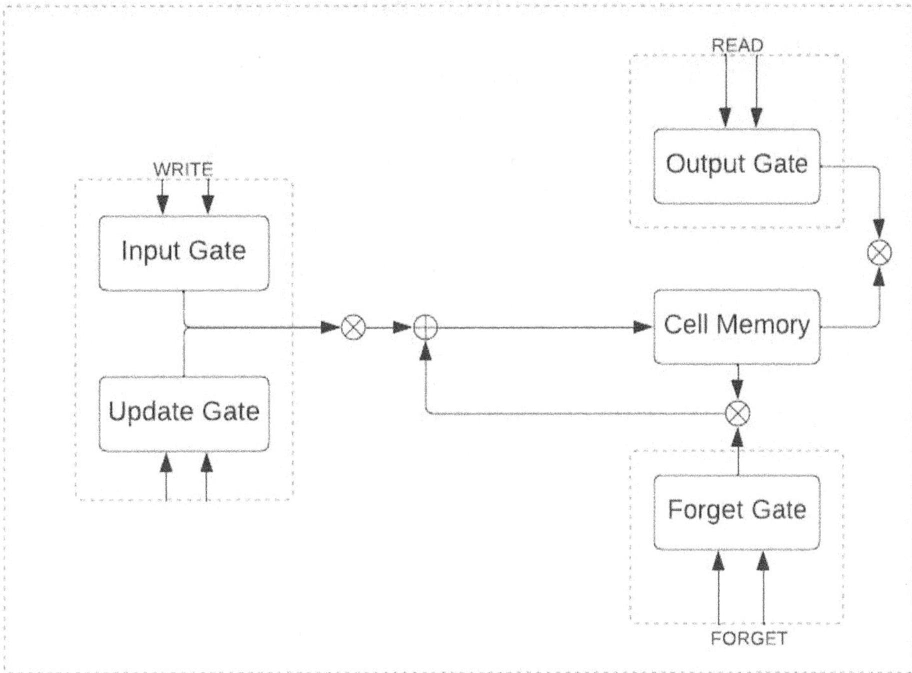

The mean squared error, or loss, between the expected value and the actual value is decreased by the training procedure. The network weights are iteratively changed based on the training data in this optimisation strategy. The learning rate for each network parameter is maintained distinct and modified when new knowledge is acquired. The model can be used to make predictions once it has been trained. Given an input pollutant concentration, the model predicts the amount of contaminants for the following hour. The forecast for the nth hour, which is required to anticipate the levels of contaminants for the next five hours, is obtained from the forecast for the n-1th hour as input.

PROPOSED METHODOLOGY

Accidental environmental pollution, especially air pollution, has been the subject of a lot of recent research. The biological ecosystem of the planet would slowly deteriorate if environmental issues are disregarded in the name of progressive transformation. Accidents involving air pollution have been occurring quite frequent lately, harming both the economy and human life. Air pollution control must be analysed to determine the exact extent of the damage to understand the fundamentals of pollution. Utilizing suitable economic, legal, technical, and other methods, air pollution assessment aims to identify and quantify the scope and amount of environmental pollution damage.

The primary objective of air pollution early warning systems is to alert relevant agencies when the air quality satisfies the early warning requirement. The resource, the contaminant, and the region of influence are all taken into account by an exhaustive pollution warning system. Meteorological factors can forecast urban air pollution episodes to give an early warning. Air pollution forecasts should therefore contain emergency warnings as a critical part of the overall emergency system, together with management for prevention and control and emergency measures, given the fact that urban air pollution occurrences are on the rise.

Whether Particulate Matter is the target pollutant or not, approaches based on artificial neural networks outperform other methods in air pollution forecasting research employing machine learning. It is only natural that research on air pollution forecasting has recently shifted focus to deep learning techniques due to the success of deep learning approaches in a range of other application fields. The upcoming iteration of Recurrent Neural Networks (RNN) used in deep learning is called Long Short-Term Memory (LSTM) because of its optimised architecture for swiftly identifying patterns in sequential input. The ability to learn and recall across lengthy periods, as well as the lack of dependence on input from pre-specified window-lagged observations, are advantages of this type of network. The purpose of this work is to investigate the possibilities of an LSTM-based strategy for predicting daily $PM_{2.5}$ concentrations and to construct a general prediction model that could be applied to any specific region in India. Although this research focused on predicting $PM_{2.5}$ concentrations, it can be used to a greater spectrum to anticipate additional air contaminants.

Algorithm: The basic algorithm and flow of the forecasting model are as follows:

Step 1: Import and preprocess the data: This entails loading the air quality dataset and turning it into a more suitable format, by using methods such as altering time intervals, deleting missing values, or normalizing the features, before splitting the dataset into training and testing data.

Step 2: Designing the LSTM Model: In this step, we select the amount of LSTM layers, the amount of hidden units in each layer, and the input and output dimensions that are going to be used, based on the requirements of the study, and then tweak these variables to optimize model performance. This also involves defining the activation functions, regularization, and optimization techniques.

Step 3: Training the Model: This includes configuring the hyperparameters, initializing the LSTM model, and feeding the training data into the network to propagate the outputs over time. Following that, the loss function between the predicted and ground truth values would be computed and minimized using backpropagation. The model is then tested against the training data, and the parameters are tweaked to increase accuracy.

Step 4: Validating the Model: The testing data is used to forecast air quality values and assess the model's performance. Metrics such as the mean squared error and the R2 score can be computed. For this study, we are going to be using the root mean squared error (RMSE) metric. The model parameters, such as learning rate, dropout, and batch size, are then tweaked to increase the model's performance.

Dataset Description: The weather and pollution data each hour for five years in Beijing, China, were captured and used as the dataset for this study. The data includes the date and time, the level of PM2.5 pollution, and weather details including dew point, temperature, pressure, wind speed, wind direction, and the total number of snowfall and rain hours.

- No: serial number for the row
- year: the year on which data was collected in that particular row
- month: the month on which data was collected in that particular row
- day: the day on which data was collected in that particular row
- hour: the hour at which data was collected in that particular row
- pm2.5: concentration of Particulate Matter 2.5

Figure 2. Flowchart for representing air quality forecasting and predicting $PM_{2.5}$ pollutant

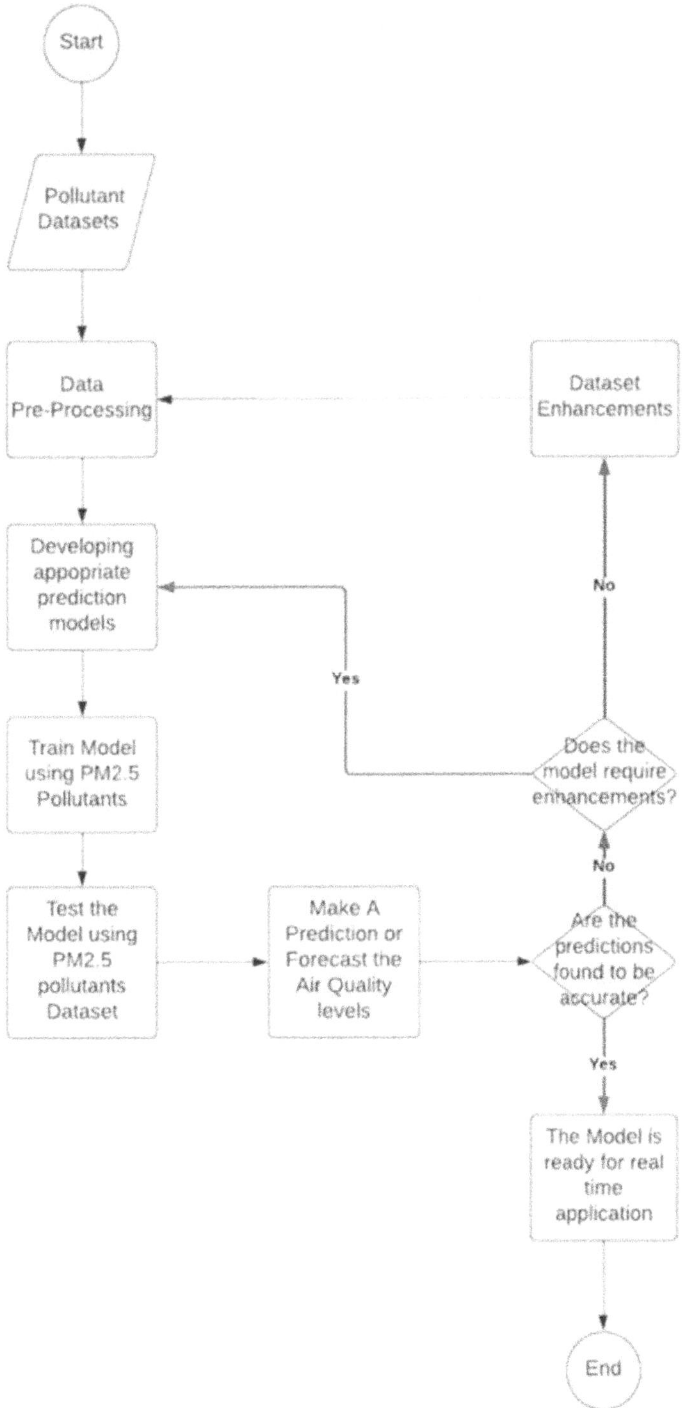

Figure 3. Flowchart for the multivariate LSTM model

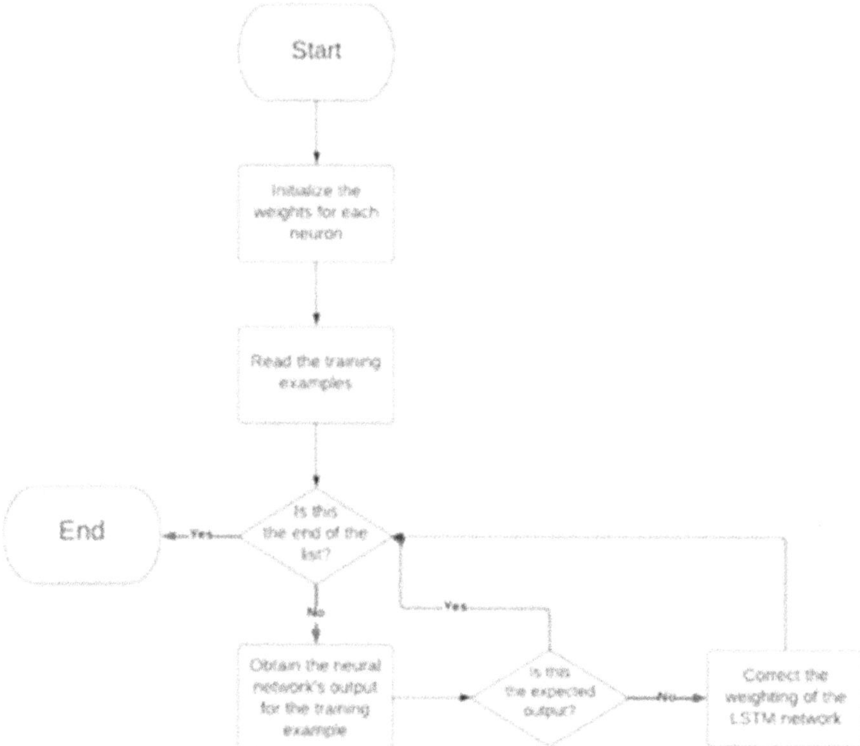

Figure 4. Flowchart for training and testing the LSTM model

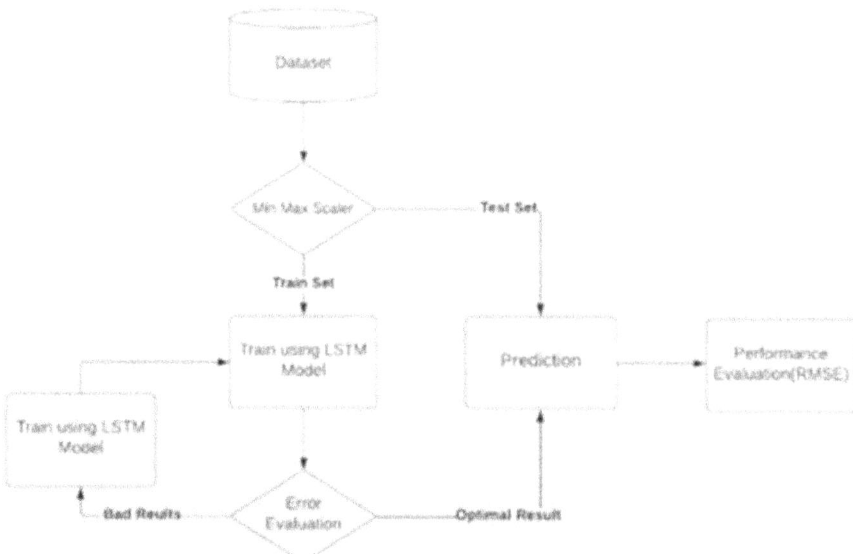

- DEWP: Dew Point recorded at the moment of data collection
- TEMP: Temperature
- PRES: Pressure
- cbwd: overall wind direction
- Iws: cumulative wind speed
- Is: cumulative hours of snow
- Ir: cumulative hours of rain

Using this information, we can construct a forecasting problem where we determine the level of pollution for the next hour while taking into account the weather and pollution from the previous hours. The dataset was split into training and testing stages, with the training phase using the data from the first three years and the testing phase using the data from the final year.

Figure 5. Sample data consisting of pollution levels dewpoint, temperature, wind speed, and snowfall

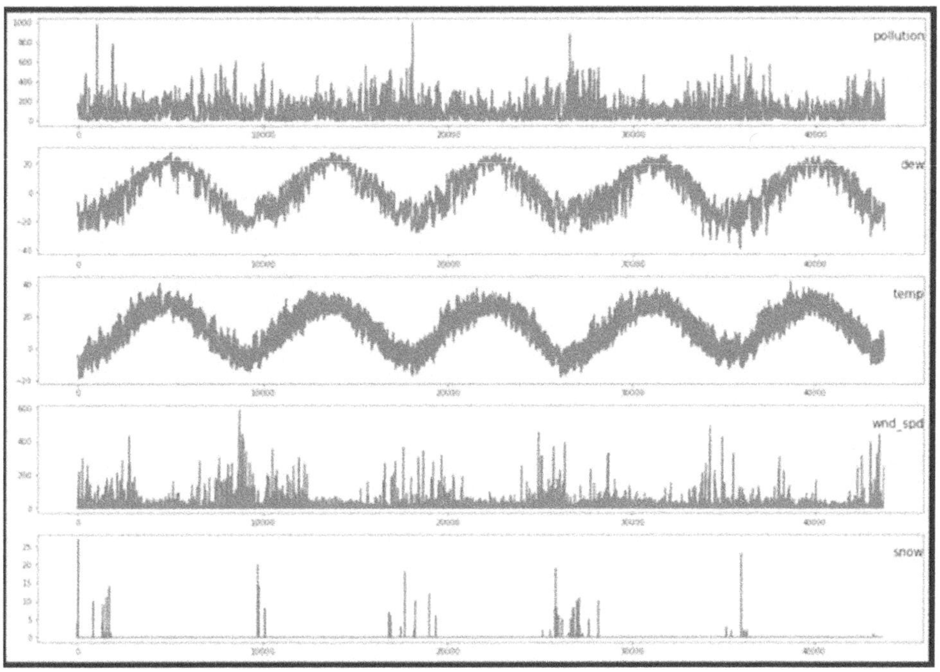

Firstly, since this is a real-world data set it needs to be pre-processed. So the dataset is first scaled using the MinMaxScaler function. This function converts features by scaling each them to a predetermined range. Each feature is scaled and translated separately by this estimator so that it falls within the predetermined range of the training set, in this case between zero and one, where feature range = min, max. This is further reframed and converted to supervised learning data. Finally, the unwanted data and columns are dropped to produce the required dataset. The data in our dataset is not stationary. This suggests that there is a time-dependent structure in the data. Seasonality and trend can be unique features of time-varying distributions, and as a result, the mean and variance of the series fluctuate, making it difficult to model their behaviour properly. Static data is simpler to model and will almost certainly result in more accurate forecasts. The trend can be taken from the observations and then used for forecasting. It can then be scaled back to the original value for prediction. After that, training data and testing data are separated from the pre-processed dataset. Since the dataset contains time series data, LSTM was used as the foundation for the base model, which was then gradually improved.

MODEL DESCRIPTION

The base model has about 64 LSTM units with epochs=50 and one dense layer is added as well. The LSTM cell boosts long-term memory even more effectively because it allows for the learning of extra characteristics. Consequently, it is the best recurrent neural network for prediction, especially when the data exhibits a longer-term trend.

Then the data is inversely scaled for better visual interpretation. This is done mainly for the testing and validation data. To further improve upon the existing base model, another model is created which has more LSTM units and an extra dense layer to fine tune the parameters. It was found that from a total of 288129 parameters there were at least 128 non-trainable parameters. The new model has 256 LSTM memory units, with a dense layer of 64 memory units and a dropout rate of 0.25. The Dropout function is added to prevent overfitting of the new model to the dataset. The new model is then fitted using the test data and training data.

Then, we converted the test data into supervised data from time-series form, for passing it through the neural network trained model for the final testing phase. The final iteration of the model was created with lesser dense layers (32 units) to observe any change to the value loss as well as RMSE value. This model is created to explore possibilities to improve upon the previously created models by observing changes made to the number of units and layers in the model.

EXPERIMENTAL RESULTS

It is found that the base LSTM model can predict the future values of the pollution pretty well for the testing and validation data which is shown in Fig.6. The RMSE value for the testing data is low as well (approximately=2.58 units). This shows that the base model forms a strong foundation to develop better and more optimized models. The new model with 256 LSTM memory units and a dense layer of 64 units had shown significantly better results compared to the base model. As you can see from the plotted graphs in Fig.7, the model prediction values are better fitted to the testing and validation data. The RMSE value is also lower than the base model(approximately = 2.45 units). This indicates a significant improvement in the functioning of the model. Finally, in the third model that was created with lesser dense layers, it was found that the model had significantly dropped in performance, and the prediction values are found to be overfitting the testing data in comparison to the previous two models as shown in Fig.8. It also had a significantly higher RSME value (approximately=12.49) which indicates that the model has a lower loss value of errors.

Table 1. Comparative analysis of the three models after training and testing

Model	Number of LSTM Units	Number of Dense Layers	Dropout Function	Root Mean Squared Error(RMSE)
Base Model	64	1	0	2.5857236
Modified Model 1	256	2 (64 units)	0.25	2.4526236
Modified Model 2	256	2 (32 units)	0.25	2.585187

Figure 6. Graphs showing the accuracy and loss values of base LSTM model

Figure 7. Graphs showing the accuracy and loss values of modified LSTM with 256 memory units

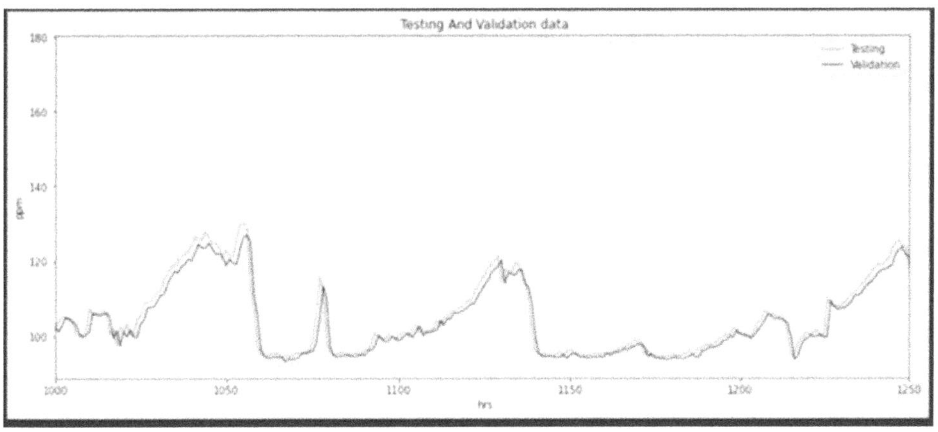

CONCLUSION

This study looked at the issue of predicting $PM_{2.5}$ concentrations on an hourly basis for a period of five years using air quality indicators like temperature, wind speed, pollution levels, dew point, rainfall, and snowfall. To perform these predictions, a base model was created using Multivariate LSTM techniques. This technique was chosen since we are dealing with time series data. The base LSTM model consisted of 64 LSTM memory units and one dense layer. From this base model, we created two more models

Figure 8. Graphs showing the accuracy and loss values of modified LSTM with 32 memory units

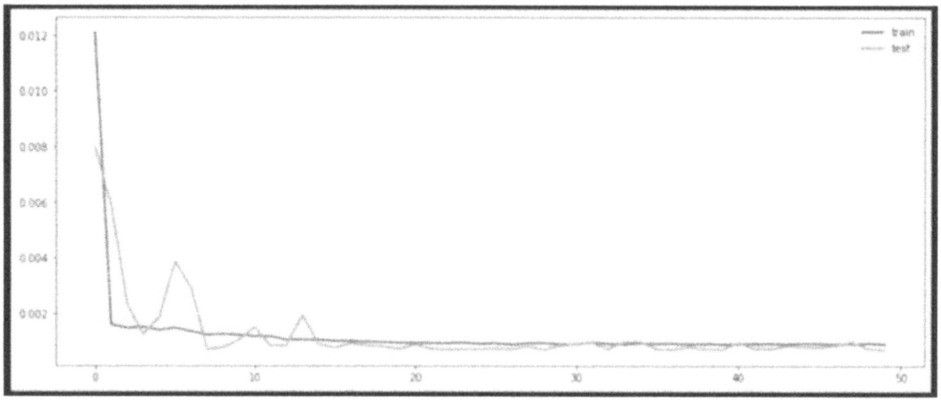

with 256 LSTM memory units and a dropout rate of 0.25. The main difference between these two models was the number of dense layer units it consisted of. The second model had 64 dense layer units while the third model had 32 dense layer units. The three models were then trained and tested using the preprocessed data, which was divided into training data that used the first 3 years of raw data and testing data which used the final years raw data. Among the three models, the second model proved to be more promising with the lowest RMSE value as well as the best quality of fitting between the testing data and validation data. This indicates that if the second model was used in a real time scenario, it would provide predictions with relatively higher accuracy. Although the RMSE for the testing data is rather acceptable, the training data and testing data lose of errors differ significantly, which suggests some overfitting. By employing appropriate techniques, such as regularisation in dense layers, adding to and adjusting dropout layer parameters, and other measures, we can attempt to lessen this. In order to improve forecast accuracy, more research is needed to investigate alternative methodologies (hybrid or ensemble). The implementation and integration of deep learning model variations that contain progressive learning, as well as the introduction of self-identifying techniques for model identification, will probably be the focus of future development.

REFERENCES

Arsov, M., Zdravevski, E., Lameski, P., Corizzo, R., Koteli, N., Gramatikov, S., Mitreski, K., & Trajkovik, V. (2021, February 10). Multi-Horizon Air Pollution Forecasting with Deep Neural Networks. *Sensors (Basel)*, *21*(4), 1235. doi:10.339021041235 PMID:33578633

Belavadi, S. V., Rajagopal, S. R. R., & Mohan, R. (2020). Air Quality Forecasting using LSTM RNN and Wireless Sensor Networks. *Procedia Computer Science*, *170*, 241–248. doi:10.1016/j.procs.2020.03.036

Božnar, M. Z., Mlakar, P., Grašič, B., Gršić, Z., Hettrich, S., Mancini, F., Patryl, L., & Thiessen, K. M. (2022, June 1). Modelling air pollution around nuclear power plants: Validation of dispersion models using tracer data. *Journal of Radiological Protection*, *42*(2), 020519. doi:10.1088/1361-6498/ac7a6f PMID:35775740

Chen, Y., Cui, S., Chen, P., Yuan, Q., Kang, P., & Zhu, L. (2021, March 11). An LSTM-based neural network method of particulate pollution forecast in China. *Environmental Research Letters*, *16*(4), 044006. doi:10.1088/1748-9326/abe1f5

Cordova, C. H., Portocarrero, M. N. L., Salas, R., Torres, R., Rodrigues, P. C., & López-Gonzales, J. L. (2021, December 20). Air quality assessment and pollution forecasting using artificial neural networks in Metropolitan Lima-Peru. *Scientific Reports*, *11*(1), 24232. doi:10.103841598-021-03650-9 PMID:34930975

Ejohwomu, O. A., Shamsideen Oshodi, O., Oladokun, M., Bukoye, O. T., Emekwuru, N., Sotunbo, A., & Adenuga, O. (2022). Modelling and Forecasting Temporal PM2.5 Concentration Using Ensemble Machine Learning Methods. *Buildings*, *12*(1), 46. doi:10.3390/buildings12010046

Guo, Q., He, Z., Li, S., Li, X., Meng, J., Hou, Z., Liu, J., & Chen, Y. (2020). Air Pollution Forecasting Using Artificial and Wavelet Neural Networks with Meteorological Conditions. *Aerosol and Air Quality Research*, *20*(6), 1429–1439. doi:10.4209/aaqr.2020.03.0097

Haq, M. A. (2021). SMOTEDNN: A Novel Model for Air Pollution Forecasting and AQI Classification. *Computers, Materials & Continua*, *71*(1), 1403–1425. doi:10.32604/cmc.2022.021968

Isaev, E., Ajikeev, B., Shamyrkanov, U., Kalnur, K., Maisalbek, K., & Sidle, R. C. (2022). Impact of Climate Change and Air Pollution Forecasting Using Machine Learning Techniques in Bishkek. *Aerosol and Air Quality Research*, *22*(3), 210336. doi:10.4209/aaqr.210336

Kaya, K., & Gündüz Öğüdücü, U. (2020, February 25). Deep Flexible Sequential (DFS) Model for Air Pollution Forecasting. *Scientific Reports*, *10*(1), 3346. doi:10.103841598-020-60102-6 PMID:32098977

Kothandaraman, D., Praveena, N., Varadarajkumar, K., Madhav Rao, B., Dhabliya, D., Satla, S., & Abera, W. (2022, June 27). Intelligent Forecasting of Air Quality and Pollution Prediction Using Machine Learning. *Adsorption Science and Technology*, *2022*, 1–15. doi:10.1155/2022/5086622

Limouni, T., Yaagoubi, R., Bouziane, K., Guissi, K., & Baali, E. H. (2022). Univariate and Multivariate LSTM Models for One Step and Multistep PV Power Forecasting. *International Journal of Renewable Energy Development*, *11*(3), 815–828. doi:10.14710/ijred.2022.43953

Michaelides, S., Paronis, D., Retalis, A., & Tymvios, F. (2017). Monitoring and Forecasting Air Pollution Levels by Exploiting Satellite, Ground-Based, and Synoptic Data, Elaborated with Regression Models. *Advances in Meteorology*, *2017*, 1–17. doi:10.1155/2017/2954010

Samal, K. K., Babu, K. S., Das, S. K., & Acharaya, A. (2019). Time Series based Air Pollution Forecasting using SARIMA and Prophet Model. *Proceedings of the 2019 International Conference on Information Technology and Computer Communications*. IEEE. 10.1145/3355402.3355417

Shokouhifar, M., & Ranjbarimesan, M. (2022). Multivariate time-series blood donation/demand forecasting for resilient supply chain management during COVID-19 pandemic. *Cleaner Logistics and Supply Chain*, *5*, 100078. doi:10.1016/j.clscn.2022.100078

Wu, Z., Zhao, W., & Lv, Y. (2022). An ensemble LSTM-based AQI forecasting model with decomposition-reconstruction technique via CEEMDAN and fuzzy entropy. *Air Quality, Atmosphere & Health*, *15*(12), 2299–2311. doi:10.100711869-022-01252-6 PMID:36196368

Yoo, M. (2022). An Efficient Multivariate Autoscaling Framework Using Bi-LSTM for Cloud Computing. *Applied Sciences (Basel, Switzerland)*, *12*(7), 3523. doi:10.3390/app12073523

Zakaria, N. N., Othman, M., Sokkalingam, R., Daud, H., Abdullah, L., & Abdul Kadir, E. (2019, September 22). Markov Chain Model Development for Forecasting Air Pollution Index of Miri, Sarawak. *Sustainability (Basel)*, *11*(19), 5190. doi:10.3390u11195190

Zhalehdoost, A., & Taleai, M. (2022). A Review of the Application of Machine Learning and Geospatial Analysis Methods in Air Pollution Prediction. *Pollution*, *8*(3), 904–933. doi:10.22059/poll.2022.336044.1300

Chapter 7
Dynamic Application of Unmanned Aerial Vehicles for Analyzing the Growth of Crops and Weeds for Precision Agriculture

I. Daniel Lawrence
https://orcid.org/0000-0002-8958-2515
Garuda Aerospace Private Ltd., India

R. Vijayakumar
Garuda Aerospace Private Ltd., India

J. Agnishwar
Garuda Aerospace Private Ltd., India

ABSTRACT

The resource that the global economy most heavily depends on is agricultural productivity. Nowadays, drones are widely used in a variety of civic applications, including agriculture. However, the aim of this work is to provide an overview of how unmanned aerial vehicles are used in agriculture fields. A development in drone technology for agricultural events can restructure the farm sector to become dynamic and intelligent rather than static and manual. Precision agriculture can be made more effective with the deployment of drones integrated with the proper camera, sensors, and integrating modules. The two main forces behind agriculture automation are precision agriculture and wireless sensor network. Drones, automation, and spectral cameras are all combined to increase the possibilities of managing crop yield and weed issues more effectively. The best design concept and analysis of an autonomous agricultural crop and weed management UAV are discussed. In order to meet the increasing need for food over the past few decades, farming practices have undergone a significant evolution.

DOI: 10.4018/978-1-6684-8516-3.ch007

1. INTRODUCTION

The sustainability of any economy is greatly influenced by the production of agriculture. India depends primarily on agriculture, but it is still far from adopting the most recent technologies to create high-quality farms. Unmanned aerial vehicle (UAVs) application in remote sensing, photogrammetry, and precision agriculture has already begun in developed nations. Precision farming is a concept that is currently being developed to increase productivity and agricultural yields while addressing all issues with labour and agricultural process performance that will immediately impact the agricultural production. This idea aims to conduct an ideal farming procedure. However, This UAVs' application in agriculture is still severely constrained and relies on a variety of criteria, including the weight, flying range, load capacity, construction, and cost of the UAVs themselves. Some studies looked into the systems, technologies, methods, and limitations of UAVs. The creation of a UAV necessitates the use of components and techniques including autonomous flight control, aerodynamic modeling and merging hardware and software systems. It is extremely quick and could lighten a farmer's workload. Drones have made significant advancements in recent years. They have had a big influence on agricultural operations because they have allowed farmers to save a lot of money, run their businesses more profitably, and enhance operational efficiency. Over the past few decades, agricultural drones have been the focus of extensive scholarly investigation. The use of small category remote pilot operated UAVs that weight less than 25 kg, including the payload, is the major emphasis of this study. Typically, these drones have small operating ranges and short flying times, which might vary from a minutes to hours. They are typically used for duties like agricultural monitoring, applying fertilizer, and applying pesticides. The Global agricultural drone market size forecast period 2022 to 2030 (USD Billion) is displayed Figure 1. (Aditya et al, 2016). For the past few decades, Precision farming has made use of remote sensing technologies for a different agricultural sectors, including crop phenology evaluation and yield prediction (Mulla et al, 2013) studied a list of research studies using drones for remote sensing with a variety of objectives., such as pest, pathogen, drought, and nutrient deficiency detection (Barbedo et al, 2019). Precision agriculture is used in a variable field of agricultural contexts, including pest detection and control, fertilization, irrigation, planting, and harvesting (Mahlein AK et al., 2016, Nawar S et al., 2017, Chlingaryan et al., 2018, Gautam et al., 2019, Lama et al., 2020). A variety of machine learning techniques and remote sensing technologies were used to assess agricultural yield output (Diaz-Gonzalez et al., 2022).

According to the findings, unmanned aerial vehicles (UAVs) are much more efficient than satellite systems at evaluating soil indicators and provide superior outcomes in terms of geographic resolution, information timeliness, and flexibility. Therefore, the use of UAVs for Agriculture field monitoring may improve the long-term sustainability of subsequent farming systems that must serve a population that is expanding worldwide. Drone technology integration with other technologies to optimize smart agriculture tasks like agricultural field surveillance, control and monitoring of crops and decision-making. Not only do advances in UAVs, sensor and other associated technologies, as well as their data transfer and various communication methods, need to be made in order to develop agricultural surveillance research (Alsamhi et al., 2021). To satisfy the demands of expanding food production and agricultural demand, the utilization of unmanned aerial vehicles are growing in an agriculture fields.

Figure 1. The Global agricultural drone market size forecast period 2022 to 2030 (USD Billion)

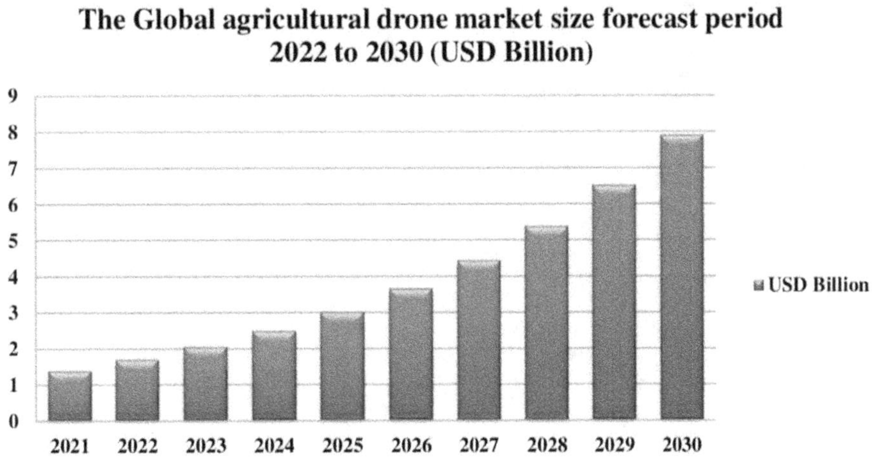

2. INSIGHT INTO MATERIALS

2.1 Unmanned Aerial Vehicle (Drone)

Unmanned aerial vehicles (UAVs) are drones that can fly and are controlled via radios. The multi-rotor UAVs may be further subdivided according to the number of rotors in its platform. Evidently, many UAV models that have been employed during the past two decades. Normal UAV designs include the hexa-copter, which have six rotors in their models. These rotors produce the lift for the Hexa-copter. The three opposite rotors of a six-rotor system rotate clockwise (CW), while the remaining three rotate counterclockwise (CCW) (CCW) (Herwitz et al., 2004, Huang et al., 2009, Kale et al., 2015, Spoorthi et al., 2017, Morey et al., 2017, Yallappa et al., 2017, Sarghini F et al., 2017). In comparison to handheld, ground-based technologies, sensing drones may be able to cover bigger areas. In comparison to manned aircraft and orbital systems, UAVs have the advancement of being able to fly at lower altitudes, which improves spatial resolution in photographs (Aasen et al., 2018). A tiny autonomous mini unmanned rotorcraft vehicle can be built using certain methods and essential parts, such as building hardware components, integrating software systems, controlling autonomous flight, and designing and implementing aerodynamic models (Cai et al., 2010). Different indices can be used by UAVs to observe the crop. In a single flight, UAVs may cover up to hectares of fields. The quadcopter is outfitted with thermal and multi-spectral cameras that are attached on its bottom in order to collect the reflectance of the plant canopy while the monitoring is being made. The camera takes picture per second and sent to the ground station, saves it in memory (Bendig et al., 2012, Colomina et al., 2014). International partners must be careful to abide by the correct laws as different nations may have different requirements for flying and registering drones (Stocker et al., 2017). Figure 2 illustrates the Agriculture Drone's block diagram.

Figure 2. Block diagram of UAV

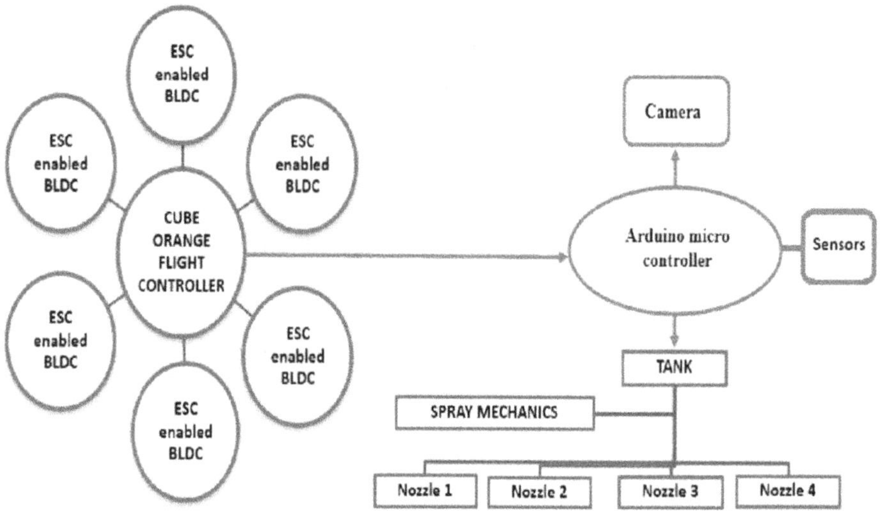

2.2 Multispectral, Hyperspectral, and RGB Sensors

Based on a thorough analysis of the literature, it has been determined that there are mostly three types of cameras deployed for weed patch recognition by UAVs: The information gathered by RGB, hyperspectral, and multispectral cameras for the purpose of identifying weeds is fairly comparable (Louargant et al., 2018). According to the spectral length and quantity of measurements they can capture, the most significant sensors now used as payload can be broadly divided into three classes are RGB (Red, Green, Blue), Hyperspectral and Multispectral sensors. An advanced technique such artificial intelligence and machine learning for object detection, phenology, diseases, and similar applications use Multispectral, hyperspectral and RGB Sensors (Manfreda et al., 2018). Apart from when employing multispectral and hyperspectral sensors, RGB sensors don't require radiometric calibration. It is the non-intrusive, labor-intensive technique that might be employed to identify stress on plants before changes become obvious to the naked eye. Near-infrared light (NIR, 700-1,400 nm) and visible light (PAR, 400-700 nm) spectral ranges are typically assessed by remote sensing equipment for crops, with the majority of studies focusing on the 400-1,000 nm region (Nansen et al., 2016). Commercial cameras with RGB sensors are the most prevalent and widely available. Because of their tremendous potential and inexpensive operating requirements, the majority of research has been concentrating on their prospective uses for years (Jannoura et al., 2015). Although an RGB sensor is only provides a small amount of spectral data. Although a multispectral produces more spectral data, A hyperspectral is often more adept than a multispectral in detecting minute variations in canopy reflectance (Yang et al., 2009). Hyper spectral sensors would need to be mounted on drones designed to carry heavier payloads because they are often larger. Furthermore, they tend to be more expensive, and data processing calls for more time and expertise, which restricts their utility for independent growers (Aasen et al., 2018). The specification of available image sensors like RGB, Hyper and multi spectral sensors are shown in Table 1.

Table 1. The specifications of Multispectral, hyperspectral and RGB cameras

The primary characteristics of multispectral sensors				
S.No	**Camera model**	**Resolution [*Mpx*]**	**Spectral bands**	**Ground sample distance**
1	Micasense RedEdge-M	1280 × 960 (1.2 Mpx per EO band)	Red, Green, Blue, Near-Infrared, Red Edge	8 (per band) at 120 m AGL
2	Micasense RedEdge-MX	1280 × 960 (1.2 Mpx per EO band)	Blue, green, red, red edge, near infrared (NIR)	8 (per band) at 120 m AGL
3	TertaCam MCAW 6	1.3	6 user selectable narrow bands (450–1000 μm)	–
The primary characteristics of hyperspectral sensors				
S.No	**Camera model**	**Lens**	**Spectral range**	**Spectral bands [number and μm]**
1	Hyperspectral camera - Rikola Ltd.	CMOS	500 μm –900 μm	**40 (10μm)**
2	Headwall Photonics - hyperspec X-series NIR	InGaAs	900 μm –1700 μm	**62 (12.9 μm)**
3	Cornirg micro HSI 410 SHARK	CCD/CMOS	400 μm –1000 μm	**300 (2μm)**
4	CUBERT	Snapshot + PAN	450 μm –995 μm	**125 (8μm)**
5	Specim - AISA KESTREL16	Push-broom	600 μm –1640 μm	**350 (3-8μm)**
RGB cameras and their major features				
S.No	**Camera model**	**Sensor type and resolution [*Mpx*]**	**Sensor format**	**Sensor Size (*mm*)**
1	Canon EOS 5d Mark IV	CMOS 30.4	Full Frame	36.0 *mm* × 24.0 *mm*
2	Sony Alpha 7R II	CMOS 42	Full Frame Mirror less	35.0 *mm* × 24.0 *mm*
3	Sony Alpha a6300	CMOS 24	Small Frame Mirror less	23.5 *mm* × 15.6 *mm*
4	Panasonic Lumix DMC GX8	CMOS 20	Small Frame Mirror less	17.3 *mm* × 13 *mm*

A significant problem is data analysis. A hyperspectral sensor often produces numerous terabytes of data per mission, which needs to be carefully stored, processed using particular software, and examined by professionals with extensive knowledge. Although multispectral data processing is now significantly faster than hyperspectral data processing (Yang et al., 2009). Hyperspectral sensors better spectral resolution makes it possible to estimate several characteristics like leaf water content or vegetarianism with greater accuracy (Suomalainen et al., 2014). The new algorithm for picture segmentation was described by Singh V in 2017. This method is employed to categories and identifies various plant leaf diseases. The MATLAB environment is used in this work to detect diseases. Low-altitude, excellent RGB photos are necessary (Singh et al., 2017). IoT technology must be employed to boost the productivity of the food demand in furtherance of fulfil the rising food demand of the expanding global population in light of the steadily decreasing amount of arable land (Ayaz et al., 2019). Multi- and hyperspectral sensors take pictures of things with spatial information in several wavebands. Spatial features or patterns depends on how close an object is to the sensor. Because of this, compared to platforms that operate at low altitudes like drones, satellite photos have lower spatial resolution (Lee et al., 2017). A pre-existing dataset related to the two distinct crops of wheat and rice in India was mined using data mining methods. Prior to conducting more data analysis on highly dimensional data, dimension reduction is necessary to improve Data precision and representation since superfluous characteristics diminish the efficiency of all techniques of analysis of data (Sanghvi et al., 2015). An improvement in crop health monitoring utilizing spectral imaging, which was collected with a spectral camera. By calculating crops' chlorophyll content, the spectral band were calculated to distinguish between healthy and diseased plants. It was then utilized to precisely identify the area under stress.

3. RESEARCH METHODOLOGIES

Crop output and quality can be significantly impacted by biological hazards such bacteria, viruses, weeds, fungi, and insects. All the above are most serious issue and are responsible for a notable yield loss on a global scale. The main reason for the poor productivity is that yield estimation and crop health monitoring are still done using traditional agricultural methods. The instincts and expertise of farmers are the only sources of inspiration for these techniques. As a result of the inaccurate measurement of the area under insect/pest attack in this situation, excessive insecticide and pesticide spraying may occur, which could have a negative influence on the nutrient content of crops (Shafi et al., 2019).

3.1 Crop Management

A crucial component of precision agriculture, continuous crop monitoring necessitates the long-term logging of sensor data. As a result, UAVs that have cameras and sensors can be used to keep an eye on agricultural fields. Crop monitoring can be used to locate pest hotspots, which is the first step in precision pest management. Second, aerial drones can move equipment through fields while mounting precision control systems and both technologies. Another significant benefit of using drones for precision pest control is that it may be more affordable and cause less environmental damage. Drones that use sensors could shorten the time spent searching for pests, while drones that use actuators could limit the area where pesticide sprays are required and cut the cost of doing so (Nansen et al., 2016). Plants respond physiologically to biotic stress, such as herbivory by insect pests, by changing how their leaves reflect light. Such changes can be detected by modern imaging technologies, which can be employed as noninvasive crop monitoring techniques (Iost Filho et al., 2020). The below Table 2, classify optical sensors into RGB, multispectral, and hyperspectral categories, depending on the platform they are mounted on. Reflectance is a technique used to identify water stress in crops. Plant leaves generate energy through thermal emission and fluorescence (Ishimwe et al., 2014). Diseases may seriously harm agricultural areas, reducing their ability to produce. These disorders often spread because of plants and pests. Pesticides and fertilizers are used to get rid of insects and pests in order to solve this problem and improve crop quality (Mogili et al., 2018). UAVs assist in finding land that is ideal for farming and build a detailed agricultural database to support the to improve the quality of the crop, pests and insects are eliminated using pesticides and fertilizers (Hoummaidi et al., 2021). The data gathered from sensors and mapping of spectral images taken by a drone are used to keep track on crop health. Some of the real time trails are shown in detail in Table 2.

Table 2. RGB, multispectral, and hyperspectral categories, depending on the platform

Author	Implementation Details	Techniques	Remarks
Duana et al., 2017.	implementation of NDVI technology as well as the use of multispectral cameras on UAVs to monitor the crop.	Normalized Difference Vegetation Index(NDVI)	Multi-spectral
Reinecke et al., 2017.	Crop growth and yield are monitored using both the UVIRCO camera on the drone and Aerobotics.	Georeference Orthomosaics, NDIV	UVIRCO corona camera, GPS
Kerkech et al., (2018).	Deep learning and UAV RGB images are used to find ESCA illness in vineyard field.	Deep learning of CNN	UAV images, YUV color shapes, vegetation indices
Tejas S. Kabra et al.(2017)	The quadcopter makes it easier to spot problems in agricultural fields.	Georeference	UAV images, GPS

The quadcopter may be used to cultivate crops both indoors and outside. A quad copter is an autonomous aircraft that uses an Android device to spray fertilizer and pesticides. Bluetooth technology is used for real-time operation in the communication between the quad copter and android device. This technology is utilized to lessen issues in the agricultural fields while also increasing crop yield (Kedari et al., 2016). An electrostatic spray technology with a hexa rotor UAV was developed to decrease the loss of pesticide (Yanliang et al., 2017). The discharge and pressure rate of a designed hexacopter mounted sprayer, spray uniformity and liquid reduction, drop concentration and sizes, as well as field and laboratory evolutions, are all investigated (Yallappa et al., 2017). To spray a vast area for this purpose, heavy payload UAVs are needed. Through the use of a PWM controller, the pesticide application system mounted to a UAV performs more effectively (Sarghini et al., 2017, Sarghini et al., 2017). Crop-dusting drones have several advantages over conventional methods, including the ability to transport huge storage reservoirs, the ability to be handled securely and even on their own, and the ability to be maintained for a far lower cost. The most affordable way of crop dusting to below 50 ha is, in the opinion of experts, the use of a drone. Additionally, compared to conventional technology, it is believed that drones can perform aerial spraying up to five times faster (H.Burema et al., 2016, H.Burema at al., 2017). Drones are perfectly suited for agricultural tasks like crop dusting. Spraying crops with powdered fertilizer, pesticides, or insecticides, particularly from the air, is known as crop-dusting. Crops require appropriate fertilizer and pesticide spraying in order to maintain yields (S. Spoorthi et al., 2017, M.M. Vihari et al., 2018). Intelligent agricultural management has been presented, which includes the idea of using drones to spray pesticides, fertilizers, etc. The notion of using pesticides and fertilizers in the area is encouraged, and it ensures that there will be no adverse effects on the farmers. When a disease is detected manually or through technology, a drone in the shape of a quad copter has been developed and put into operation with advancement to spray pesticides for immediate response (Mustafi et al., 2021). In order to boost agricultural efficiency and create the approach for smart farming, that agricultural areas may be partially modernised or fully modernised of agricultural farmlands can be accomplished and approved by the community of farmers. Agricultural drone and Agricultural Field Mapping, Sensing and Actuation by Drone are represented in Figure 3 and Figure 4 respectively.

Figure 3. Agricultural drone

Figure 4. Agricultural field mapping, sensing, and actuation by drone

3.2 Sustainable Weed Management

A rise in global temperatures, an increase in atmospheric carbon dioxide, and unpredictable moisture are the primary impacts of a changing climate. These effects make weed management in agricultural practices difficult (Chauhan et al., 2020). One of the most detrimental abiotic variables in agriculture, weeds significantly reduces yields across the globe. Crop output and quality can be significantly impacted by biological hazards such bacteria, viruses, weeds, fungi, and insects. Weeds are the most serious issue among them and are responsible for a notable yield loss on a global scale. To make quick progress in this subject, it is essential to have a technical and scientific understanding of the methods for determining and the accessible technology. The employment of UAVs and machine learning algorithms allows to the accurate detection of weeds in farmland, which may increase weed control sustainability (Maes et al., 2019). Reconciling agricultural yield with biodiversity preservation is one of agriculture's major concerns globally. A varied weed ecosystem reduced productivity losses in the winter wheat industry (Adeux et al, 2019). The SSWM strategy, the combination of known and new technologies will considerably enhance the long-term viability of weed control. To get rid of weeds, utilise certain algorithms. Examples include using mechanical methods, spraying pesticides, or autonomous weeding robots (Talaviya et al., 2020). To accomplish this purpose, a specialized weed picture dataset must be created. This method must inevitably depend on a dataset of images collected in certain experimental fields (Wspanialy., 2020). UAVs are able to fly over large areas of a field, delivering the photographic data needed to identify weed patches. Different technologies, such as neural networks and object-based image analysis, are used to process these photos (Maes et al., 2019). These weed species are common around the world and can be a significant danger to agricultural production. Therefore, the deployment of UAVs in tandem with image processing techniques is help to control weed efficiently that impact with crops while also providing important environmental advantages (Hunter et al., 2020). The various observation of weeds is given in Table 3.

Table 3. The different types of weed observations for different crops

Crop	Weed (common name)	Weed (Scientific name)	Type of Camera used	Results
Zea mays (Corn)	Babysbreath	Gypsophila paniculata	RGB	Identify different weeds
Hordeum vulgare (Barley)	Bunchgrass	Phalaris minor	RGB	Identify different weeds
Lens esculenta (Lentil)	Egyptian crowfoot grass	Dactyloctenium aegyptium	Hyperspectral	Identify different weeds
Pisum sativum (Peas)	False amaranth	Digera arvensis	Hyperspectral	Identify different weeds
Phaseolus vulgaris (Beans)	Awnless barnyard grass	Echinochloa colona	Hyperspectral	Identify different weeds
Carthamus tinctorius (Safflower)	Ragwort	Jacobaea vulgaris (Senecio jacobaea)	Multispectral	Discriminate weeds in pastures
Cicer arietinum (Chickpea)	Bufel Grass	Cenchrus ciliaris	RGB	Discriminate two different weeds
Triticum spp. (Wheat)	Spinifex	Triodia sp	RGB	Discriminate two different weeds

continues on following page

Table 3. Continued

Crop	Weed (common name)	Weed (Scientific name)	Type of Camera used	Results
Triticosecale (Triticale)	Kochia	Bassia scoparia	Hyperspectral	Discriminate glyphosate and dicamba resistant genotypes from 12sensitive genotypes
Triticosecale (Triticale)	Marestail	Conyza canadensis	RGB	Comparison of cereal genotypes
Triticosecale (Triticale)	Common lambsquarters	Chenopodium album	Hyperspectral	Identify different weeds
Beta vulgaris (Beet root)	Weeds		Multispectral	Discriminate crop Vs weeds
Beta vulgaris (Beet root)	Weeds		Multispectral	Discriminate crop Vs weeds
Beta vulgaris (Beet root)	Ryegrass	Lolium multiforum	Multispectral	Discriminate crop Vs weeds
Beta vulgaris (Beet root)	Thistle	Cirsium arvense	Multispectral	Discriminate crop Vs weeds
Cicer arietinum (Chickpea)	Weeds		Hyperspectral	Discriminate crop Vs weeds
Glycine max (soybean)	Palmer amaranth	Amaranthus palmeri	RGB	Assessment of crop injury from dicamba
Glycine max (soybean)	Barnyardgrass	Echinochloa crus-galli		
Glycine max (soybean)	Large crabgrass	Digitaria sanguinalis		
Heliathus annuus (Sunflower)	Pigweed	Amaranthus blitoides	RGB	Discriminate crop Vs weeds
Heliathus annuus (Sunflower)	Mustard	Sinapis arvensis	Multispectral	Discriminate crop Vs weeds
Heliathus annuus (Sunflower)	Bindweed	Convolvulus arvensis L	Multispectral	Discriminate crop Vs weeds
Heliathus annuus (Sunflower)	Lambsquarters	Chenopodium album L	Multispectral	Discriminate crop Vs weeds
Hordeum vulgare (Barley)	Thistle	Cirsium arvense	RGB	Discriminate crop Vs weeds
Lactuca sativa (Lettuce)	Common groundsel	Senecio vulgaris	Multispectral	Discriminate crop Vs weeds
Lactuca sativa (Lettuce)	Shepherd's purse	Capsella bursa pastoris	Multispectral	Discriminate crop Vs weeds
Lactuca sativa (Lettuce)	Sow thistle	Sonchus spp	Multispectral	Discriminate crop Vs weeds
Sorghum spp. (Millet)	Amaranth	Amaranthus macrocarpus	Hyperspectral	Discriminate crop Vs weeds
Sorghum spp. (Millet)	Pigweed	Portulaca oleracea	Hyperspectral	Discriminate crop Vs weeds
Sorghum spp. (Millet)	Barnyard grass	Echinochloa crus-galli	Hyperspectral	Discriminate crop Vs weeds
Sorghum spp. (Millet)	Mallow	E. colona	Hyperspectral	Discriminate crop Vs weeds
Sorghum spp. (Millet)	Nut grass	Malva spp.	Hyperspectral	Discriminate crop Vs weeds

continues on following page

Table 3. Continued

Crop	Weed (common name)	Weed (Scientific name)	Type of Camera used	Results
Sorghum spp. (Millet)	Fat Hen	Cyperus rotundus Chenopodium album	Hyperspectral	Discriminate crop Vs weeds
Triticum spp (Wheat)	Wild oat	Avena sterilis	Multispectral	Discriminate crop Vs weeds
Triticum spp (Wheat)	Canarygrass	Phalaris canariensis	Multispectral	Discriminate crop Vs weeds
Triticum spp (Wheat)	Ryegrass	Lolium rigidum	Multispectral	Discriminate crop Vs weeds
Triticum spp (Wheat)	Wild oat	Avena fatua	Hyperspectral	Discriminate crop Vs weeds
Triticum spp (Wheat)	Canarygrass	Phalaris canariensis	Multispectral	Discriminate crop Vs weeds
Triticum spp (Wheat)	Ryegrass	Lolium rigidum	Multispectral	Discriminate crop Vs weeds
Vitis vinifera (Grape)	Bermuda grass	Cynodon dactylon	RGB	Discriminate crop Vs weeds
Zea mays (Maize)	Weeds		Multispectral	Discriminate crop Vs weeds
Zea mays (Maize)	Common lambs quarters	Chenopodium album	Multispectral	Discriminate monocotyledons (crops) Vs dicotyledons (weeds)
Zea mays (Maize)	Thistle	Cirsium arvense	Multispectral	Discriminate crop Vs weeds
Zea mays (Maize)	Common lambsquarters	Chenopodium album	Multispectral	Discriminate crop Vs weeds
Zea mays (Maize)	Thistle	Cirsium arvense	Multispectral	Discriminate crop Vs weeds
Zea mays (Maize)	Mat amaranth	Amaranthus blitoides	Multispectral	Discriminate crop Vs weeds
Zea mays (Maize)	Johnsongrass	Sorghum halepense	Multispectral	Discriminate crop Vs weeds
Zea mays (Maize)	Palmer amaranth	Amaranthus palmeri	Hyperspectral	Discriminate glyphosate resistant from glyphosate sensitive weeds
Reddy KN et al., 2014, Garcia-Ruiz et al., 2015, Tiwari O et al., 2019, Sandino J et al., 2018, Scherrer et al., 2019, Svensgaard et al., 2019, De Castro et al., 2017, Fawakherji et al., 2019.				

4. ARTIFICIAL INTELLIGENCE AND MACHINE LEARNING IN PRECISION AGRICULTURE

Artificial Intelligence (AI) and machine learning (ML) are cutting-edge technologies in the agricultural industry. Technology and equipment driven by AI and ML have enhanced the current agriculture sector. This technology has increased real-time agricultural monitoring, harvesting, processing, and selling. The goal of agricultural robot development is to provide high-value AI and ML solutions for the aforementioned sector. The agricultural sector is at risk as a result of the growing world population, but AI and ML may be able to provide a much-needed solution. Due to the AI-based technology solutions, farmers can now produce more with less resources while also improving the quality of their product and reducing the amount of time it takes for their harvests to be sold. 75 million linked devices will be used by farmers by 2020 (Panpatte, 2018). AI is a tool for creativity that models how human intellect and ability are processed by technology, mostly computers, robots, and digital equipment. The use of AI technology may be found in a number of industries that have had the quickest development in recent years, including healthcare, retail, intelligent process automation, and marketing. One of the main tenets of AI, ML enables individuals to operate productively. To learn from datasets and make predictions and choices based on that learning, machine learning (ML) uses statistical and mathematical techniques (Traore et al, 2017). Additionally, the implementation and applications of AI have enormous benefits that have revolutionize the agri-food industry. First off, AI provides more effective process for growing, harvesting, and selling. It also emphasizes spotting flawed crops and increasing the chance of growing healthy harvests. Finally, AI is being used for weather forecasting and 98% accurate disease and pest identification (Sujatha et al 2021). The advanced AI and ML-based technologies has additional benefits for the agri-food supply chain, including lower costs associated with employee training, shorter problem-solving times, fewer human errors, less need for human intervention, and automated decision-making that is accurate, and robust at right time.

5. CONCLUSION

Smart farming will spread to every corner of the globe during the coming years. The use of UAVs in intelligent agriculture has considerable promise. The affordability and simplicity of using UAVs for smart farming, however, may be significant factors in encouraging farmers to employ them. The usage of agricultural drones is one of the main advancements for increasing production in Indian agriculture. An important factor in raising crop output is the necessity to monitor the crops in agri fields. In the end, one or more sensing drones that can identify pest hotspots will connect with another drone that can actuate and dispense agrochemicals precisely where they are needed. These drones may also return to their base of operations on their own, requiring no further human intervention to refuel.

- This research paper's key benefit and purpose is that drones can be used by farmers to spray pesticides and preserve their crops while being controlled by someone operating from a secure location.
- This review briefly examines the use of UAVs for crop, weed monitoring and sprinkling setup.
- The sprinkling structure is used GPS coordinates to automatically go to the diseased regions and spray the insecticides there.
- The design of a drone for RGB (Red, Green, Blue), Multispectral sensors, Hyperspectral sensors has been addressed.
- The area has advanced from creating effective UAV systems to adding advanced methods into the construction of agricultural drones.
- Drones can be used for efficient crop health monitoring and environmental damage prevention; our farmers will be able to procure more yields.

ACKNOWLEDGMENT

This work was partially supported by Garuda Aerospace Research and Development Program under "Precision Agriculture development".

REFERENCES

Aasen, H., & Bolten, A. (2018). Multi-temporal high-resolution imaging spectroscopy with hyperspectral 2D imagers – From theory to applicaton. *Remote Sensing of Environment, 205*, 374–389. doi:10.1016/j.rse.2017.10.043

Aasen, H., Honkavaara, E., Lucieer, A., & Zarco-Tejada, P. (2018). Quantitative remote sensing at ultra-high resolution with UAV spectroscopy: A review of sensor technology, measurement procedures, and data correction workflows. *Remote Sensing (Basel), 10*(7), 1091. doi:10.3390/rs10071091

Adeux, G., Vieren, E., Carlesi, S., Bàrberi, P., Munier-Jolain, N., & Cordeau, S. (2019). Mitigating crop yield losses through weed diversity. *Nature Sustainability, 2*(11), 1018–1026. doi:10.103841893-019-0415-y

Aditya, S., Natu., K., & S., C. (2016). Adoption and Utilization of Drones for Advanced Precision Farming: A Review. *International Journal on Recent and Innovation Trends in Computing and Communication, 4*, 563 - 565.

Alsamhi, S. H., Afghah, F., Sahal, R., Hawbani, A., Al-qaness, M. A. A., Lee, B., & Guizani, M. (2021). Green internet of things using UAVs in B5G networks: A review of applications and strategies. *Ad Hoc Networks, 117*, 102505. doi:10.1016/j.adhoc.2021.102505

Ayaz, M., Ammad-Uddin, M., Sharif, Z., Mansour, A., & Aggoune, E. H. M. (2019). Internet-of-Things (IoT)-based smart agriculture: Toward making the fields talk. *IEEE Access : Practical Innovations, Open Solutions, 7*, 129551–129583. doi:10.1109/ACCESS.2019.2932609

Barbedo, J. G. A. (2019). A review on the use of unmanned aerial vehicles and imaging sensors for monitoring and assessing plant stresses. *Drones (Basel), 3*(2), 40. doi:10.3390/drones3020040

Bendig, J., Bolten, A., & Bareth, G. (2012). Introducing a low-cost mini-UAV for thermal-and multispectral-imaging. *The International Archives of the Photogrammetry, Remote Sensing and Spatial Information Sciences, 39*(B1), 345–349. doi:10.5194/isprsarchives-XXXIX-B1-345-2012

Burema, H., & Filin, A. (2016). *Aerial Farm Robot System for Crop Dusting, Planting, Fertilizing and Other Field Jobs.* United States Patent US9382003B2.

Cai, G., Chen, B. M., & Lee, T. H. (2010). An overview on development of miniature unmanned rotorcraft systems. *Frontiers of Electrical and Electronic Engineering in China, 5*(1), 1–14. doi:10.100711460-009-0065-3

Chauhan, B. S. (2020). Grand challenges in weed management. *Frontiers in Agronomy, 1*, 3. doi:10.3389/fagro.2019.00003

Chlingaryan, A., Sukkarieh, S., & Whelan, B. (2018). Machine learning approaches for crop yield prediction and nitrogen status estimation in precision agriculture: A review. *Computers and Electronics in Agriculture, 151*, 61–69. doi:10.1016/j.compag.2018.05.012

Colomina, I., & Molina, P. (2014). Unmanned aerial systems for photogrammetry and remote sensing: A review. *ISPRS Journal of Photogrammetry and Remote Sensing*, *92*, 79–97. doi:10.1016/j.isprsjprs.2014.02.013

De Castro, A. I., Peña, J. M., Torres-Sánchez, J., Jiménez-Brenes, F., & López-Granados, F. (2017). mapping Cynodon dactylon in vineyards using UAV images for site-specifc weed control. *Advances in Animal Biosciences*, *8*(2), 267–271. doi:10.1017/S2040470017000826

Diaz-Gonzalez, F. A., Vuelvas, J., Correa, C. A., Vallejo, V. E., & Patino, D. (2022). Machine learning and remote sensing techniques applied to estimate soil indicators – review. *Ecological Indicators*, *135*, 108517. doi:10.1016/j.ecolind.2021.108517

Duan, T., Chapman, S. C., Guo, Y., & Zheng, B. (2017). Dynamic monitoring of NDVI in wheat agronomy and breeding trials using an unmanned aerial vehicle. *Field Crops Research*, *210*, 71–80. doi:10.1016/j.fcr.2017.05.025

El Hoummaidi, L., Larabi, A., & Alam, K. (2021). Using unmanned aerial systems and deep learning for agriculture mapping in Dubai. *Heliyon*, *7*(10), e08154. doi:10.1016/j.heliyon.2021.e08154 PMID:34703924

Fawakherji, M., Potena, C., Bloisi, D. D., Imperoli, M., Pretto, A., & Nardi, D. (2019). UAV image based crop and weed distribution estimation on embedded GPU boards. *Communications in Computer and Information Science*, *1089*, 100–108. doi:10.1007/978-3-030-29930-9_10

Garcia-Ruiz, F. J., Wulfsohn, D., & Rasmussen, J. (2015). Sugar beet (Beta vulgaris L.) and thistle (Cirsium arvensis L.) discrimination based on feld spectral data. *Biosystems Engineering*, *139*, 1–15. doi:10.1016/j.biosystemseng.2015.07.012

Gautam, P. V., Kushwaha, H. L., Kumar, A., & Kumar, D. (2019). Mechatronics application in precision sowing: A review. *International Journal of Current Microbiology and Applied Sciences*, *8*(4), 1793–1807. doi:10.20546/ijcmas.2019.804.208

Herwitz, S. R., Johnson, L. F., Dunagan, S. E., Higgins, R. G., Sullivan, D. V., Zheng, J., Slye, R. E. (2004) Imaging from an unmanned aerial vehicle: agricultural surveillance and decision support. *Computers and electronics in agriculture, 44*(1).

Huang, J.-R., Sun, J.-Y., Liao, H.-J., & Liu, X.-D. (2015a). Detection of brown planthopper infestation based on SPAD and spectral data from rice under different rates of nitrogen fertilizer. *Precision Agriculture*, *16*(2), 148–163. doi:10.100711119-014-9367-4

Huang, Y., Hoffmann, W. C., Lan, Y., Wu, W., & Fritz, B. K. (2009). Development of a spray system for an unmanned aerial vehicle platform. *Applied Engineering in Agriculture*, *25*(6), 803–809. doi:10.13031/2013.29229

Hunter, J. E. III, Gannon, T. W., Richardson, R. J., Yelverton, F. H., & Leon, R. G. (2020). Integration of remote-weed mapping and an autonomous spraying unmanned aerial vehicle for site-specific weed management. *Pest Management Science*, *76*(4), 1386–1392. doi:10.1002/ps.5651 PMID:31622004

Iost Filho, F. H., Heldens, W. B., Kong, Z., & de Lange, E. S. (2020). Drones: Innovative technology for use in precision pest management. *Journal of Economic Entomology, 113*(1), 1–25. doi:10.1093/jee/toz268 PMID:31811713

Ishimwe, R., Abutaleb, K., & Ahmed, F. (2014). Applications of Thermal Imaging in Agriculture—A Review. *Advances in Remote Sensing, 3*(3), 128–140. doi:10.4236/ars.2014.33011

Jannoura, R., Brinkmann, K., Uteau, D., Bruns, C., & Joergensen, R. G. (2015). Monitoring of crop biomass using true colour aerial photographs taken from a remote controlled hexacopter. *Biosystems Engineering, 129*, 341–351. doi:10.1016/j.biosystemseng.2014.11.007

Kabra, T. S., Kardile, A. V., Deeksha, M. G., Mane, D. B., Bhosale, P. R., & Belekar, A. M. (2017). Design, Development & Optimization of a Quad-Copter for Agricultural Applications. *International Research Journal of Engineering and Technology, 04*(7).

Kale, S. D., Khandagale, S. V., Gaikwad, S. S., Narve, S. S., & Gangal, P. V. (2015). Agriculture Drone for Spraying Fertilizer and Pesticides. Published in *International journal of Advanced Research in Computer science and Software engineering, 5*(12), 804- 807.

Kedari, S., Lohagaonkar, P., Nimbokar, M., Palve, G., & Yevale, P. (2016). Quadcopter-A Smarter Way of Pesticide Spraying. *Imperial Journal of Interdisciplinary Research, 2*(6).

Kerkech, M., Hafiane, A., & Canals, R. (2018). Deep leaning approach with colorimetric spaces and vegetation indices for vine diseases detection in UAV images. *Computers and Electronics in Agriculture, 155*, 237–243. doi:10.1016/j.compag.2018.10.006

Lama, G. F. C., Errico, A., Francalanci, S., Solari, L., Preti, F., & Chirico, G. B. (2020). of flow resistance models based on field experiments in a partly vegetated reclamation channel. *Geosciences, 10*(2), 47. doi:10.3390/geosciences10020047

Lee, H., Moon, A., Moon, K., & Lee, Y. *Disease and pest prediction IoT system in orchard: A preliminary study*. In *Proceedings of the 2017 Ninth International Conference on Ubiquitous and Future Networks (ICUFN)*, Milan, Italy. 10.1109/ICUFN.2017.7993840

Louargant, M., Jones, G., Faroux, R., Paoli, J. N., Maillot, T., Gée, C., & Villette, S. (2018). Unsupervised classification algorithm for early weed detection in row-crops by combining spatial and spectral information. *Remote Sensing (Basel), 10*(5), 1–18. doi:10.3390/rs10050761

Maes, W. H., & Steppe, K. (2019). Perspectives for remote sensing with unmanned aerial vehicles in precision agriculture. *Trends in Plant Science, 24*(2), 152–164. doi:10.1016/j.tplants.2018.11.007 PMID:30558964

Maes, W. H., & Steppe, K. (2019). Perspectives for remote sensing with unmanned aerial vehicles in precision agriculture. *Trends in Plant Science, 24*(2), 152–164. doi:10.1016/j.tplants.2018.11.007 PMID:30558964

Mahlein, A. K. (2016). Plant disease detection by imaging sensors—Parallels and specific demands for precision agriculture and plant phenotyping. *Plant Disease, 100*(2), 241–251. doi:10.1094/PDIS-03-15-0340-FE PMID:30694129

Manfreda, S., McCabe, M. E., Miller, P. E., Lucas, R., Madrigal, V. P., & Mallinis, G. (2018). On the use of unmanned aerial systems for environmental monitoring. *Remote Sensing (Basel), 10*(4), 641. doi:10.3390/rs10040641

Mogili, U. R., & Deepak, B. B. V. L. (2018). Review on application of drone systems in precision agriculture. *Procedia Computer Science, 133*, 502–509. doi:10.1016/j.procs.2018.07.063

Morey, N. S., Mehere, P. N., & Hedaoo, K. (2017). Agriculture Drone for Fertilizers and Pesticides Spraying. *International Journal for engineering applications and Technology.*

Mulla, D. J. (2013). Twenty five years of remote sensing in precision agriculture: Key advances and remaining knowledge gaps. *Biosystems Engineering, 114*(4), 358–371. doi:10.1016/j.biosystemseng.2012.08.009

Mustafi, S., Ghosh, P., Roy, K., Dan, S., Mukherjee, K., & Nath Mandal, S. (2021). Drones for Intelligent Agricultural Management. In P. Krause & F. Xhafa (Eds.), *IoT-based Intelligent Modelling for Environmental and Ecological Engineering. Lecture Notes on Data Engineering and Communications Technologies* (Vol. 67). Springer. doi:10.1007/978-3-030-71172-6_4

Nansen, C. (2016). The potential and prospects of proximal remote sensing of arthropod pests. *Pest Management Science, 72*(4), 653–659. doi:10.1002/ps.4209 PMID:26663253

Nansen, C., & Elliott, N. (2016). Remote sensing and reflectance profiling in entomology. *Annual Review of Entomology, 61*(1), 139–158. doi:10.1146/annurev-ento-010715-023834 PMID:26982438

Nawar, S., Corstanje, R., Halcro, G., Mulla, D., & Mouazen, A. M. (2017). Delineation of soil management zones for variable-rate fertilization: A review. *Advances in Agronomy, 143*, 175–245. doi:10.1016/bs.agron.2017.01.003

Panpatte, D. G. (2018). *Artificial Intelligence in Agriculture: An Emerging Era of Research. Intutional Science.* CANADA.

Reddy, K. N., Huang, Y., Lee, M. A., Nandula, V. K., Fletcher, R. S., Thomson, S. J., & Zhao, F. (2014). Glyphosate-resistant and glyphosate-susceptible Palmer amaranth(Amaranthus palmeri S. Wats.): Hyperspectral refectance properties of plants and potential for classifcation. *Pest Management Science, 70*(12), 1910–1917. doi:10.1002/ps.3755 PMID:24497403

Reinecke, M., & Prinsloo, T. (2017, July). The influence of drone monitoring on crop health and harvest size. *International Conference on Next Generation Computing Applications.* IEEE. 10.1109/NEXTCOMP.2017.8016168

Salnikov, V., Filin, A., & Burema, H. Hybrid Airship-Drone Farm Robot System for Crop Dusting, Planting, Fertilizing and Other Field Jobs, United States Patent US9852644B2, Dec. 26, 2017.

Sandino, J., & Gonzalez, F. A Novel Approach for invasive weeds and vegetation surveys using UAS and artifcial intelligence. 2018 23rd Int Conf Methods Models Autom Robo. 2018.

Sanghvi, Y., Gupta, H., Doshi, H., Koli, D., Ansh, A., & Gupta, U. Comparison of Self organizing maps and Sammon's mapping on agricultural datasets for precision agriculture. In Proceedings of the 2015. International Conference on Innovations in Information, Embedded and Communication Systems (ICIIECS), Coimbatore, India, 19–20 March 2015; pp. 1–5. 10.1109/ICIIECS.2015.7193206

Sarghini, F., & De Vivo, A. (2017). Analysis of preliminary design requirements of a heavy lift multi rotor drone for agricultural use. *Chemical Engineering Transactions*, *58*, 625–630.

Sarghini, F., & De Vivo, A. (2017). Interference analysis of a heavy lift multi rotor drone flow field and transported spraying system. [30]. *Chemical Engineering Transactions*, *58*, 631–636.

Sarghini, F., & De Vivo, A. (2017). Analysis of preliminary design requirements of a heavy lift multi rotor drone for agricultural use. *Chemical Engineering Transactions*, *58*, 625–630.

Scherrer, B., Sheppard, J., Jha, P., & Shaw, J. A. (2019). Hyperspectral imaging and neural networks to classify herbicide-resistant weeds. *Journal of Applied Remote Sensing*, *13*(4), 1. doi:10.1117/1.JRS.13.044516

Shafi, U., Mumtaz, R., García-Nieto, J., Hassan, S. A., Zaidi, S. A. R., & Iqbal, N. (2019). Precision agriculture techniques and practices: From considerations to applications. *Sensors (Basel)*, *19*(17), 3796. doi:10.339019173796 PMID:31480709

Singh, V., & Misra, A. K. (2017). Detection of plant leaf diseases using image segmentation and soft computing techniques. *Information Processing in Agriculture*, *4*(1), 41–49. doi:10.1016/j.inpa.2016.10.005

Spoorthi, S., & Manasa, V. K. (2017). Freyr drone: pesticide/fertilizers spraying drone – an agricultural approach. International Conference on Computing and Communications Technologies, (pp. 252–255). IEEE. 10.1109/ICCCT2.2017.7972289

Spoorthi, S., Shadaksharappa, B., Suraj, S., & Manasa, V. K. (2017). Freyr drone: Pesticide/fertilizers spraying drone-an agricultural approach. *IEEE 2nd International Conference on In Computing and Communications Technologies (ICCCT - 2017),* (pp. 252-255). IEEE. 10.1109/ICCCT2.2017.7972289

Stöcker, C., Bennett, R., Nex, F., Gerke, M., & Zevenbergen, J. (2017). Review of the current state of UAV regulations. *Remote Sensing (Basel)*, *9*(5), 459. doi:10.3390/rs9050459

Sujatha, R., Chatterjee, J. M., Jhanjhi, N., & Brohi, S. N. (2021). Performance of deep learning vs machine learning in plant leaf disease detection. *Microprocessors and Microsystems*, *80*, 103615. doi:10.1016/j.micpro.2020.103615

Suomalainen, J., Anders, N., Iqbal, S., Roerink, G., Franke, J., Wenting, P., Hünniger, D., Bartholomeus, H., Becker, R., & Kooistra, L. (2014). A lightweight hyperspectral mapping system and photogrammetric processing chain for unmanned aerial vehicles. *Remote Sensing (Basel)*, *6*(11), 11013–11030. doi:10.3390/rs61111013

Svensgaard, J., Jensen, S. M., Westergaard, J. C., Nielsen, J., Christensen, S., & Rasmussen, J. (2019). Can reproducible comparisons of cereal genotypes be generated in feld experiments based on AV imagery using RGB cameras? *European Journal of Agronomy*, *106*, 49–57. doi:10.1016/j.eja.2019.03.006

Talaviya, T., Shah, D., Patel, N., Yagnik, H., & Shah, M. (2020). Implementation of artificial intelligence in agriculture for optimisation of irrigation and application of pesticides and herbicides. *Artificial Intelligence in Agriculture*, *4*, 58–73. doi:10.1016/j.aiia.2020.04.002

Tiwari, O., Goyal, V., Kumar, P., & Vij, S. (2018). An experimental set up for utilizing convolutional neural network in automated weed detection. Proc 2019 4th Int Conf Internet Things Smart Innov Usages. IEEE.

Traore, B. B., Kamsu-Foguem, B., & Tangara, F. (2017). Data mining techniques on satellite images for discovery of risk areas. *Expert Systems with Applications*, *72*, 443–456. doi:10.1016/j.eswa.2016.10.010

Vihari, M. M., & Nelakuditi, U. R. (2018). IoT based unmanned aerial vehicle system for agriculture applications, in: International Conference on Smart Systems and Inventive Technology, (pp. 26–28). IEEE. 10.1109/ICSSIT.2018.8748794

Wspanialy, P., Brooks, J., & Moussa, M. (2020). *An image labeling tool and agricultural dataset for deep learning*. arXiv preprint arXiv:2004.03351.

Yallappa, D., Veerangouda, M., Maski, D., Palled, V., & Bheemanna, M. (2017, October) Development and evaluation of drone mounted sprayer for pesticide applications to crops. *IEEE Global Humanitarian Technology Conference (GHTC) 2017 IEEE* (pp. 1-7). IEEE. 10.1109/GHTC.2017.8239330

Yallappa, D., Veerangouda, M., Maski, D., Palled, V., & Bheemanna, M. (2017, October). Development and evaluation of drone mounted sprayer for pesticide applications to crops. *IEEE Global Humanitarian Technology Conference (GHTC) 2017 IEEE* (pp. 1-7). IEEE. 10.1109/GHTC.2017.8239330

Yang, C., Everitt, J. H., Bradford, J. M., & Murden, D. (2009a). Comparison of airborne multispectral and hyperspectral imagery for estimating grain sorghum yield. *Transactions of the ASAE. American Society of Agricultural Engineers*, *52*, 641–649.

Yang, Z., Rao, M. N., Elliott, N. C., Kindler, S. D., & Popham, T. W. (2009). Differentiating stress induced by greenbugs and Russian wheat aphids in wheat using remote sensing. *Computers and Electronics in Agriculture*, *67*(1-2), 64–70. doi:10.1016/j.compag.2009.03.003

Yanliang, Z., Qi, L., & Wei, Z. (2017). Design and test of a six-rotor unmanned aerial vehicle (UAV) electrostatic spraying system for crop protection. *International Journal of Agricultural and Biological Engineering*, *10*(6), 68–76. doi:10.25165/j.ijabe.20171006.3460

Chapter 8
YOLO-IP:
An Efficient and Robust Deep Learning Framework to Detect Insect Pests for Agricultural Applications

Rashmiranjan Nayak
https://orcid.org/0000-0002-1380-701X
National Institute of Technology, Rourkela, India

Prashant Dwivedi
National Institute of Technology, Rourkela, India

Umesh Chandra Pati
https://orcid.org/0000-0001-9805-2543
National Institute of Technology, Rourkela, India

ABSTRACT

The food security of any country may be jeopardized due to improper management of agricultural insect pests. Accurate pest detection and efficient pest control strategies must be employed in time to grow healthy crops for achieving food security in a country and worldwide. Hence, developing efficient and robust techniques to detect agricultural insect pests using computer vision approaches is one of the essential steps for timely managing insect pests. This chapter presents a short survey of deep learning-based object detection techniques focusing on insect pest detection and associated insect pest image datasets. Subsequently, a transfer learning-based custom You Only Look Once (YOLOv5) model is developed using the publicly available dataset IP102 for detecting agricultural insect pests with the help of computer vision approaches. The hyperparameters of the proposed insect pest detector are optimized using the genetic algorithm-based hyperparameter evolution method. The performance metrics of the proposed insect pest detector are found to be promising.

DOI: 10.4018/978-1-6684-8516-3.ch008

INTRODUCTION

It has been observed that from the post-green revolution decades to the early twentieth century, crop losses due to agricultural insect pests have decreased from 13.6% to 10.8% (Dhaliwal, Jindal, & Mohindru, 2015). India is mostly an agriculture-based economy. However, the farmers and residents in India are dealing with agriculture pests (such as rice leaf roller, grub, aphids, peach borer, red spider, and so on) in their fields as well as house gardens, respectively (Wu, Zhan, Lai, Cheng, & Yang, 2019). Food security in the country is jeopardized due to agricultural insect pests, and agriculture in India suffers a total annual crop loss of approximately US$36 billion (Dhaliwal, Jindal, & Mohindru, 2015). Since the post-green revolution decades to the early twentieth century, crop losses due to agricultural insect pests in India have decreased from 23.3% to 15.7% (Dhaliwal, Jindal, & Mohindru, 2015). Accurate pest detection and efficient pest control strategies must be employed in time to grow healthy crops for achieving food security in the country and worldwide. Hence, developing efficient and robust techniques to detect agricultural insect pests using computer vision approaches is one of the essential steps for timely managing insect pests.

Motivation

Due to the country's rapidly expanding population, the Indian government must ensure food security for its people. However, insect pests play a critical role in destroying healthy crops, making it challenging to manage agriculture-based product production. Farmers are spraying pesticides (chemicals) in an unregulated quantity to combat the problem of insect pests and associated diseases. It harms the crop, soil, water, and the health of the consumers. The majority of the time, pesticides (chemicals) are chosen using a labor-intensive human-centric approach. Farmers identify the insect pest in the crop using manual inspection, which mostly leads to wrong classification of the insect pests due to the high resemblances among the various insect pests. Subsequently, incorrect selection of pesticide (chemical) happens, resulting in severe crop and soil damage as well as a significant financial loss for the farmer. So, the use of the required quantity of pesticides (chemicals) in the crop is dependent on the desired information about the insect pest in the crop, such as where the insect pest is widely distributed (location) and what categories they belong to (classification). Further, agricultural pests are of various sizes ranging from tiny, small, and large. Usually, the insect pests reside over the body or leaf of the crop as overlapping clusters. Usually, there is a lack of proper visibility in the crop field due to the heavily populated plants in the fields. Hence, key research challenges of insect pest detection include high variation in the object sizes, complex background, occlusion, and low illumination (Ahmad, et al., 2022). Hence, these research challenges make object detection even more challenging for insect pest detection applications. Due to the availability of massive computational facilities and large datasets, a data-driven technique like deep learning-based computer vision approaches can efficiently classify and detect agricultural insect pests (Domingues, Brandão, & Ferreira, 2022; Saleem, Potgieter, & Arif, 2021). These are the major motivating factors for carrying out the research on this topic.

Contributions

The key contributions of the research work are briefly outlined as follows.

- A short survey of the Deep Learning (DL)-based object detection techniques focusing on insect pest detection is presented.
- A brief discussion about the publicly available image datasets for detecting agricultural insect pests is covered.
- A transfer learning-based custom You Only Look Once (YOLOv5) model is developed using the publicly available dataset IP102 for detecting agricultural insect pests with the help of computer vision approaches.
- The proposed insect detector is improved by optimizing its hyperparameters using the Genetic Algorithm-based hyperparameter evolution method.
- A brief comparative analysis of the proposed insect pest detector with state-of-the-art techniques is presented to validate its effectiveness.

Organization

The remainder of the chapter is structured as follows. A short survey of deep learning-based object detection techniques focusing on insect pest detection is presented in Section 2. Section 3 briefly discusses the publicly available image datasets for detecting agricultural insect pests. Section 4 proposes a transfer learning-based YOLOv5 model using the publicly available dataset IP102 for detecting agricultural insect pests with high accuracy. Experimental results with critical evaluations are discussed in Section 5. At last, the chapter is concluded in Section 6.

RELATED WORKS

There is an increasing trend in market competition for agricultural-based products at both quality and quantity levels. Hence, smart farming is the only solution to produce the best quality products based on agriculture. Appropriate tackling of the problems associated with crop production due to agricultural insect pests is essential for yielding the best quality crops. If the presence of insect pests in crops is detected early, then timely treatment incorporating suitable pesticides in appropriate quantities may be done to minimize crop loss. Object detection in computer vision is defined as the process of the identification and localization of the objects present within an image. Object detection generates bounding boxes around the detected objects, determining where the objects are in a given scene (Sabat, Pati, & Das, 2021). In other words, object detection combines categorization and localization to identify which objects are present in an image and where they are located in the image. DL-based approaches are preferred to machine learning and traditional computer vision-based approaches for object detection due to their 1) ability to provide high accuracy, 2) ability to provide high processing speed, and 3) robustness in a complex scene. Because of the widespread use of DL approaches in object detection techniques, a number of methods for detecting (both classification and localization) agricultural insect pests have been developed. Hence, in this Section, a short survey of DL-based object detection techniques with a focus on insect pest detection is presented.

Object detectors are broadly classified into two types: single-stage detectors and two-stage detectors.

Single-Stage Object Detectors

Single-stage object detectors are based on the regression proposal-based approach. The single-stage object detectors perform the classification and localization (bounding-box regression) of the objects without using pre-generated region proposals. Single Shot Detection (SSD) (Liu, et al., 2016) and You Only Look Once (YOLO) (Redmon & Farhadi, 2018) are the widely used single-stage objection methods. One-stage detectors have higher inference speeds for object detection than two-stage detectors. Recently, a lightweight, small object detection algorithm using SSD feature fusion has been prosed to detect grain pests (Lyu, Jin, Zhen, Sun, & Xu, 2021). Further, YOLO (Redmon, Divvala, Girshick, & Farhadi, 2016) (Redmon & Farhadi, 2017) is widely used for object detection due to its fast detection capability with high accuracy. The YOLO is preferred over other DL-based object detectors, such as SSD and Region proposal-based approaches, due to its high accuracy, low false alarm, and fast response in real-time applications (Redmon & Farhadi, 2017). Prior to YOLO, object detection models localize the objects in the image by using regions. YOLO model looks at the portions of the input images where the probability of finding the object is high instead of focusing on the entire input image. Further, a single convolutional neural network (CNN) in YOLO object detectors performs object identification (prediction of class probabilities) and object localization (bounding box) (Redmon & Farhadi, 2018). Hence, a short survey of the most widely used single-stage detector, i.e., the YOLO series, is presented as follows.

YOLOv1

In YOLOv1, the object detection problem is modeled as a regression problem. Here, the input image is divided into an S × S grid, and B bounding boxes are predicted with corresponding confidence scores and C class probabilities (Redmon, Divvala, Girshick, & Farhadi, 2016). The object is detected by that grid cell where the center of an object falls. Further, Confidence is defined as $\Pr(Object) * IoU_{pred}^{truth}$ (Redmon, Divvala, Girshick, & Farhadi, 2016). Confidence is zero only when no object is present in the grid cell (Redmon, Divvala, Girshick, & Farhadi, 2016). Otherwise, the confidence score = IoU_{pred}^{truth} (Redmon, Divvala, Girshick, & Farhadi, 2016). Five predictions, such as x, y, w, h, and confidence, are made for every bounding box. Further, (x, y) denotes the box center related to the grid cell-bound, (w, h) denotes width and height pair of the whole image. For getting the confidence scores related to each box class, during test time, the conditional class probabilities (CCP) and the individual box confidence predictions are multiplied as shown in Eq. (1) (Redmon, Divvala, Girshick, & Farhadi, 2016).

$$\Pr\left(Class_i|Object\right) * \Pr\left(Object\right) * IoU_{pred}^{truth} = \Pr\left(Class_i\right) * IoU_{pred}^{truth} \quad (1)$$

GoogLeNet model inspired the network architecture of YOLOv1 for the task of classification of images. It used a total of twenty-four convolutional layers to extract features from the sample images, followed by two fully connected layers to predict the output probabilities and coordinates. The final layer of the model provides the class probabilities and the bounding box coordinates (Redmon, Divvala, Girshick, & Farhadi, 2016). The bounding box coordinates are now normalized by dividing the box width and height by the image width and height that enforce the values between 0 and 1 (Redmon, Divvala, Girshick, & Farhadi, 2016). To constrain the x and y coordinates of each bounding box between 0 and 1, they must be parametrized so that the associated offsets from the location of a specific grid cell are managed (Redmon, Divvala, Girshick, & Farhadi, 2016). The final layer in the model uses a linear activation function, while the other layers in the model use a leaky rectified linear activation function (Redmon, Divvala, Girshick, & Farhadi, 2016). Two parameters such as λc_{ord} and $\lambda no_{obj\ a}$re widely used to overcome the disadvantages of the sum of squared errors. $\lambda \mathrm{cor}_{d\ is}$ set to 5 to increase the loss of predictions of coordinates of each bounding box and $\lambda \mathrm{noo}b_{j\ is}$ set to 0.5 to decrease the loss of predictions of confidence of each bounding box that does not contain an object (Redmon, Divvala, Girshick, & Farhadi, 2016). To avoid overfitting, a dropout rate of 0.5 is used in front of the first connected layer to prevent co-adaptation among layers. For data augmentation, the image exposure and sample saturation are arbitrarily set to a factor of 1.5 in the HSV (Hue, Saturation, Value) color space (Redmon, Divvala, Girshick, & Farhadi, 2016). A translation and scaling of 20% of the authentic image size are also used for extensive data augmentation.

YOLO9000

Instead of model dropout, batch normalization is applied in each convolutional layer of the YOLO architecture to attain regularization without overfitting (Redmon & Farhadi, 2017). In YOLOv2 or YOLO9000, the first full resolution 448×448 images are used in the classification network for at least ten epochs on the ImageNet dataset. Subsequently, the filters adjusted and reset the resulting resolutions for the object detection by the trained network. Anchor boxes are used to predict bounding boxes in object detection sample images instead of a fully connected layer. The input image resolution of the YOLO9000 architecture is reduced from 448 × 448 to 416 × 416. Hence, the feature map contains an odd number of locations, resulting in a single cell in the center (Redmon & Farhadi, 2017). The distance metric used in YOLOv2 is shown in Eq. (2) (Redmon & Farhadi, 2017).

$$Dis(B,C) = 1 - IoU(B,C) \tag{2}$$

Here, B = Bounding Box and C = Centroid. An average IoU with the closest centroid is plotted by running k-means for various values of k. In YOLO9000, k=5 is selected to achieve a good tradeoff between high recall and model complexity (Redmon & Farhadi, 2017). YOLO9000 is improved upon YOLOv1 by using direct location prediction, fine-grained features, and multi-scale training. The model predicts the locations directly instead of predicting first offsets and later location coordinates related to the grid cell location to remove the model instability issue. Logistic activation is used to normalize the ground truth and constrain the values to range from 0 to 1. The 13 × 13 feature map benefits from fine-grained features to localize the tiny object in the sample image. A newly developed approach involving the addition of a passthrough layer is employed to bring the features from the early layers of the model at 26 × 26 resolution. In other words, a passthrough layer, like the identity mapping in ResNet, joins together the features of differing resolutions by stacking properties of neighboring channels rather than spatial locations. Finally, the detector uses this concatenated feature map to access the fine-grained features, which increases the performance by 1%. YOLO9000 uses multi-scale training where the input image resolution changes after every ten epochs so that the trained model becomes robust to the dynamic changes of input image resolution during inference. Joint training on classification and detection data has been proposed to strengthen YOLO (Redmon & Farhadi, 2017).

YOLOv3

Bounding boxes in YOLOv3 are predicted by utilizing dimension clusters as anchor boxes, as in YOLO9000 (Redmon & Farhadi, 2018). In YOLOv3, by using logistic regression objectness score is predicted for each bounding box. The sum of squared error loss is used during training. Four coordinates have been predicted for each bounding box by the network, such as t_x, t_y, t_w, t_h (Redmon & Farhadi, 2018). The bounding box around the objects in the image can be predicted using Equation (3)- Equation (6) (Redmon & Farhadi, 2018).

$$b_x = \sigma^* (t_x) + c_x \quad (3)$$

$$b_y = \sigma * (t_y) + c_y \quad (4)$$

$$b_w = (p_w) * \exp(t_w) \quad (5)$$

$$b_h = (p_h) * \exp(t_h) \quad (6)$$

Here, (c_x, c_y) represents the offset distances of the cell measured from the top left corner of the image and (p_w, p_h) height and width of the bounding box prior. In YOLOv3, the model predicts three boxes at each scale with four bounding boxes, one objectness prediction, and 80 class predictions. Hence, the resultant tensor is of dimensions $N \times N \times [3*(4+1+80)]$. The upsampled features (by a factor of two) are merged with the feature map taken earlier from the network using concatenation. The addition of a few extra convolutional layers to the combined feature map to extract more abstract spatial features results in a tensor that is twice the size (Redmon & Farhadi, 2018). For feature extraction, YOLOv3 employs a novel network incorporating elements from YOLOv2 and Darknet-19. This hybrid network is built with the help of shortcut connections, 3×3 and 1×1 convolutional layers. Hence, the hybrid network is larger in size than its predecessors.

YOLOv4

YOLOv4 (Bochkovskiy, Wang, & Liao, 2020) is proposed to achieve optimal speed and accuracy for object detection by using state-of-the-art Bag-of-Freebies (BoF) and Bag-of-Specials (BoS) methods such as Weighted Residual Connections (WRC), etc. CSPDarknet53 comprising of the 29 convolutional layers 3×3, 725×725 receptive fields, and 27.6M parameters, is used against CSPResNext50 to meet the requirements of the detector such as resolution (larger input network size), more layers, as well as more parameters (Bochkovskiy, Wang, & Liao, 2020). In YOLOv4, an SPP block is added to the CSPDarknet53, which greatly expands the receptive field and effectively extracts the most relevant context features without sacrificing network speed (Bochkovskiy, Wang, & Liao, 2020). YOLOv4 uses PANet for parameter aggregation across different levels of the backbone and detector. The architecture of YOLOv4 uses the anchor-based head from YOLOv3. The combination of BoF and BoS has improved object detection training and inference processes (Bochkovskiy, Wang, & Liao, 2020). The robustness of the YOLOv4 has been enhanced further with the addition of a new data augmentation technique known as Mosaic and Self-Adversarial Training (SAT). The best hyperparameters are chosen using genetic algorithms. SAM and PAN are modified, and a new method called Cross Mini-Batch Normalization (CmBN) is introduced. CmBN only collects statistics within a single batch between mini-batches.

An improved YOLOv4-based pest detector has been proposed for precision agriculture (Lippi, et al., 2021). Further, Pest-YOLO, an improved object detector using a modified CNN and YOLOv4 for multi-feature fusion, is proposed to detect agriculture pests (Tang, Chen, Qi, Zhang, & Chen, 2021). However, there is a need to explore the scalability feature of the YOLOv5 for insect pest detection applications.

PP-YOLO

PP-YOLO (PaddlePaddle-YOLO) (Long, et al., 2020) does not provide a novel detection method; instead, it combines the existing performance evaluation techniques to create an efficient detector. A few 3×3 convolutional layers at the end of ResNet50-vd have been replaced with deformable convolutional layers to maintain effectiveness and efficiency. As a result, a modified version that serves as the primary baseline for PP-YOLO is created by simply replacing the YOLOv3 backbone with ResNet50-vd-dcn (Long, et al., 2020). A pyramid of features is constructed in the detection Neck by connecting (lateral connections) the output of the backbone (C3, C4, C5 feature maps) to the FPN module. However, the detection head is the same as YOLOv3. The batch size should be large to stabilize the training of an object detection model. Hence, one can vary the training batch size from 64 to 192, as well as the learning rate and training schedule to achieve the best training results for object detection. The use of an exponential moving average to calculate moving averages of trained model parameters during training time is advantageous. A shadow parameter is created for each and every parameter T (Long, et al., 2020). The calculation of $T_{(EMA)}$ is shown in Eq. (7) (Long, et al., 2020).

$$T_{(EMA)} = \lambda * \left(T_{(EMA)}\right) + \left(1 - \lambda\right) T \tag{7}$$

Here, λ =Decay=0.998 and $T_{(EMA)}$ = shadow parameter. DropBlock is only applicable to the FPN because adding DropBlock to the backbone reduces performance (Long, et al., 2020). In contrast to YOLOv4, a new branch has been added in PP-YOLO to calculate IoU loss for improvement of the localization accuracy (Long, et al., 2020). The IoU prediction branch for the object detection model is trained using IoU aware loss. The detection confidence score is calculated during inference by simply multiplying the predicted IoU by the classification probability and objectiveness score (Long, et al., 2020). Bounding Box (BBox) centers can occasionally be found lying on the grid cell's boundary. Eq. (8) and Eq. (9) are modified as Eq. (10) and Eq. (11), respectively, to predict those centers (Long, et al., 2020).

$$x = S_f.(G_x + \sigma(P_x)) \tag{8}$$

$$y = S_f.(G_y + \sigma(Py)) \tag{9}$$

$$x = S_f.(G_x + \alpha.\sigma(P_x) - (\alpha - 1)/2) \tag{10}$$

$$y = S_f.(G_y + \alpha.\sigma(P_y) - (\alpha - 1)/2) \tag{11}$$

where, G_x, G_y are integers and $\alpha = 1.05$.

YOLOv5

YOLOv5 (Jocher, et al., 2020) is developed by Ultralytics to achieve the best accuracy and efficiency in object detection tasks. However, YOLOv5 is unique from earlier versions. It is not based on Darknet but on Pytorch instead. YOLOv5 is released in five different model sizes such as nano-size model (YOLOv5n), small-size model (YOLOv5s), medium-size model (YOLOv5m), large-size model (YOLOv5l), and extra-large size model (YOLOv5x). All the models are operated in the same manner except for the fact that the sizes are different due to the use of different numbers of layers and parameters. A generalized architecture of YOLOv5 model (Jocher, et al., 2020; Ahmad, et al., 2022) is presented in Fig. 1. The YOLOv5 network consists of three discrete components: the backbone, the neck, and the head (YOLO layer). The responsibility of the backbone layer is to extract the unique features of the input images. YOLOv5 employs the cross-stage partial network (CSPNet) with Darknet, also known as CSPDarknet53, as the network architecture's backbone. The SSP block broadens the receptive field while separating the critical features from the backbone (Ahmad, et al., 2022). Subsequently, accuracy increases, inference speed decreases, and model size decreases due to reduced model parameters. In YOLOv5, a path aggregation network (PANet) is employed as a bottleneck to increase information flow. YOLOv5 (Jocher, et al., 2020), like YOLOv3 (Redmon & Farhadi, 2018) and YOLOv4 (Bochkovskiy, Wang, & Liao, 2020), generates three distinct feature maps in order to conduct multi-level prediction, assuring the prediction of tiny to large-sized objects. The primary distinction between YOLOv3, YOLOv4, and YOLOv5 is based on the backbone architectures of the models. Darknet53 is used as the backbone network in YOLOv3, and CSPdarknet53 is used in YOLOv4. However, as the backbone, YOLOv5 (Jocher, et al., 2020) employs the efficient focus structure of CSPDarknet53, which assures forward-backward propagation optimization by reducing memory demands and the number of layer weights. The final prediction of the class chance, object location, and enclosing box dimensions are all transmitted as vectors by the model head (YOLO layer). Recently, YOLOv5 has been used to detect Insect Pests using IP-23 datasets (Ahmad, et al., 2022). The multiple bottom-up and top-down layers (newFPN) in PANet increased low-level feature propagation in the model and therefore information flows in YOLOv5.

Recently, different variants of YOLOv5-based efficient object detectors have been proposed for detecting agricultural pests (Aladhadh, et al., 2022), jute plant disease with associated pests (Li, Ahmed, Wu, & Sethi, 2022), fruit pests (Li, Wang, Jalil, & Wang, 2023), pests on the soybean crops (Verma, Tripathi, Singh, Ojha, & Saxena, 2021), forest pests with associated diseases (Verma, Tripathi, Singh, Ojha, & Saxena, 2021), cotton plant diseases with associated pests (Zhang, Ma, Hu, Li, & Li, 2022), orchard pest (Zhang, Cai, Fan, Song, & Jin, 2022), hazelnut orchards (Lippi, Carpio, Contarini, Speranza, & Gasparri, 2022). However, there is ample scope to develop further YOLO-based lightweight, robust, and efficient insect pest detectors.

Two-Stage Object Detectors

The two-stage object detectors are also known as the region proposal-based approaches. In this method, a region proposal is first generated, then the object classification is done for each region proposal. Two-stage detectors usually have better accuracy but are slower than one-stage detectors (Jiao, et al., 2019). Widely used Region-Proposal-based object detectors are Region-based Convolutional Neural Networks (R-CNN) (Girshick, Donahue, Darrell, & Malik, 2014), Fast-RCNN (Girshick, 2015), and Faster-RCNN (Ren, He, Girshick, & Sun, 2015).

Figure 1. The generalized architecture of YOLOv5 model
(Jocher, et al., 2020; Ahmad, et al., 2022)

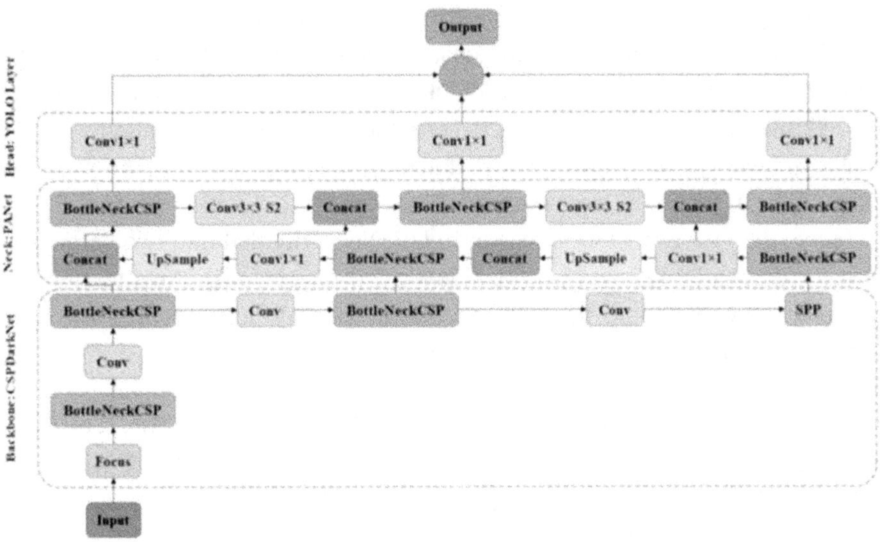

Table 1. Comparative analysis of single-stage YOLO-based object detection models

Reference	Model	Architecture	Advantages	Disadvantages	Remarks
(Redmon, Divvala, Girshick, & Farhadi, 2016)	YOLOv1	Inspired by GoogLeNet, YOLO VGG-16	Extremely fast, Trained directly on full images	Unusual aspect ratios, More localization errors	Whole image divides into S × S grid cell and it is simple to construct.
(Redmon & Farhadi, 2017)	YOLO9000	Darknet-19	mAP increases due to batch normalization and increased input size.	YOLO9000 struggles to model some categories like "sunglasses" or "swimming trunks."	YOLO9000 still requires more accuracy.
(Redmon & Farhadi, 2018)	YOLOv3	Darknet-53	It provides good accuracy as compared to YOLOv1 and YOLOv2.	Extra accuracy is achieved at the cost of speed.	YOLOv3 is a little bigger network than YOLOv2.
(Bochkovskiy, Wang, & Liao, 2020)	YOLOv4	Backbone: CSPDarknet53, Neck: SPP, PAN, Head: YOLOv3	It provides very high fps with competitive accuracy.	Low recognition accuracy.	The model may be trained and deployed on a standard GPU with 8-16 GB of VRAM.
(Long, et al., 2020)	PP-YOLO	ResNet, ResNet50-vd-dcn, FPN	It achieves a superior mix of effectiveness (45.2% mAP) and efficiency (72.9 FPS).	Low performance	The model combines various existing tricks to improve the model without increasing the model computational complexity.
(Ahmad, et al., 2022)	YOLOv5	Backbone: CSPDarknet, Neck: PANet, Head: YOLO Layer	YOLOv5 (27MB) is 88% the size of YOLOv4 (244 MB). In addition, YOLOv5 is 180% faster than YOLOv4 (140 FPS vs 50 FPS).	• YOLOv5 has almost the same accuracy that of accurate as YOLOv4.	The scalability models of YOLOv5 provide good accuracy and online performance as compared to previous versions of YOLO.

Recently, a new backbone based on improved ResNet-101 has been proposed for the Faster RCNN network (Zhao, et al., 2022). This is because the ResNet-101 can overcome the difficulty faced VGG-16 network in extracting features for the tiny objects. However, the enhancement in the feature extraction ability of the ResNet-101 is achieved at the cost of the explosion of hyperparameters and increased computational cost. The aggregated residual block replaces the actual residual block with 64 paths for convolution, making it simple for the network to learn the features of small and tiny objects. The hyperparameters are also reduced from 278 K to 271 K. There is more loss in feature information due to missing some grids due to the use of a 1×1 convolutional kernel in ResNet with a stride of 2 for down-sampling operations. Hence, the 1×1 convolutional layer in DPeNet is divided into two parts: an average pooling layer with a stride of 2 and a 1×1 convolutional layer with a stride of 1. An attention module consisting of global pooling, full connecting, sigmoid function layer, and the residual block is used to know the feature information learned by the insect pest detection model.

Further, in nature, agriculture insect pests vary in size from tiny to moderate to large. It is advisable to reduce the image size to extract information about small and tiny pests. This is because low-resolution images with reduced size have much feature information (Adarsh, Rathi, & Kumar, 2020). Hence, a research work (Zhao, et al., 2022) uses images of reduced sizes having the resolution of 225 × 300, 750 × 750, and 1125 × 1500. Here, 195 images out of 325 images are used for training, 97 images out of 325 images for validation, and 33 images out of 325 images for testing. Subsequently, 100 images of resolution 225 × 300 could be generated from one original image, 12 images of resolution 750 × 750 could be generated from one original image, and four images of resolution 1125 × 1500 could be generated from one original image. Therefore, there are a total of 22815 insect pest images for the train set, 11349 insect pest images for validation, and 3861 insect pest images for the test set.

Recently, various variants of RCN, such as custom RCN (Dalai & Senapati, 2019), improved mask FRCNN (Deepika & Arthi, 2022), Faster RCN (Deng, Mao, Zeng, Zeng, & Wei, 2022), anchor-free FRCN (Jiao, Dong, Zhang, Xie, & Wang, 2020), Faster RCN and mask RCN (Lin, Chang, & Chen, 2020), augmentation with FRCN (Patel & Bhatt, 2021), multi-scale super-resolution enabled RCN (Teng, Zhang, Dong, Zheng, & Liu, 2022), sampling-balanced region proposal network (Wang, et al., 2021), are used to develop efficient pest detectors. In general, object detection techniques have been advanced significantly for video surveillance applications. However, there is relatively less progress in agricultural insect pest detection using state-of-the-art object detection techniques. A lightweight and robust object detector is needed to detect agricultural insect pests efficiently.

PUBLICLY AVAILABLE IMAGE DATASETS FOR DETECTING AGRICULTURAL INSECT PESTS

Many agricultural diseases are caused by insect pests, which harm crops and slow the production of agricultural products. DL-enabled computer vision approaches can early detect insect pests that harm healthy crops, which protects the crops from further damage caused by diseases. However, image datasets of various insect pests that can harm crops are required for developing DL-based insect pest detectors. This Section presents a brief discussion of the publicly available image datasets for detecting agricultural insect pests.

IP23 Dataset

A very new agriculture insect pest dataset, also called IP-23, with 23 different insects has been developed for obtaining the detection performance of YOLOv5, YOLOR, YOLOv3, and YOLO-Lite (Ahmad, et al., 2022). The IP-23 dataset is created by collecting images from different search engines and databases like Google, Baidu, Kaggle, Lostock, Dream, Bing, and Flickr using the internet. The image dimension is resized to 640 × 640 (Ahmad, et al., 2022). Due to the less number of image samples, data augmentation like rotation, horizontal flipping, hue, blur, and saturation has been done so that the number of samples could be increased.

IP102 Dataset

The IP102 dataset is a large-scale benchmark dataset for recognizing insect pests (Wu, Zhan, Lai, Cheng, & Yang, 2019). There are 75222 insect pest images in 102 categories, including 19000 insect pest annotated images with bounding boxes. There are 737 samples in each class in the IP102 dataset. The IP102 dataset is divided into two sections: field crops (FC) and economic crops (EC). Rice, Corn, Wheat, Beet, and Alfalfa are the five superclasses in FC, while Vitis, Citrus, and Mango are the three superclasses in EC. The IP102 dataset has 102 classes in total with a data split ratio of 6:1:3:: train:validation:test. There are 45095 insect images for training, 7508 for validation, and 22619 for testing, as mentioned in Table 2. Here, the 'Class' specifies the number of sub-classes within the corresponding superclass. (Wu, Zhan, Lai, Cheng, & Yang, 2019)

Table 2. Training, validation, and testing sets with corresponding imbalance ratio of the IP102 dataset

	Superclass	Class	Training	Validation	Test	Imbalance ratio
Field Crops	Rice	0014	005,043	00,843	002,531	006.4
	Corn	0013	008,404	01,399	004,212	027.9
	Wheat	0009	002,048	00,340	001,030	005.2
	Beet	0008	002,649	00,441	001,330	015.4
	Alfalfa	0013	006,230	01,037	003,123	010.7
Economic Crops	Vitis	0016	010,525	01,752	005,274	074.8
	Citrus	0019	004,356	00,725	002,192	017.6
	Mango	0010	005,840	00,971	002,927	061.7
IP-102	Field Crops	0057	024,602	04,098	012,341	039.4
	Economic Crops	0045	020,721	03,448	010,393	080.8
	IP102	0102	045,095	07,508	022,619	080.8

Pest 24

Pest24 (Wang, et al., 2020), a large-scale multi-target standardized data set of agricultural pests, is offered for training pest detection deep learning models. The Pest24 dataset comprises 25,378 field pest annotated images corresponding to the 24 dominant classes of agricultural pests in China. It contains insect pest images having densely distributed, very small-scale, highly correlated objects.

Insect Pest Dataset Dedicated to Tea Plants

Another insect pest dataset having ten categories with 40 to 70 samples (Images) for each category which mainly harms tea plants, has been published (Deng, Wang, Han, & Yu, 2018). Here, some images have been taken from DSLR Camera and a few from IPM images, Dave's Garden, etc. Images have scale, position, viewpoint, and lighting conditions variations. This dataset includes ten classes of insect pests, such as Locusta migratoria, Parasa lepida, etc.

Insect Pest Dataset by (Wang, Lin, Ji, & Liang, 2012)

A small inspect pest dataset comprised of nine classes with 25 insect pest images for each is proposed (Wang, Lin, Ji, & Liang, 2012). The dataset includes nine different classes of insect pests, such as Coleoptera, Heteroptera, etc.

Insect Pest Dataset by (Xie, et al., 2018)

This insect pest dataset was introduced in 2018 by (Xie, et al., 2018). The dataset is comprised of 4500 insect pest images having insects of crops including corn, soybean, wheat, and canola. It includes 40 distinct insect pest species, such as Dolycoris baccarum (Linnaeus), Lycorma delicatula (White), etc. The resolution of the images is 200 × 200.

Three important factors, such as the relative scale of the pests, the number of instances present in the input images, and object adhesion, affect the detection accuracy of the developed insect pest detector (Wang, et al., 2020). Though a few image datasets for detecting agricultural insect pests are available in the public domain, only the IP102 insect pest dataset is most suitable for developing the DL-based insect pest detectors due to its diversity in the classes with a large volume of data.

PROPOSED METHODOLOGY

The proposed DL model is a transfer learning-based customized object detector. It is custom trained on the IP102 agricultural insect pest dataset using the YOLOv5 model trained on the COCO dataset as the source model. The end-to-end development of the proposed model can be represented in Fig. 2 and is explained using the following steps.

Figure 2. Block diagram of the end-to-end pipeline of the proposed insect pest detector

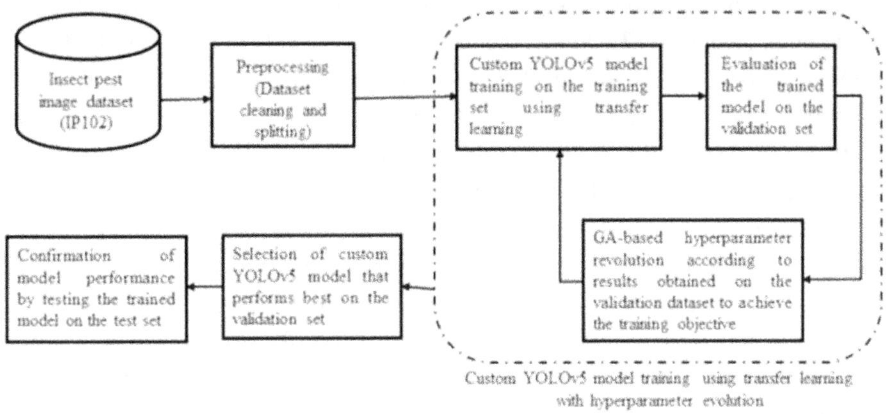

Dataset Collection and Preprocessing

The proposed model is trained on one of the state-of-art publicly available insect pest datasets, i.e., the IP102 dataset (Wu, Zhan, Lai, Cheng, & Yang, 2019). The IP102 dataset comprises around 18980 images of 102 different classes of agricultural insect pests collected for insect pest classification as well as localization purposes. However, the original IP102 dataset contains some images with no annotations and some annotations having no images. Hence, the downloaded IP102 image dataset is cleaned by filtering out the images having no annotation files and annotations having no images. The total number of pest images with annotations files for all 102 classes of insect pests is 12580. Originally, the annotation files of the images were available with .xml extensions. However, the annotations must be with .txt extensions for training the model using the YOLO framework. Hence, the .xml annotation files are converted to .txt annotation files with proper class mapping using Python script. Subsequently, all the images with corresponding annotations are split into three subsets: train, validation, and test in the ratio of 0.7:0.2:0.1. Hence, the total number of images available in the train, validation, and test folders are 8806, 2516, and 1258, respectively. Similarly, the same number of respective annotation files are also available in the train, validation, and test folders.

Custom Model Training Using Transfer Learning

Deep learning techniques are data-driven approaches. When the training data or data distribution changes, it is necessary to rebuild the model from scratch. Collecting large training datasets to build any deep learning model from scratch is usually impossible or too expensive (Shao, Zhu, & Li, 2014). However, deep learning provides model reusability. Hence, knowledge transfer or transfer learning is applied to address the issues associated with small datasets. Transfer learning involves the utilization of the weights of the successfully trained model on a big dataset as the initial weights for another new problem of the related field having a smaller dataset (Shorten & Khoshgoftaar, 2019). In other words, transfer learning is useful for solving new problems by applying prior knowledge where the source domain and target domain must be matched (Pan & Yang, 2010). Specifically, transfer learning is very beneficial in image processing as many image datasets share better-learned low-level spatial characteristics. In this case, collecting and preparing a large dataset of insect pests is a tedious, time-consuming, and expensive task. Hence, the proposed insect pest detector is a custom model trained on the IP102 dataset using transfer learning, where weights of the YOLOv5m model (Jocher, et al., 2020) trained on the COCO dataset are used as the initial weights. Usually, a YOLO-based object detector is composed of several parts, such as inputs, backbones, necks, and heads. The selected resolution of the input insect pest images is 640 × 640 (Height×Width). A new CSP-Darknet53 deep neural architecture is used as the backbone of the YOLOv5 object detector. An improved feature fusion and transmission networks, i.e., Spatial Pyramid Polling - Fast (SPPF) and Cross Stage Partial Connections- Path Aggregation Network (CSP-PAN), are used in the neck of the YOLOv5 architecture. Further, the YOLOv5 uses the same YOLOv3 head as its head for multi-scale object detection. Data augmentation techniques such as mosaic, copy-paste, random affine, mixup, augment HSV, and random horizontal flip are applied during training to improve the robustness of the pest detector. The training strategy of the proposed YOLOv5-based pest detector includes multi-scale training, auto anchor generation, exponential moving average, mixed precision, and evolution of hyperparameters. The loss function of YOLOv5 comprises three parts such as class loss, objectness loss, and location loss. Finally, the objectness losses of the top three prediction layers (P3, P4, and P5) of the YOLOv5 model are balanced using various weights of 4.0, 1.0, and 0.4, respectively.

Hyperparameter Evolution of the Transfer Learning-Based Custom YOLOv5 Model

The hyperparameters of the custom YOLOv5 insect pest detector are further optimized using Genetic Algorithm (GA)-based hyperparameter evolution. Hyperparameters are the key control parameters that control various aspects of the training and, subsequently, the overall model performance. However, finding optimal values for the hyperparameters for particular model development is challenging. Usually, traditional hyperparameter tuning methods, such as grid searches, are not suitable for tuning the hyperparameters of complex deep neural networks. Due to the involvement of high-dimensional search spaces, unknown correlations between dimensions, and costly evaluation of the fitness function at each point, grid searches rapidly become intractable. Hence, GA is found to be a suitable candidate for the searches of the hyperparameters during hyperparameter evolution (Jocher, et al., 2020). The hyperparameters are appropriately initialized with default values of the YOLOv5l model trained on the COCO dataset before applying the hyperparameter evolution. The hyperparameter evolution is aimed at maximizing the fitness function. In other words, the fitness function involves the values meant to be maximized. In YOLOv5, the default fitness function combines key performance metrics such as mAP@0.5 (contributing 10% of the weight) and mAP@0.5:0.95 (contributing 90% of the weight) with the absence of Precision and Recall. The key genetic operators of the GA are crossover and mutation. A mutation with a probability of 0.80 and a variance of 0.040 is used to produce new offspring from a combination of the best parents from all previous generations (Jocher, et al., 2020). Finally, the hyperparameters are evolved from the base scenario as per the defined fitness function to develop an optimized YOLOv5-based custom insect pest detector.

EXPERIMENTAL RESULTS ANALYSIS

The experimental results of the proposed insect pest detector can be analyzed as follows.

Experimental Setup

The experimental setup used to develop the proposed model is mentioned in Table 3.

Table 3. Experimental Setup

Hardware setup		Software setup	
CPU	Intel Xeon Silver 4214 (64-bit, 12 cores)	Operating system	Ubuntu 20.04 LTS
RAM	DDR4 (64 GB)	Programming language	Python 3.8
GPU	NVIDIA RTX A4000 (16 GB)	Deep learning framework	Pytorch 1.8.2 LTS
Storage	512 GB (SSD) + 2 TB (HDD)	CUDA compatibility	11.1

Training of Custom YOLOv5 Model Using Transfer Learning With GA-Based Hyperparameter Evolution

The proposed model is custom trained on insect pest dataset IP102 using transfer learning. Here, the YOLOv5m model trained on the COCO dataset is used as the source model whose knowledge is transferred to the target (proposed) model using transfer learning-based training. The resolution of the input insect pest images is selected as 640 × 640 (Height × Width) during training. The batch size of the input image is kept at 64 during training due to resource constraints imposed by the hardware training platform. The model is trained using Stochastic Gradient Descent optimizer with a learning rate of 0.009 with parameter groups 79 weight (decay=0.0), 82 weight(decay=0.0005), and 82 bias. The model is trained for 1000 epochs with a patience of 100 for early stopping criteria. The model is optimized using GA-based hyperparameter evolution with fitness function comprising of key performance metrics such as mAP@0.5 (contributing 10% of the weight) and mAP@0.5:0.95 (contributing 90% of the weight) with the absence of Precision and Recall. Important optimized hyperparameters are lr0=0.009, lrf=0.0009, momentum = 0.937, weight decay = 0.0005. The model is comprised of 291 layers, 21279459 parameters, 21279459 gradients, and is of 49.5 GFLOPs. The auto-anchors are selected as 2.99 anchors/target, with 1.000 Best Possible Recall (BPR), which fit the given dataset well. Three components of the overall loss function, i.e., box loss, object loss, and class loss, are plotted for the standard performance metric levels, such as mAP50 and mAP50-95, as shown in Fig. 3. From Fig. 3, it is evident that the insect pest detector achieves excellent detection accuracies, i.e., mAP@0.5 of 99.36% and mAP@0.5:0.95 of 93.51%. Further, all the loss curves converge properly, and the validation loss is less than the training loss. Hence, it may be concluded that the model is best fitted.

Quantitative Result Analysis

Key performance metrics that are widely used for the quantitative analysis of object detectors are discussed with appropriate results as follows.

Figure 3. Loss plots of the proposed model for training and validation sets of the IP102 dataset

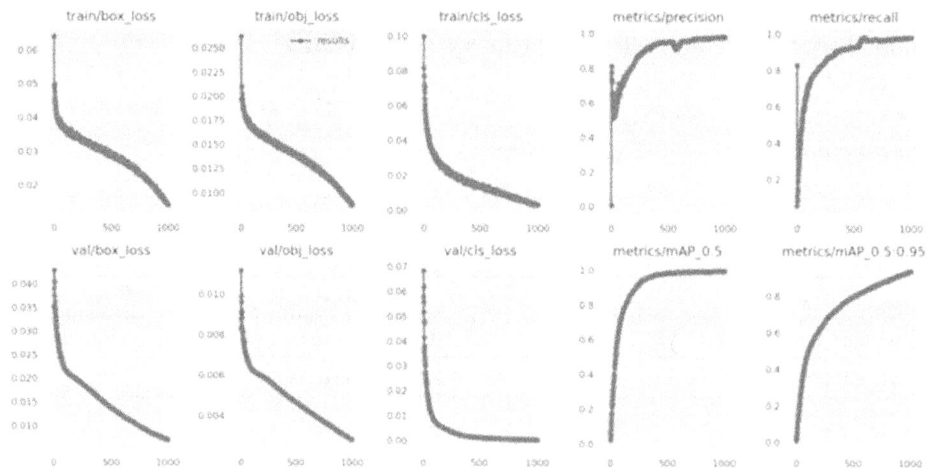

Intersection Over Union (IOU)

IoU is one of the basic metrics used to calculate the distance from the expected bounding box (ground truth) to the actual bounding box (predicted) by the object detector (Ahmad, et al., 2022). IoU is the ratio of the intersection of the overlapped area (actual and anticipated bounding box) to the total area (the union of the actual and anticipated bounding box). Usually, mAP@IoU = 0.5, and mAP@IoU = 0.5:0.95 is used for location evaluation of the predicted bounding boxes. For the IoU value > 0.5, the object detection result of the classifier is found as True Positive (TP), and for the IoU value < 0.5, the object detection result of the classifier is found as False Positive (FP). Similarly, when the predicted results of the object detector indicate the wrong prediction with respect to the given ground truth, then it is known as False Negative (FN) (Ahmad, et al., 2022).

Mean Average Precision (mAP)

It is one of the key performance metrics for any object detector and its calculation is truly dependent upon the area covered by the precision-recall curve of the object detector. Mathematically, mAP can be as the Average Precision for each class and subsequent average over the number of classes as defined in Equation (12),

$$mean\ Average\ Precision = mAP = \frac{1}{N}\sum_{i=1}^{N} AP_i Ni \tag{12}$$

where, N number of classes, and AP_i is the Average Precision corresponding to class i.

Confusion Matrix

It is used to present the predicted labels and corresponding true labels in matrix form to evaluate the model performance (Das, Meher, Panda, & Abraham, 2019). Further, a few secondary performance metrics, such as accuracy, Precision, recall, etc., can be calculated by using the obtained confusion matrix for the particular object detector. The confusion matrix of the proposed object detector for the IP102 dataset is presented in Fig. 4. The model provides good accuracy with minimal wrong predictions, as evident from the major diagonal elements and off-diagonal elements of the confusion matrix.

Precision

It can be defined as the ratio of the TP of the classifier to the sum of the TP and FP of the classifier, as represented in Equation (13).

$$Precision = \frac{True_{positive}}{True_{positive} + False_{positive}} \tag{13}$$

The precision curve (P curve) of the proposed model is presented in Fig. 5. It can be observed that the precision value increases with an increase in the model confidence. The precision value is 1.00, with a confidence level of 0.966.

Recall

It is simply the ratio of the TP to the sum of the correct result (TP) and incorrect result (FN) of the classifier, as defined in Equation (14). The recall curve (R curve) of the proposed model is presented in Fig. 6.

$$Recall = \frac{TP\left(True_{positive}\right)}{TP\left(True_{positive}\right) + FN\left(False_{negative}\right)} \tag{14}$$

PR Curve

The PR curve is a two-dimensional plot between Precision (Y-axis) and Recall (X-axis). The PR of the proposed insect pest detector is presented in Fig. 7. The area under the PR curve provides a summary measure of the performance of the developed insect pest detector. The PR curve depicts the accuracy of 0.994 mAP at the threshold of 0.5 for all 102 classes.

F_1 Score

$F_1 Score$ is a derived performance metric that relates both Precision and Recall, as mentioned in Equation (15).

$$F_1Score = 2\times\left\{\frac{Precission\times Recall}{(Precsion+Recall)}\right\} \quad (15)$$

The F_1Score of the proposed model plotted for different confidence levels is presented in Fig. 8. It can be observed from Fig. 7 that the optimized F_1Score is 0.98 at the confidence level 0.630.

Qualitative Result Analysis

A few widely used important results for the qualitative analysis of object detectors are discussed with appropriate results as follows. The visual results of the proposed insect pest detector for the train set of the IP102 dataset are presented in Fig. 9. Here, the annotated images of the input training batch 0 of the IP102 dataset are shown where class ids such as 0, 1, 2, …, 101 are indicated for the corresponding insect pest classes. Further, class-predicted results (only class levels) for various insect pest images of batch 0 of the validation set are shown in Fig. 10. Similarly, class-predicted levels with confidence scores for various insect pest images of batch 0 of the validation set are shown in Fig. 11. From the validation results, it is visually evident that insect pests are correctly detected with high detection accuracy.

Figure 4. Confusion matrix of the proposed insect pest detector for the IP102 dataset

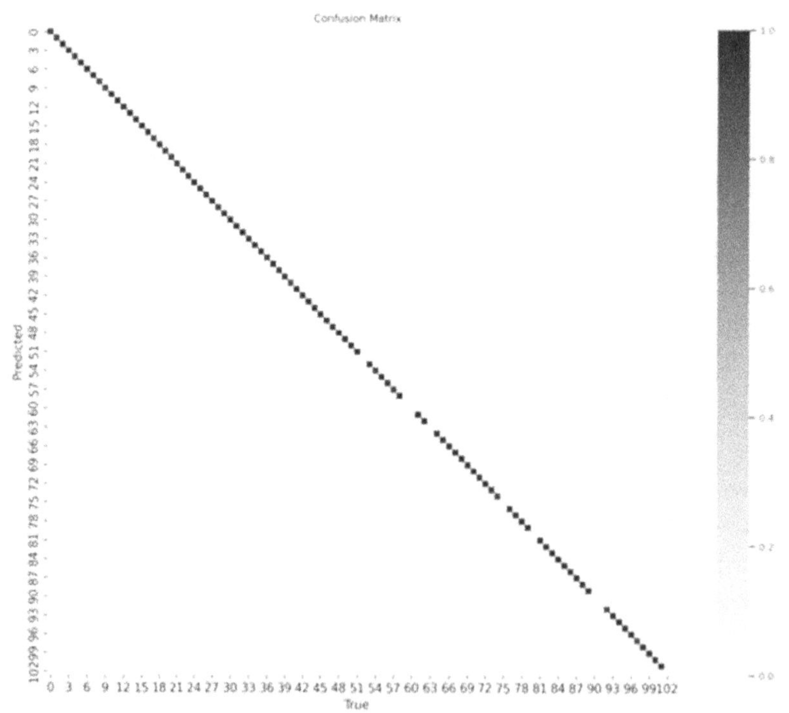

Figure 5. Precision curve of the proposed insect pest detector for the IP102 dataset

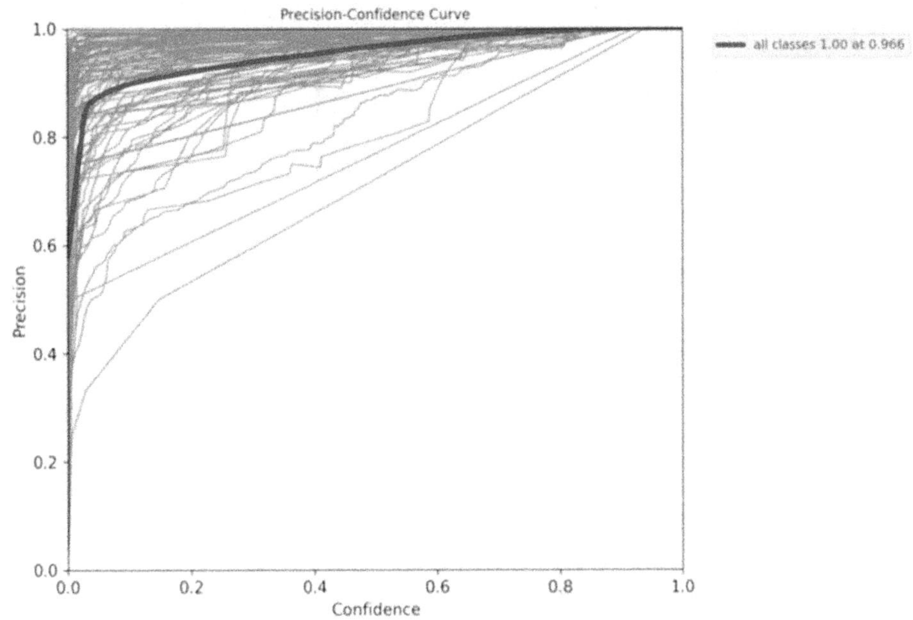

Figure 6. Recall curve of the proposed insect pest detector for the IP102 dataset

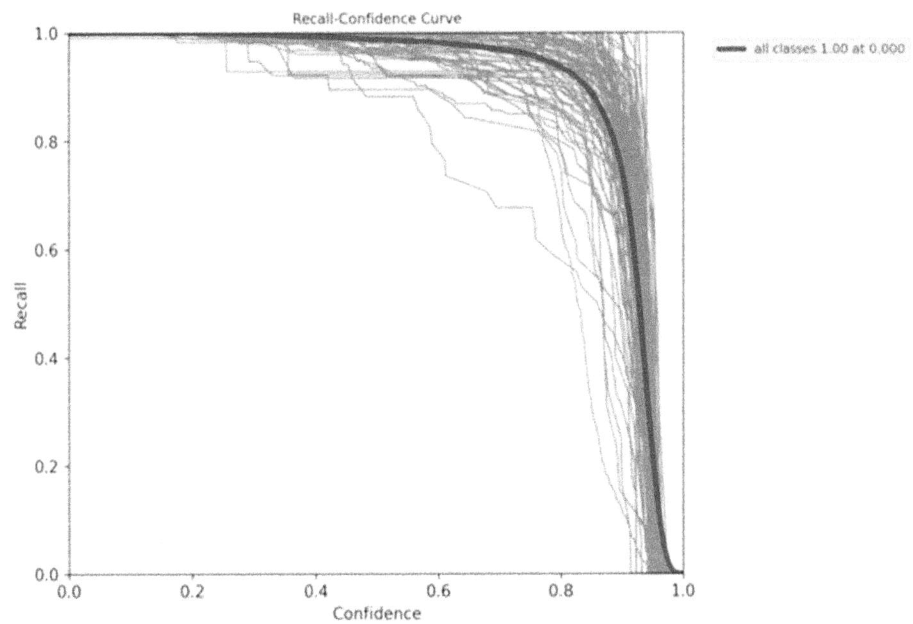

Figure 7. PR curve of the proposed insect pest detector for the IP102 dataset

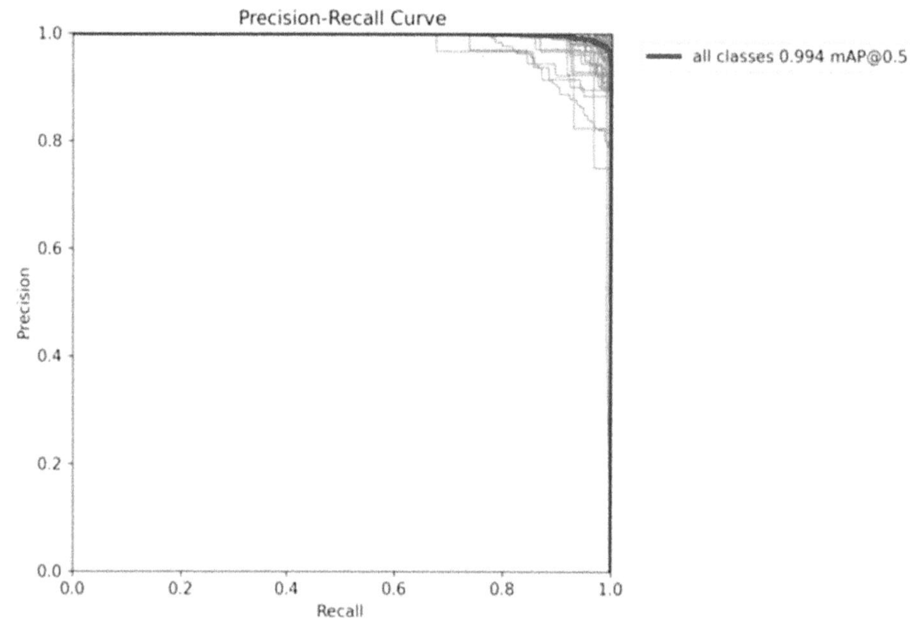

Figure 8. F_1-Confidence curve of the proposed insect pest detector for the IP102 dataset

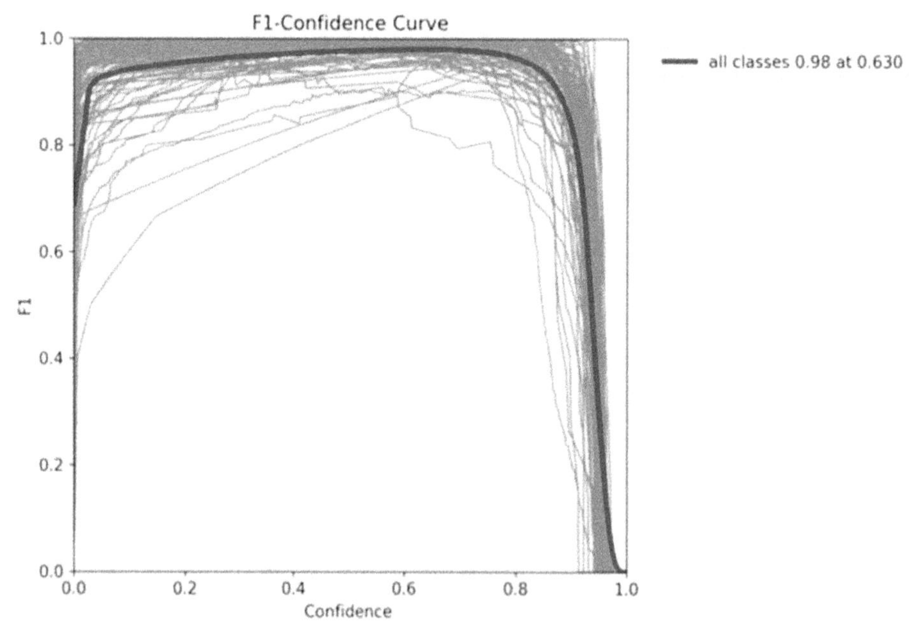

Figure 9. Visual results corresponding to input Batch 0 during training of the proposed insect pest detector for the IP102 dataset

Figure 10. Visual results showing the class-predicted levels corresponding to Batch 0 during validation of the proposed insect pest detector for the IP102 dataset

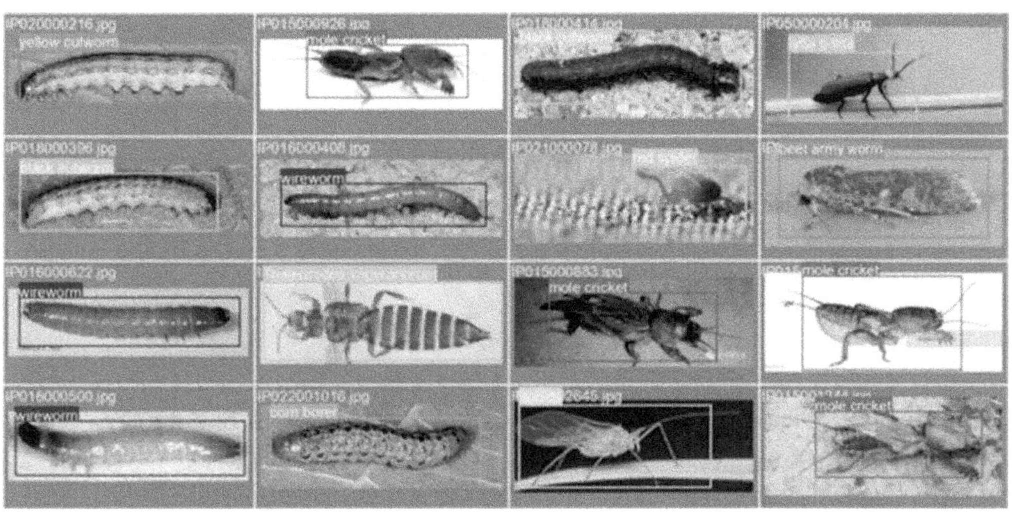

Figure 11. Visual results corresponding showing the class-predicted levels with confidence scores of Batch 0 during validation of the proposed insect pest detector for the IP102 dataset

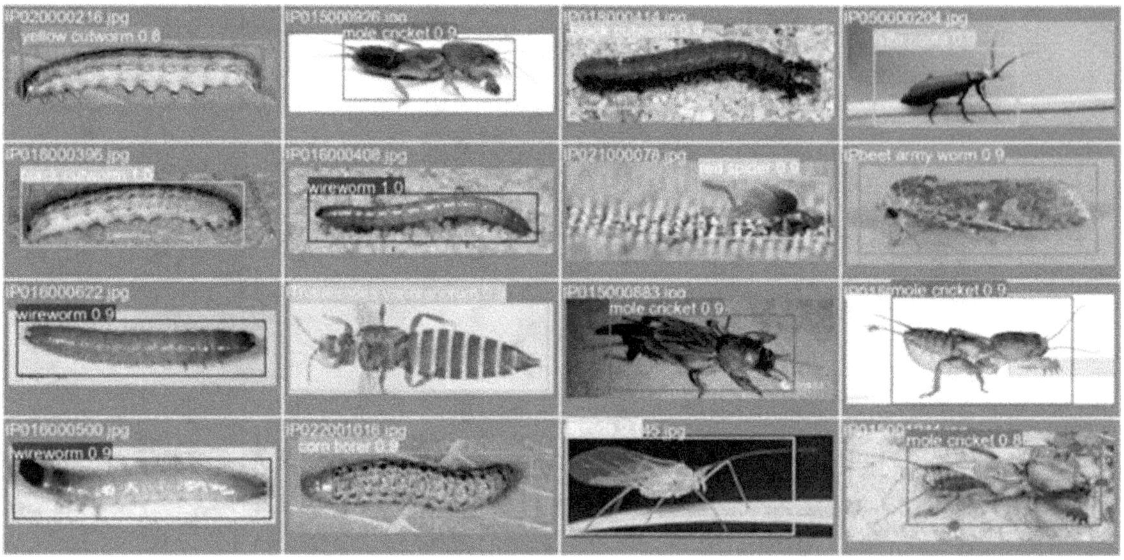

Comparative Analysis of the Proposed Insect Pest Detector

A brief comparative analysis of the proposed insect pest detector with key state-of-the-art techniques is presented in Table 4. It is also evident that the proposed model performs better as compared to the few important latest works due to GA-based hyperparameter evolution.

Table 4. Comparative Analysis

Ref.	Model	Dataset	mAP@0.5	FPS
(Zhang, Cai, Fan, Song, & Jin, 2022)	YOLOv5 with GhostNet backbone	Orchard pest dataset	0.9898	6.99
(Aladhadh, et al., 2022)	YOLOv5m	Own medium-scale pest detection dataset	0.8450	10.16
(Zhang, Huang, Sun, & Wu, 2022)	AgriPest-YOLO	Pest24	0.7130	11.36
Proposed model	Transfer learning-based custom YOLOv5 model optimized with the GA-based hyperparameter evolution	IP102	**0.99362**	**18.5**

CONCLUSION

DL-based single-stage object detectors like YOLO series object detectors are found to be effective in insect pest detection due to their high accuracy and speed. This work proposes an improved YOLOv5-based object detection model for detecting agricultural insect pests. The robustness of the proposed model is achieved by applying state-of-the-art image augmentation techniques. Further, the transfer learning-based custom YOLOv5m model is improved by applying GA-based hyperparameter evolution. The accuracy of the proposed insect pest detector is found to be promising. The best results of the proposed insect pest detector are mAP@0.5 = 0.99362, mAP@0.5:0.95=0.9351, best precision value = 0.98234, and best recall value = 0.98125. Future works may be the development of more efficient, robust, and lightweight single-stage insect pest detectors for mobile applications. Subsequently, the developed model can be deployed in mobile phones as mobile applications to facilitate the widespread use of this technology.

REFERENCES

Adarsh, P., Rathi, P., & Kumar, M. (2020). YOLO v3-Tiny: Object Detection and Recognition using one stage improved model. *2020 6th International Conference on Advanced Computing and Communication Systems (ICACCS)* (pp. 687–694). IEEE. 10.1109/ICACCS48705.2020.9074315

Ahmad, I., Yang, Y., Yue, Y., Ye, C., Hassan, M., Cheng, X., Wu, Y., & Zhang, Y. (2022). Deep learning based detector YOLOv5 for identifying insect pests. *Applied Sciences (Basel, Switzerland)*, *12*(19), 10167. doi:10.3390/app121910167

Aladhadh, S., Habib, S., Islam, M., Aloraini, M., Aladhadh, M., & Al-Rawashdeh, H. S. (2022). An Efficient Pest Detection Framework with a Medium-Scale Benchmark to Increase the Agricultural Productivity. *Sensors (Basel)*, *22*(24), 1–17. doi:10.339022249749 PMID:36560117

Bochkovskiy, A., Wang, C.-Y., & Liao, H.-Y. M. (2020). Yolov4: Optimal speed and accuracy of object detection. *arXiv preprint arXiv:2004.10934*. doi:/arXiv.2004.10934 doi:10.48550

Dalai, R., & Senapati, K. K. (2019). An Intelligent Vision based Pest Detection System Using RCNN based Deep Learning Mechanism. *2019 International Conference on Recent Advances in Energy-Efficient Computing and Communication, ICRAECC 2019*, (pp. 1-4). IEEE. 10.1109/ICRAECC43874.2019.8995072

Das, P. K., Meher, S., Panda, R., & Abraham, A. (2019). A review of automated methods for the detection of sickle cell disease. *IEEE Reviews in Biomedical Engineering*, *13*, 309–324. doi:10.1109/RBME.2019.2917780 PMID:31107662

Deepika, P., & Arthi, B. (2022). Prediction of plant pest detection using improved mask FRCNN in cloud environment. *Measurement: Sensors*, *24*, 100549. doi:10.1016/j.measen.2022.100549

Deng, F., Mao, W., Zeng, Z., Zeng, H., & Wei, B. (2022). Multiple Diseases and Pests Detection Based on Federated Learning and Improved Faster R-CNN. *IEEE Transactions on Instrumentation and Measurement*, *71*, 1–11. doi:10.1109/TIM.2022.3201937

Deng, L., Wang, Y., Han, Z., & Yu, R. (2018). Research on insect pest image detection and recognition based on bio-inspired methods. *Biosystems Engineering*, *169*, 139–148. doi:10.1016/j.biosystemseng.2018.02.008

Dhaliwal, G. S., Jindal, V., & Mohindru, B. (2015). Crop losses due to insect pests: Global and Indian scenario. *Indian Journal of Entomology*, *77*(2), 165–168. doi:10.5958/0974-8172.2015.00033.4

Domingues, T., Brandão, T., & Ferreira, J. C. (2022). Machine Learning for Detection and Prediction of Crop Diseases and Pests: A Comprehensive Survey. *Agriculture*, *12*(9), 1–23. doi:10.3390/agriculture12091350

Girshick, R. (2015). Fast r-cnn. *Proceedings of the IEEE international conference on computer vision*, (pp. 1440–1448). doi:/arXiv.1504.08083 doi:10.48550

Girshick, R., Donahue, J., Darrell, T., & Malik, J. (2014). Rich feature hierarchies for accurate object detection and semantic segmentation. *Proceedings of the IEEE conference on computer vision and pattern recognition*, (pp. 580–587). IEEE. 10.1109/CVPR.2014.81

Jiao, L., Dong, S., Zhang, S., Xie, C., & Wang, H. (2020). AF-RCNN: An anchor-free convolutional neural network for multi-categories agricultural pest detection. *Computers and Electronics in Agriculture*, *174*, 105522. doi:10.1016/j.compag.2020.105522

Jiao, L., Zhang, F., Liu, F., Yang, S., Li, L., Feng, Z., & Qu, R. (2019). A Survey of Deep Learning-Based Object Detection. *IEEE Access : Practical Innovations, Open Solutions*, *7*, 128837–128868. doi:10.1109/ACCESS.2019.2939201

Jocher, G., Stoken, A., & Borovec, J., Christopher, S., Changyu, L., & Rai, P. (2020). ultralytics/yolov5: v3.1 - Bug Fixes and Performance Improvements. *ultralytics/yolov5: v3.1 - Bug Fixes and Performance Improvements*. Zenodo. doi:10.5281/zenodo.4154370

Li, D., Ahmed, F., Wu, N., & Sethi, A. I. (2022). YOLO-JD: A Deep Learning Network for Jute Diseases and Pests Detection from Images. *Plants*, *11*(7), 1–13. doi:10.3390/plants11070937 PMID:35406915

Li, K., Wang, J., Jalil, H., & Wang, H. (2023). A fast and lightweight detection algorithm for passion fruit pests based on improved YOLOv5. *Computers and Electronics in Agriculture*, *204*, 107534. doi:10.1016/j.compag.2022.107534

Lin, T. L., Chang, H. Y., & Chen, K. H. (2020). The pest and disease identification in the growth of sweet peppers using faster r-cnN and mask r-CNn. *Journal of Internet Technology*, *21*, 605–614. doi:10.3966/160792642020032102027

Lippi, M., Bonucci, N., Carpio, R. F., Contarini, M., Speranza, S., & Gasparri, A. (2021). A YOLO-based pest detection system for precision agriculture. *2021 29th Mediterranean Conference on Control and Automation. Med (New York, N.Y.)*, *2021*, 342–347. doi:10.1109/MED51440.2021.9480344

Lippi, M., Carpio, R. F., Contarini, M., Speranza, S., & Gasparri, A. (2022). A Data-Driven Monitoring System for the Early Pest Detection in the Precision Agriculture of Hazelnut Orchards. *IFAC-PapersOnLine*, *55*(32), 42–47. doi:10.1016/j.ifacol.2022.11.112

Liu, W., Anguelov, D., Erhan, D., Szegedy, C., Reed, S., Fu, C.-Y., & Berg, A. C. (2016). Ssd: Single shot multibox detector. *Computer Vision–ECCV 2016: 14th European Conference, Amsterdam, The Netherlands, October 11–14, 2016. Proceedings*, *14*(Part I), 21–37. doi:10.1007/978-3-319-46448-0_2

Long, X., Deng, K., Wang, G., Zhang, Y., Dang, Q., & Gao, Y. (2020). PP-YOLO: An effective and efficient implementation of object detector. *arXiv preprint arXiv:2007.12099*. doi:/arXiv.2007.12099 doi:10.48550

Lyu, Z., Jin, H., Zhen, T., Sun, F., & Xu, H. (2021). Small Object Recognition Algorithm of Grain Pests Based on SSD Feature Fusion. *IEEE Access : Practical Innovations, Open Solutions*, *9*, 43202–43213. doi:10.1109/ACCESS.2021.3066510

Pan, S. J., & Yang, Q. (2010). A survey on transfer learning. *IEEE Transactions on Knowledge and Data Engineering*, *22*(10), 1345–1359. doi:10.1109/TKDE.2009.191

Patel, D., & Bhatt, N. (2021). Improved accuracy of pest detection using augmentation approach with Faster R-CNN. *IOP Conference Series. Materials Science and Engineering*, *1042*(1), 012020. doi:10.1088/1757-899X/1042/1/012020

Redmon, J., Divvala, S., Girshick, R., & Farhadi, A. (2016). You only look once: Unified, real-time object detection. *Proceedings of the IEEE conference on computer vision and pattern recognition*, (pp. 779–788). IEEE. 10.1109/CVPR.2016.91

Redmon, J., & Farhadi, A. (2017). YOLO9000: better, faster, stronger. *Proceedings of the IEEE conference on computer vision and pattern recognition*, (pp. 7263–7271). IEEE. 10.48550/arXiv.1612.08242

Redmon, J., & Farhadi, A. (2018). Yolov3: An incremental improvement. *arXiv preprint arXiv:1804.02767*, 1-6. doi:/arXiv.1804.02767 doi:10.48550

Ren, S., He, K., Girshick, R., & Sun, J. (2015). Faster r-cnn: Towards real-time object detection with region proposal networks. *Advances in Neural Information Processing Systems*, *28*, 1–14. doi:10.48550/arXiv.1506.01497

Sabat, N. K., Pati, U. C., & Das, S. K. (2021). A Short Survey on Real-Time Object Detection and Its Challenges. *Advances in Systems, Control and Automations: Select Proceedings of ETAEERE 2020*, (pp. 93–100). Springer. 10.1007/978-981-15-8685-9_9

Saleem, M. H., Potgieter, J., & Arif, K. M. (2021). *Automation in Agriculture by Machine and Deep Learning Techniques: A Review of Recent Developments*. *22*, 2053–2091. Springer. doi:10.1007/s11119-021-09806-x

Shao, L., Zhu, F., & Li, X. (2014). Transfer learning for visual categorization: A survey. *IEEE Transactions on Neural Networks and Learning Systems*, *26*(5), 1019–1034. doi:10.1109/TNNLS.2014.2330900 PMID:25014970

Shorten, C., & Khoshgoftaar, T. M. (2019). A survey on image data augmentation for deep learning. *Journal of Big Data*, *6*(1), 1–48. doi:10.118640537-019-0197-0

Tang, Z., Chen, Z., Qi, F., Zhang, L., & Chen, S. (2021). Pest-YOLO: Deep Image Mining and Multi-Feature Fusion for Real-Time Agriculture Pest Detection. *2021 IEEE International Conference on Data Mining (ICDM). 2021-December*, pp. 1348–1353. IEEE. 10.1109/ICDM51629.2021.00169

Teng, Y., Zhang, J., Dong, S., Zheng, S., & Liu, L. (2022). MSR-RCNN: A Multi-Class Crop Pest Detection Network Based on a Multi-Scale Super-Resolution Feature Enhancement Module. *Frontiers in Plant Science*, *13*, 1–12. doi:10.3389/fpls.2022.810546 PMID:35310676

Verma, S., Tripathi, S., Singh, A., Ojha, M., & Saxena, R. R. (2021). Insect Detection and Identification using YOLO Algorithms on Soybean Crop. *IEEE Region 10 Annual International Conference, Proceedings/TENCON, 2021-December*, (pp. 272–277). IEEE. 10.1109/TENCON54134.2021.9707354

Wang, J., Lin, C., Ji, L., & Liang, A. (2012). A new automatic identification system of insect images at the order level. *Knowledge-Based Systems, 33*, 102–110. doi:10.1016/j.knosys.2012.03.014

Wang, Q. J., Zhang, S. Y., Dong, S. F., Zhang, G. C., Yang, J., Li, R., & Wang, H. Q. (2020). Pest24: A large-scale very small object data set of agricultural pests for multi-target detection. *Computers and Electronics in Agriculture, 175*, 105585. doi:10.1016/j.compag.2020.105585

Wang, R., Jiao, L., Xie, C., Chen, P., Du, J., & Li, R. (2021). S-RPN: Sampling-balanced region proposal network for small crop pest detection. *Computers and Electronics in Agriculture, 187*, 106290. doi:10.1016/j.compag.2021.106290

Wu, X., Zhan, C., Lai, Y.-K., Cheng, M.-M., & Yang, J. (2019). Ip102: A large-scale benchmark dataset for insect pest recognition. *Proceedings of the IEEE/CVF conference on computer vision and pattern recognition*, (pp. 8787–8796). IEEE. 10.1109/CVPR.2019.00899

Xie, C., Wang, R., Zhang, J., Chen, P., Dong, W., Li, R., Chen, T., & Chen, H. (2018). Multi-level learning features for automatic classification of field crop pests. *Computers and Electronics in Agriculture, 152*, 233–241. doi:10.1016/j.compag.2018.07.014

Zhang, W., Huang, H., Sun, Y., & Wu, X. (2022). AgriPest-YOLO: A rapid light-trap agricultural pest detection method based on deep learning. *Frontiers in Plant Science, 13*, 1–16. doi:10.3389/fpls.2022.1079384 PMID:36589124

Zhang, Y., Cai, W., Fan, S., Song, R., & Jin, J. (2022). Object Detection Based on YOLOv5 and GhostNet for Orchard Pests. *Information (Basel), 13*(11), 548. doi:10.3390/info13110548

Zhang, Y., Ma, B., Hu, Y., Li, C., & Li, Y. (2022). Accurate cotton diseases and pests detection in complex background based on an improved YOLOX model. *Computers and Electronics in Agriculture, 203*, 107484. doi:10.1016/j.compag.2022.107484

Zhao, N., Zhou, L., Huang, T., Taha, M. F., He, Y., & Qiu, Z. (2022). Development of an automatic pest monitoring system using a deep learning model of DPeNet. *Measurement, 203*, 111970. doi:10.1016/j.measurement.2022.111970

Chapter 9
A Deep Learning-Based Vector Autoregressive-Gated Recurrent Unit Hybrid Model for Long-Term Forecasting of Weather Parameters for Smart Farms

Naba Krushna Sabat
National Institute of Technology, Rourkela, India

Rashmiranjan Nayak
https://orcid.org/0000-0002-1380-701X
National Institute of Technology, Rourkela, India

Umesh Chandra Pati
https://orcid.org/0000-0001-9805-2543
National Institute of Technology, Rourkela, India

Santos Kumar Das
National Institute of Technology, Rourkela, India

ABSTRACT

Agriculture is inextricably linked to the environment. Climate change has an effect directly on agricultural activities. India gets severely impacted if there is a loss of yields, which affects human lives. Hence, monitoring climate and its impact on the agricultural field is essential for a country like India. This chapter proposes a novel deep learning-based hybrid vector autoregressive–gated recurrent unit model (VAR-GRU model) for weather forecasting involving the four important weather parameters such as temperature, pressure, humidity, and wind speed for the cities of Bengaluru and temperature, pressure, dew point, and wind speed for the cities of Dongsi. The effectiveness of the proposed VAR-GRU model is proven by comparing its performance metrics (MAE, MSE, RMSE, and R^2 Score) with that of other baseline models such as LSTM, VAR, GRU, and another hybrid VAR-LSTM model. The outcomes of this research work can help in increasing crop yields by utilizing the weather forecasting results in smart farming applications.

DOI: 10.4018/978-1-6684-8516-3.ch009

INTRODUCTION

Agriculture is an important industry in the world. It plays a significant role in global economic activity and supplies the majority of food in the global food sector. However, this industry is completely dependent on the environment and its climate. Agriculture relies mostly on external environmental factors such as rainfall and sunlight for plant growth as well as development. The other climatology parameters, such as temperature, humidity, and wind, are also intricately tied to agriculture and have an impact on the growth of the crops (Vining, 1990). In recent years, Extreme Weather Events (EWE) such as heavy rainfall causing floods and higher temperatures causing droughts directly affect the agricultural sector (Cogato, Meggio, De Antoni Migliorati, & Marinello, 2019). This EWE results in a lack of food security, which puts human lives in a dangerous situation (Harvey, et al., 2018). Large populations in countries like India and China depend on a thriving agricultural sector to guarantee there is never a national food shortage. However, agricultural productivity declines when temperatures increase, resulting in unfavorable outcomes. Hence, it is very necessary to make a more accurate and exact prediction of the meteorological parameter.

Several methodologies such as numerical, and statistical methods that include Autoregressive Integrated Moving Average (ARIMA) (Shivhare, Rahul, Dwivedi, & Dikshit, 2019), Seasonal Autoregressive Integrated Moving Average (SARIMA) (Narasimha Murthy, Saravana, & Vijaya Kumar, 2018), Fb-Prophet (Toharudin, et al., 2020), Vector Moving Average (VMA), Vector Autoregressive Moving Average (VARMA) (Kadiyala & Kumar, 2014), etc., are adopted to predict the climatology parameter. However, these models are not able to predict long data sequences. Hence, researchers are more focused on machine learning (ML) models like support vector regression (SVR) (Hasan, Nath, & Rasel, 2015), xg-boost (Zheng & Wu, 2019), Support Vector Machine (SVM) (Zendehboudi, Baseer, & Saidur, 2018), Random Forest (RF) (Lahouar & Slama, 2017), Linear regression (LR) (Anusha, Chaithanya, & Reddy, 2019), etc. With the advancement of artificial intelligence and computational platforms, Deep Learning (DL) models such as Long Short Term Memory (LSTM) (Qing & Niu, 2018), Bi-directional LSTM (Bi-LSTM) (Peng, Zhang, Zhou, & Nazir, 2021), Convolutional Neural Network (CNN) (Lawal, Rehman, Alhems, & Alam, 2021), Neural Bias Expansion Analysis for Time Series (NBEATS) (Sabat, Nayak, Pati, & Das, 2022), Gated Recurrent Unit (GRU) (Darmawan, Yuliana, & Hadi, 2022), Deep Neural Network (DNN) (Ren, et al., 2021), Bidirectional GRU (Bi-GRU) (Liu, Gan, Chen, & Shu, 2023), became popular and widely used for predicting climatology parameters. The statistical approaches are appropriate for regression issues, but deep learning models are appropriate for complex data sets and long-term forecasting. Hence, this study is being examined and intended to create a hybrid model that includes both statistical and deep learning models. The climatic parameter is a multivariate parameter, which means one parameter is dependent on another. For example, temperature predictions are based on modeling historical time series data that includes not just temperatures but also humidity, wind, and other variables. Hence, a multivariate regressive model of ARIMA called Vector Auto-regressive (VAR) and a DL model GRU are combined to get more accurate and long-term forecasting results.

Motivation

Many environmental factors cause food shortages in the agriculture sector, which directly impacts the economy. One of the most important factors is the environmental climate and its variation. Over the past decade, it has been observed that the global temperature has been rising progressively each year compared to the previous decade. This temperature rise significantly influences agriculture, resulting in less food supply to meet demand. As a result, weather monitoring is one method that helps in better planning for producing both quality and quantity food in response to demand. DL models are the most powerful technique for extracting features from historical data and forecasting the future. Hence, it has been decided to propose a hybrid model for long-term forecasting of crucial weather parameters such as temperature, pressure, dewpoint, humidity, and wind speed.

Contributions

The important contributions can be outlined as follows.

- A novel deep-learning-based hybrid model, i.e., Vector Autoregressive–Gated Recurrent Unit (VAR-GRU) model, is proposed for time-series forecasting of the four important weather parameters of the cities of Bengaluru and Dongsi.
- The proposed model is validated by comparing it with the essential state-of-the-art (SOTA) methods in terms of Mean Absolute Error (MAE), Mean Square Error (MSE), Root Mean Square Error (RMSE), and R^2-score.
- Further, the results of the baseline models have been presented to ensure clarity in the resulting improvement of the proposed model.

Organization

The subsequent section of the chapter is laid out as follows. The related work is briefed in Section 2. The problem formulation for weather prediction modeling is presented in Section 3. The collection and preprocessing of data are highlighted in Section 4. Subsequently, Section 5 describes the proposed framework. Performance analysis using different evaluation metrics and experimental result analysis are discussed in Sections 6 and 7, respectively, followed by a conclusion in Section 8.

RELATED WORKS

Usually, time series forecasting techniques may be of either univariate or multivariate categories. When the forecasting model uses only one dependent variable for modeling, irrespective of the other availability of other input parameters, then it is known as a univariate model. The univariate model simplifies the forecasting process by ignoring the correlation factors among the input variables at the cost of model robustness. Similarly, when the forecasting model uses multiple dependent variables for modeling, then it is known as a multivariate model. Multivariate models explore the correlation factors among the input parameters to increase the efficiency of the model at the cost of model complexity. Boardly, the related works to forecast the time series meteorological parameters may be classified into four categories such as 1) based on statistical techniques, 2) based on machine learning techniques, 3) based on deep learning techniques, and 4) based on hybrid techniques.

Based on Statistical Techniques

Statistical techniques use classical statistical techniques such as ARIMA, SARIMA, SARIMAX, Fb-Prophet, VAR, VMA, and VARMA, etc., for time series forecasting. The ARIMA model was first proposed by Box & Jenkins for predicting time series data (Box, Jenkins, & Reinsel, 1970). An ARIMA model is proposed for forecasting temperature (minimum and maximum temperature range) and rainfall of the ancient city of Varanasi, India, for the period 1951 to 2015 (Shivhare, Rahul, Dwivedi, & Dikshit, 2019). Recently, the visibility of the atmosphere at the Hang Nadim Batam airport in Indonesia has been predicted using the ARIMA model (Salman & Kanigoro, 2021). The SARIMA model is proposed by Peng Chen et al. to forecast the temperature for the city of Nanjing, China (Chen, Niu, Liu, Jiang, & Ma, 2018). Rainfall pattern prediction using the SARIMA model for the northeast region of India for the years 1951 to 2014 is performed (Narasimha Murthy, Saravana, & Vijaya Kumar, 2018). This study has revealed that the rainfall pattern in the northeast region will be unaffected for the next three years. Another efficient model, Fb-prophet is developed to forecast the temperature for Myintkyina, Myanmar, by utilizing the daily average temperature from the period January 2010 to December 2017 (Oo & Sabai, 2020).

Further, key multivariate statistical models such as VAR, VMA, and VARMA are used to forecast the water quality parameters as well as their associated Water Quality Index (WQI) of river Ganga, India (Tejoyadav, Nayak, & Pati, 2022). In this study, the VAR method provides higher accuracy than other models by utilizing the high correlation factors among the water quality parameters. Usually, statistical techniques provide good prediction results for stationary and linear data. However, the efficiency of the statistical models significantly degrades when the input is nonlinear and complex in nature. The effectiveness of the conversion process involving nonstationary data to stationary data acts as a bottleneck in effective time series forecasting. These models are susceptible to structural changes in the form of trends or seasonality.

Based on Machine Learning (ML) Techniques

Machine learning models such as RF, LR, SVM, and SVR are used for prediction to short out the aforesaid problem. A RF model is proposed to estimate the wind power of Sidi Daoud, Tunisia. The accuracy of wind power forecasts is significantly enhanced in comparison to predictions made using traditional neural networks (Lahouar & Slama, 2017). A time series pollutant parameter, PM 2.5, in Tehran, Iran, is explored using both random forest and xgboost ML models (Zamani Joharestani, Cao, Ni, Bashir, & Talebiesfandarani, 2019). The authors investigated the time-series pollutant parameter from January 2015 to December 2018, and xgboost provides good accuracy and prediction than the RF model. The SVM model is used to predict the atmospheric temperature of Cambridge university from 2003 to 2008 (Radhika & Shashi, 2009). The authors compared the SVM result with Multi-Layer Perception (MLP) result and observed that SVM provides consistently outperforms with a lower mean squared error of 7.07 for the order of 10. Artificial neural network (ANN) is a robust and popular ML method used for classifying and predicting meteorological data because of its nonlinear function estimation and optimization problems. The authors (Shank, Hoogenboom, & McClendon, 2008) discussed the use of ANNs for estimating the dew point temperature for the next one to twelve hours using historical meteorological data. The primary objective of this research is to develop a dew point temperature prediction throughout the state using three-layer backpropagation ANNs and temperature data gathered over a three-year period from 20 stores in Georgia. However, the ML models are inefficient in handling nonstationary, trend and seasonality, and corrupted or missing time series data during prediction. On the other hand, ML models can experience model overfitting due to the limited size of the dataset.

Based on Deep Learning (DL) Techniques

The DL models overcome the limitations of traditional ML models in time series analysis by leveraging deep neural networks with multiple layers. This allows for capturing complex, nonlinear patterns in time series data. Additionally, DL models can handle large amounts of data and handle inputs of various sizes and resolutions, providing a more comprehensive analysis of time series data. Furthermore, these models are capable of handling missing or corrupted data points and are better suited for multivariate time series analysis. These characteristics make deep learning models a powerful tool for time series forecasting, classification, and anomaly detection. A deep learning architecture with multiple LSTM models combined to perform a specific task is called a cascaded LSTM network and is proposed to predict the dew point weather parameter (Al Sadeque & Bui, 2020). The weather data of Saskatoon Diefenbaker International Airport from the period December 2012 to October 2019 is considered for this research. This study concluded that the prediction result is superior to the traditional LSTM model. The authors (Karevan & Suykens, 2018) suggested a modified version of the most popular deep learning model LSTM, called the spatiotemporal stacked LSTM (ST sLSTM) model for temperature prediction. The authors obtained the LSTM model to improve the performance accuracy while the spatiotemporal property of the data is considered. The DL models such as LSTM, Bi-LSTM, GRU, Bi-GRU, and NBEATS are applied and evaluated for average temperature forecasting from the period of 2015 to 2022 for the capital city Bhubaneswar, Odisha, and it is concluded that the Bi-GRU model obtained the lowest error metrics compared to the SOTA models (Sabat, Nayak, Pati, & Das, 2022).

Although deep learning models provide accurate prediction, they have challenges in data interpretation, model generalization, model complexity, etc. No generic deep-learning model predicts weather patterns in different regions or under different conditions. On the other hand, if the data is unavailable or is of poor quality, it can negatively impact the accuracy of the predictions. These limitations highlight the importance of using hybrid deep learning methods, which can mitigate some of these issues and provide more accurate and robust weather predictions.

Based on Hybrid Techniques

The hybrid model includes a combination of statistical, ML, and DL models. These models help to predict both short-range as well as long-range weather prediction. A novel hybrid deep learning approach that uses CNN and an attention-based BiLSTM deep neural network is proposed to enhance the wind speed prediction accuracy of the National Wind Technology Center (Han, et al., 2022). For various heights (10 m, 20 m, and 30 m), the projected wind speed is compared with SOTA models, and it is determined that the suggested model gives superior performance (Han, et al., 2022). Hybrid deep learning models such as CNN-LSTM and CNN-GRU are proposed to predict the water quality (WQ) of an aquaculture farm at Kollam, Kerala, India (Haq & Harigovindan, 2022). The study investigated CNN-LSTM outperformed CNN-GRU in terms of computation time and prediction accuracy. An ensemble deep hybrid model CNN-GRU-SVR is proposed by (Kogekar, Nayak, & Pati, 2021) to forecast the water quality parameters and water quality index of river Ganga. The air quality of Shijiazhuang, Hebei Province, is predicted using the CNN-BiNLSTM-attention mechanism from the period January 2020 to September 2020 (Wang, Li, Wang, Wang, & Sun, 2022). The proposed model improves the complexity of the model and model accuracy. The methods that have been used in the most recent years to make predictions about time-series data are briefly outlined in Table 1.

Table 1. Approaches used for time series data analysis

Authors	Modeling approach	Model used	Considered time series parameter
Nikita Shivhare, et al. (Shivhare, Rahul, Dwivedi, & Dikshit, 2019)	Statistical	ARIMA	Temperature and rainfall
Narasimha Murthy, et al. (Narasimha Murthy, Saravana, & Vijaya Kumar, 2018)	Statistical	SARIMA	Rainfall
T.Toharudin, et al. (Toharudin, et al., 2020)	Deep learning and statistical	LSTM and Fb-Prophet	Air temperature
A. Kadiyala, et al. (Kadiyala & Kumar, 2014)	Statistical	VARMA	CO, CO_2 (indoor air quality)
N. Hasan, et al. (Hasan, Nath, & Rasel, 2015)	Statistical	SVR	Rainfall
H. Zheng, et al. (Zheng & Wu, 2019)	Machine learning	xgboost	Wind power
A. Zendehboudi, et al. (Zendehboudi, Baseer, & Saidur, 2018)	Machine learning	SVM	Solar and wind power
A. Lahouar, et al. (Lahouar & Slama, 2017)	Machine learning	Random Forest	Wind power
N. Anusha, et al. (Anusha, Chaithanya, & Reddy, 2019)	Machine learning	Multiple linear regression	Rainfall
X. Qing, et al. (Qing & Niu, 2018)	Deep learning	LSTM	Hourly solar irradiance
T. Peng, et al. (Peng, Zhang, Zhou, & Nazir, 2021)	Deep learning	Bi-LSTM	Hourly solar radiation
A. Lawal, et al. (Lawal, Rehman, Alhems, & Alam, 2021)	Deep learning	CNN and BiLSTM	Wind speed
N. K. Sabat et al. (Sabat, Nayak, Pati, & Das, 2022)	Deep learning	LSTM, N-BEATS, GRU	Temperature
B. Liu et al. (Liu, Gan, Chen, & Shu, 2023)	Hybrid	CNN-Bi-GRU	Temperature

PROBLEM FORMULATION

The weather parameters are time-dependent and can be treated as time series data. Further, the time series data consist of both linear and nonlinear data as expressed in Eq. (1).

$$A_t = l_t + nl_t + \theta_t \tag{1}$$

Here, l_t represents the linear time series data, nl_t represents nonlinear data, and θ_t is the error term. The linear data can be predicted using a regression model, while a deep learning model can capture the nonlinear time series data over a long period of time. Hence, a hybrid deep learning model is required to provide a suitable prediction outcome (Y_t) that incorporates both linear and nonlinear data as expressed in Eq. (2).

$$Y_t = A_{t1} + A_{t2} + A_{t3} + \ldots\ldots + A_{tn} \tag{2}$$

Here, $A_{t1}, A_{t2}, A_{t3}, \ldots\ldots, A_{tn}$ are different individual time series weather parameters.

DATA COLLECTION AND PREPROCESSING

Countries like India and China are seeing dramatic changes to their meteorological data because of their rapid industrial, construction, and population growth. Ultimately, it affects agricultural food. Hence, the present research compiles the weather data for two important global cities, i.e., Bengaluru, India, and Dongsi, China. The present research work used two real-world datasets in experimentation and the proposed best model for the most suitable weather data prediction. It is also useful to examine whether the proposed model retains its ability to predict even though the location of the meteorological data is changed.

Bengaluru Data Description

The data set for the city of Bengaluru, India, is collected from the open-source website Kaggle (Soneji, 2020). The dataset is comprised of many weather parameters like temperature, pressure, humidity, dew point, wind direction, UV index, perception, heat index, wind speed, etc., for the duration starting from January 2009 to January 2020. However, data for the four most relevant parameters, such as temperature, wind speed, humidity, and pressure, for the duration 1st March 2013 to 28th February 2017 is considered in this research work.

Dongsi Data Description

The data set for the city of Dongsi, China, is collected from the open-source UCI machine learning repository (Zhang, et al., 2017). It has 12 distinct metrics with 35064 hourly samples, including temperature, dew point, pressure, wind direction, PM 2.5, carbon monoxide, nitrogen oxide, wind speed, and many more. The data is available for the period of 1st March 2013 to 28th February 2017. However, the four most significant parameters, such as temperature, wind speed, dew point, and pressure, are considered for this study.

Data Preprocessing

In time series forecasting, one of the most important steps is data preprocessing. The raw data obtained is not good enough for use in making accurate predictions. Most of the data acquired contain errors such as missing values, abnormal data, distorted data, or noisy data. Malfunctioning sensors, natural disasters, and computer malfunctions, etc., might have caused these errors. The prediction accuracy increases if the erroneous data are recovered before the training of the model. Suitable data transformation techniques such as normalization and outlier removal have been applied to improve the data quality. The hourly collected data for both the mentioned cities are transformed to daily average data for dimensionality reduction. Following that, the linear interpolation approach is employed to fill in the missing value.

Before training, the data are standardized and adjusted so that they fit the model using a standard scalar library. The whole data set is split into a training and testing set with proportions of 8:2, respectively.

PROPOSED FRAMEWORK

The proposed deep learning-based hybrid model is developed using VAR and GRU models. Block diagram representation of the proposed model is illustrated in Fig. 1. Initially, raw weather data is preprocessed to make them suitable for the modeling. The preprocessing process involves outlier removal, data imputation, standardization, etc. Subsequently, the preprocessed data is organized as feature vectors for developing the VAR-GRU-based hybrid model. Finally, the trained hybrid model can be used for forecasting weather parameters. An overview of the models used in the proposed work and baseline models are discussed as follows.

Figure 1. Proposed deep hybrid VAR-GRU model

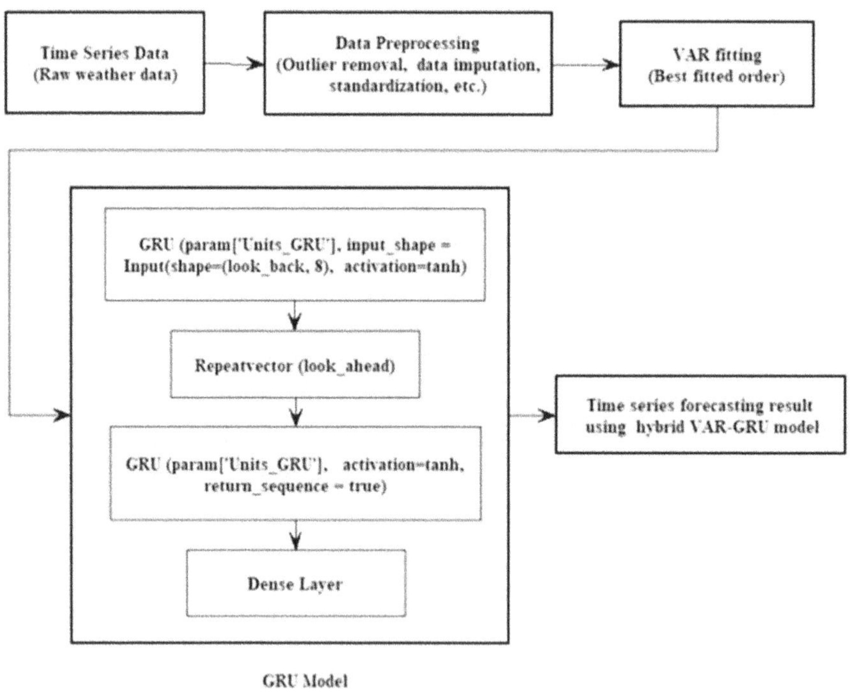

VAR Model

The time series weather parameters are multivariate in nature, and the statistical model VAR is used for multivariate data prediction. VAR is an extension of a univariate statistical model called ARIMA. The VAR model correlates present data for a specific parameter with historical data for that parameter and other related parameters. Let two-time series variables i_1 and i_2 are correlated with each other. The forecasted value for i_1 and i_2 for the time duration, t using the VAR model can be expressed as Eq. 3 and Eq. 4, considering the lag value p=1 (Ouhame & Hadi, 2019),

$$i_{1,t}=b_1+m_{11}i_{1(t-1)}+m_{12}i_{2(t-1)}+e_{1,t-1} \tag{3}$$

$$i_{2,t}=b_2+m_{21}i_{1(t-1)}+m_{22}i_{2(t-1)}+e_{2,t-1} \tag{4}$$

where, m_{11} and m_{12} are the (m x m) coefficient matrices, b_1 and b_2 are the constants, e_1 and e_2 are the error terms. The selection of order value p for the VAR model is an important task. The order value may be selected from any of the criteria, such as the Akaike Information Criterion (AIC), Bayesian Information Criterion (BIC), or Hannan-Quinn Criterion (HQC). However, the AIC criterion is applied to choose an appropriate order value in this study.

GRU Model

GRU is an advancement of the Recurrent Neural Network (RNN) network. It is quite similar to the LSTM network. However, GRU uses two gates instead of the three used by the LSTM network. The data flow inside the model is managed by two gates, the reset gate and the update gate, respectively. These gates utilize the sigmoid function, which causes their outputs to range from zero to one. The reset gate decides how much information about the prior state is retained after it has been cleared. In a similar vein, an update gate would control the extent to which the new state is simply a duplicate of the previous state. Fig. 2 depicts the GRU model architecture.

Figure 2. GRU model architecture

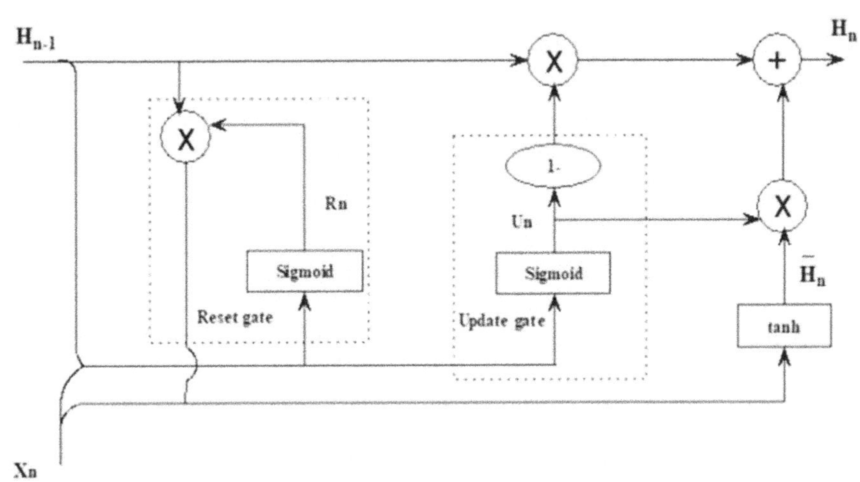

The standard GRU computation for n^{th} state is processed in accordance with reset (R_n), update (U_n), hidden (H_n), and new hidden state ($\bar{H}_n$) is expressed as in Eq. 5 (Zhou, Xu, Zeng, & Meng, 2019).

$$
\begin{aligned}
R_n &= \sigma(w^R.[H_{n-1}, X_n]) + b^R \\
U_n &= \sigma(w^U.[H_{n-1}, X_n]) + b^U \\
\bar{H}_n &= \tanh(w^H.[R_n * H_{n-1}, X_n]) + b^H
\end{aligned}
\quad (5)
$$

$$H_n = (1 - U_n) * H_{n-1} + U_n * \overline{H_n}$$

Here, X_n serves as the input, while H_{n-1} is the outputs of prior hidden state given to the GRU model. The parameters w^R, w^U, w^H and b^R, b^U, b^H are the weights of bias of reset gate, update gate, and hidden states, respectively.

Baseline Models

The proposed model is compared with a few baseline models, which are briefly explained as follows.

LSTM Model

The LSTM is a sub-class of RNN model used for processing sequential data such as time series, text, speech, and video. Traditional RNN model faces gradient vanishing issues during training. Hence, the LSTM model is designed with memory cells and gates that control the flow of information in the model and handle the problem of gradients vanishing issue. The model uses gates to determine what information to keep and what to discard from the memory cells, allowing it to retain important information over a longer period of time as compared to traditional RNNs.

Vector Autoregressive - Long Short-Term Memory Model

Vector Autoregressive-Long Short-Term Memory (VAR-LSTM) is a hybrid model that combines the strengths of two popular models in time series analysis: VAR and LSTM. VAR models the relationships between a set of time series variables, while LSTMs can capture nonlinear and long-range dependencies in sequential data. The VAR-LSTM model combines these two models to handle multivariate time series data. The VAR part models the relationships between variables, and the LSTM part captures the temporal dependencies within each variable. This hybrid model has been used in various applications such as weather forecasting, crop-yield forecasting, energy consumption prediction, traffic flow prediction, etc.

Proposed Model

An efficient hybrid model, i.e., the VAR-GRU model, is proposed. VAR-GRU is a type of time-series prediction hybrid model that combines two techniques: VAR and GRU. The VAR models are used for predicting time-series data based on past values, while GRU is a type of RNN designed to capture long-term dependencies in sequential data. In the VAR-GRU model, the VAR component captures the dependencies between past values, and the GRU component captures the long-term dependencies in the sequence. The combination of these two techniques allows for improved prediction accuracy on time-series data compared to traditional time-series models.

The proposed model's flowchart is depicted in Fig. 3. The training dataset is fitted with the VAR model by considering the lowest AIC score. The output features of the VAR model are smoothed or augmented versions of raw data, which is used as training data for the GRU model. Meanwhile, the raw training data is used as validation data set for the GRU model. This is called a two-step training procedure, i.e., the first step involves using the fitted values produced by VAR, while the second step uses the raw data to fine-tune the model using the validation process. This helps to improve the forecast result by ensuring a better-learned model. Further, the proposed model uses the grid search technique to optimize the hyperparameter until the desired level of performance is attained. Upon achieving satisfactory model performance, the model is tested using a test data set and subsequently utilized to predict the time series weather data.

METRICS FOR PERFORMANCE EVALUATION

This research article primarily focused on evaluating error measures for regression tasks to assess the forecasting performance of all the baseline and proposed models. A few key performance metrics are briefed as follows.

MAE

It quantifies the mean absolute deviation between observed and estimated values. Mathematical expression for the MAE is expressed in Eq. (6) (Wu, et al., 2021). It is commonly used in regression analysis to determine the accuracy of a model's predictions. The smaller the MAE, the more accurate the model's predictions are. MAE is easy to understand and interpret, but it does not provide information about the distribution of errors.

$$\mathrm{MAE}=\frac{1}{S}\sum_{t=1}^{S}\left|Y_t\right| \tag{6}$$

MSE

It measures the mean of the squared deviations between observed values and estimated values. Mathematical expression for the MSE is expressed in Eq. (7) (Wu, et al., 2021). It is commonly used in regression analysis to measure the predictive accuracy of a model. The smaller the MSE, the more accurate the model's predictions are. MSE is sensitive to outliers and can be affected by the scale of the data. It also penalizes large errors more than MAE, as the errors are squared.

$$\mathrm{MSE}=\frac{1}{S}\sum_{t=1}^{S}\left(Y_t\right)^2 \tag{7}$$

Figure 3. Flow chart of proposed deep hybrid VAR-GRU model

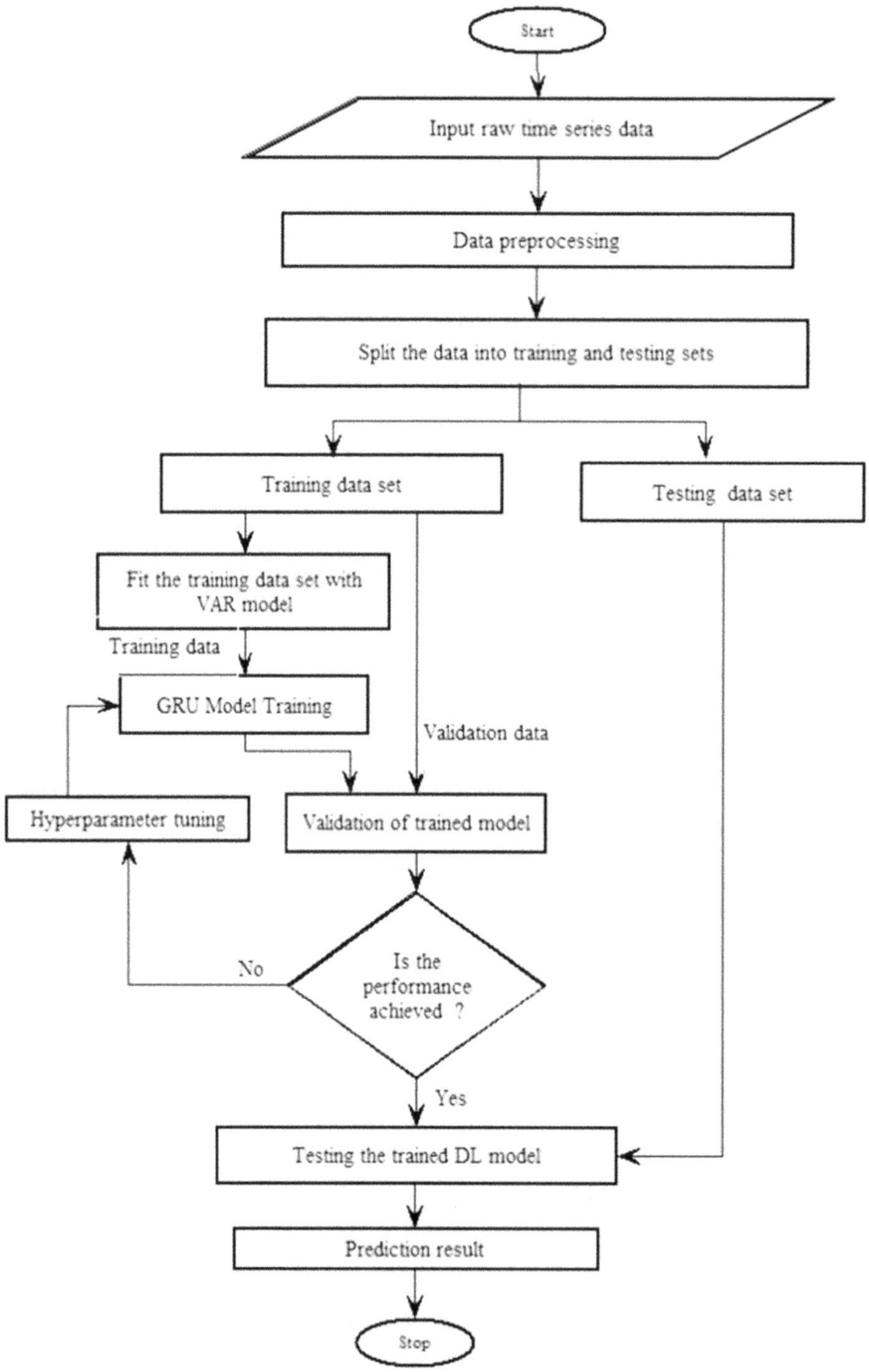

RMSE

It measures the mean of the squared deviations between observed values and estimated values and then takes the square root of that value. The mathematical formulation for the RMSE is expressed in Eq. (8) (Wu, et al., 2021). It is commonly used in regression analysis to determine the accuracy of a model's predictions. The smaller the RMSE, the more accurate the model's predictions are. The RMSE is in the same unit as the target variable, which makes it easier to interpret. However, it also suffers from the same weaknesses as MSE, in that it is sensitive to outliers and can be affected by the scale of the data. It also penalizes large errors more than MAE, as the errors are squared.

$$\text{RMSE}=\sqrt{\frac{1}{S}\sum_{t=1}^{S}(Y_t)^2} \tag{8}$$

R^2 Score

The R^2 Score is a performance metric that quantifies the proportion of the variance present in the dependent variable that can be predicted from the independent variable(s). Mathematical expression for the R^2 Score is expressed in Eq. (9) (Wu, et al., 2021). It ranges from zero to one, where one means the model explains all the variability in the target variable, and zero means the model does not explain any of the variability in the target variable. In general, a higher R-squared value indicates a better fit, that is, a model that better explains the variability of the response data around its mean. However, a high R-squared value can also indicate that the model is overfitting the data, which means it's not generalizing well to new data.

$$R^2\ \text{Score}=1-\frac{\Sigma(B_t-\widehat{B}_t)^2}{\Sigma(B_t-\overline{B}_t)^2} \tag{9}$$

Here, in Eq. (6) – Eq. (9), Y_t is expressed as the deviation between the observed value B_t and estimated value $\widehat{B}_t$, i.e. $Y_t = B_t - \widehat{B}_t$. The number of samples is denoted as S, and t defines the time period. The term $\overline{B}_t$, stands for the mean actual value.

RESULT AND DISCUSSION

The proposed deep learning-based hybrid model (VAR-GRU model) is developed for the prediction of the temperature parameter for the cities of Bengaluru, India, and Dongsi, China. Subsequently, the experimental results as discussed in the following steps.

Evaluation of Data Stationary

It is always required to analyze whether the raw data used for the prediction is stationary. The stationary data has statistical properties like mean, variance, and covariance that are time-invariant. The data stationary check can be made in different ways, such as the Augmented Dickey-Fuller test (ADF) (Cheung & Lai, 1995), Kwiatkowski-Phillips-Schmidt-Shin test (KPSS) test (Kwiatkowski, Phillips, Schmidt, & Shin, 1992), etc. However, ADF test is chosen in this study to ensure that the data are stationary. The ADF test provides information about the data and calculates the p-value; if the p-value is less than or equal to 0.05, then the null hypothesis may be rejected, the data have no unit root, and it is stationary. Fig. 4 and Fig. 5 illustrate the ADF test of temperature data for both the cities Dongsi and Bengaluru, respectively. The p-value of temperature data for both cities has been observed to be less than 0.05 ($p<0.05$). The other studied weather parameters are also tested with the ADF test, and all are observed to be stationary.

Figure 4. Augmented Dickey-Fuller test of temperature data for the city of Dongsi

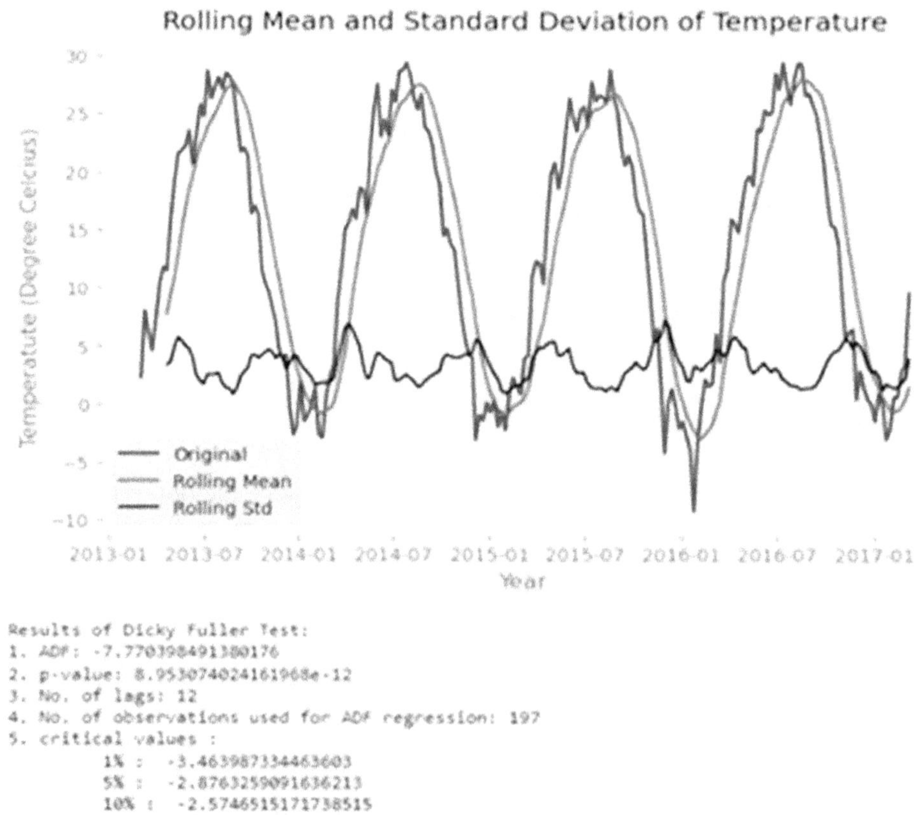

Figure 5. Augmented Dickey-Fuller test of temperature data for the city of Bengaluru

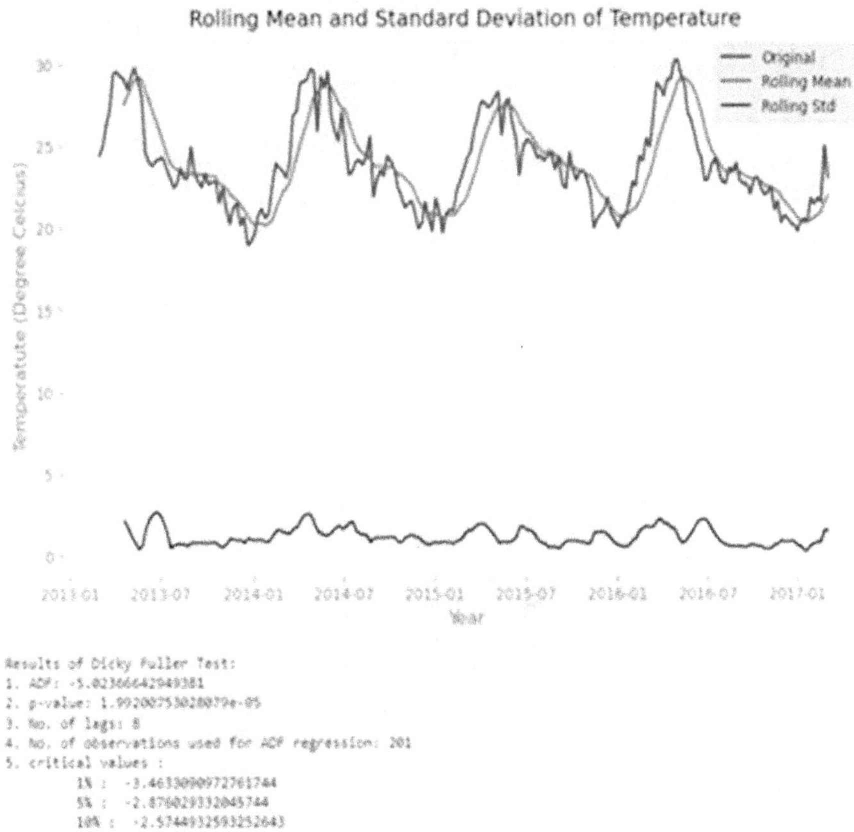

Trend Analysis Test

In order to make accurate predictions based on time series data, trend analysis is a crucial step that must be completed. In this particular investigation, the Mann–Kendall trend analysis test (Mondal, Kundu, Mukhopadhyay, & others, 2012) is used in order to determine the general tendencies of several meteorological parameters. The trend analysis of the meteorological parameters and their respective statistical values obtained from the Mann-Kendall test for both the cities Bengaluru and Dongsi, are described in Table 2. The trend shows the meteorological parameters having an increasing, decreasing, or no trend. The h-value indicates true if there is a trend, which may be increasing or decreasing; otherwise, it indicates false. The p-value and z-value are the significant test and normalized test statistics acquired from the Mann-Kendall test.

Correlation Test

The correlation test provides a numerical representation of the extent to which the variables are connected to one another. It is used to analyze the correlation coefficient between two or more variables which helps to know about the change in one variable when the other one does. Fig. 6 and Fig. 7 represent the correlation matrix between the weather parameters for the studied cities.

Table 2. Weather parameters trend analysis for the city of Bengaluru and Dongsi

Parameters	Bengaluru				Dongsi			
	Trend	**h-value**	**p-value**	**z-value**	**Trend**	**h-value**	**p-value**	**z-value**
Temperature	Decreasing	True	2.2204	-8.1862	Decreasing	True	1.9482	-4.2707
Pressure	Increasing	True	0.0	10.4381	Increasing	True	2.0341	6.7035
Wind speed	Decreasing	True	0.0189	-2.3471	No trend	False	0.4542	-0.7483
Humidity	No trend	False	0.1050	1.6208	Not available	Not available	Not available	Not available
Dew point	Not available	Not available	Not available	Not available	Decreasing	True	0.0007	-3.3527

Figure 6. A climatological parameter correlation matrix for the city of Dongsi

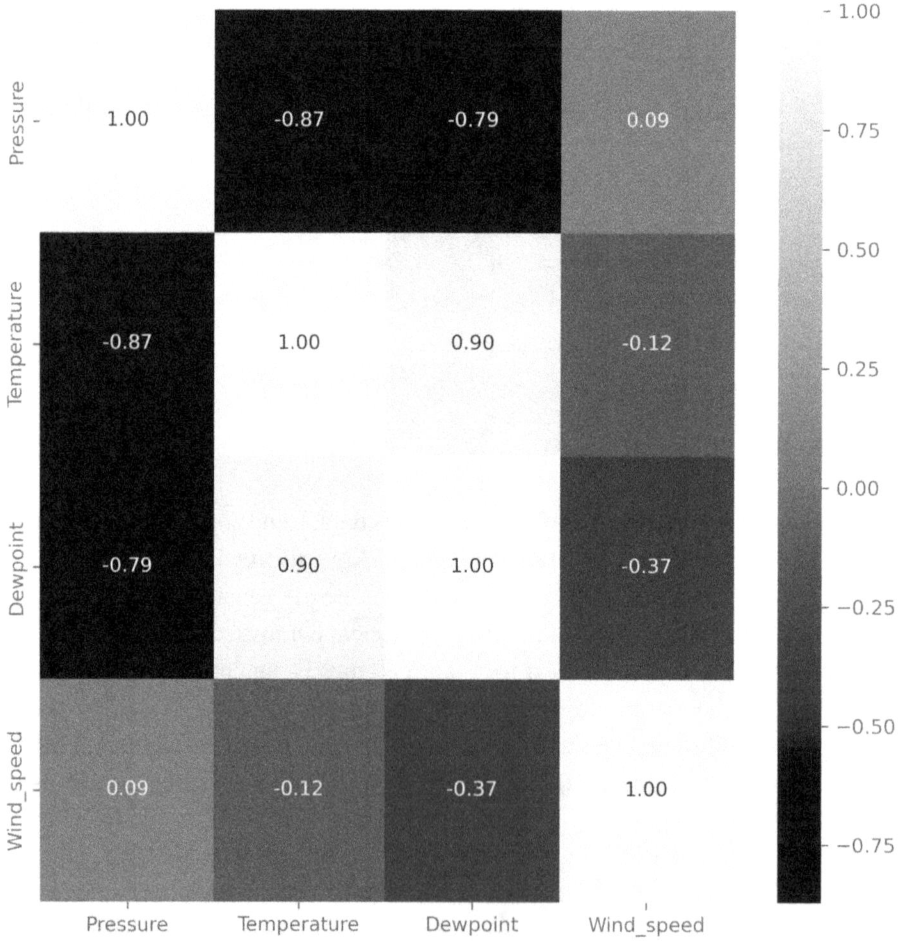

Figure 7. A climatological parameter correlation matrix for the city of Bengaluru

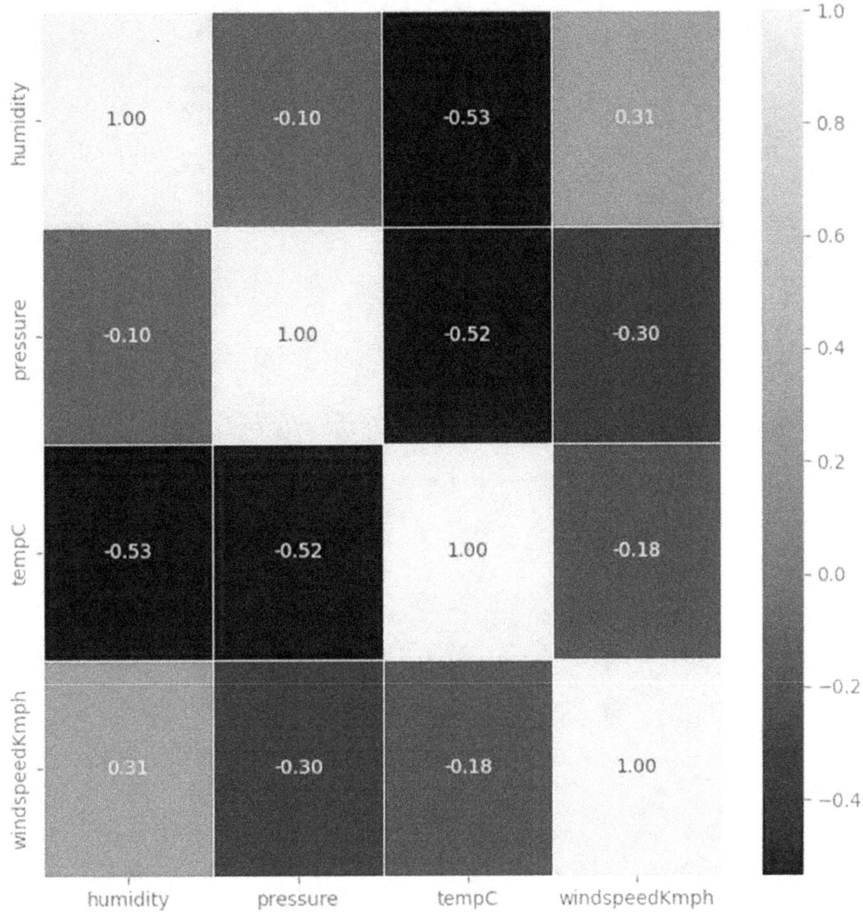

Selection of Statistic Model

The article uses a multivariate statistical model (VAR), which has been chosen based on its p-value. The parameter p-value of the VAR model is selected from the AIC test. AIC quantifies how well the predicted and true likelihood values are fitted with the model (Akaike, 1974). The minimum value of AIC is considered the best-fit parameter value for the VAR model. Table 3 describes the order value (p) and the AIC of studied cities for the VAR model. Fig. 8 and Fig. 9 depict the lags order selection of the VAR model using the AIC method.

Table 3. Best fitted order value for VAR model

Model→	VAR	
Cities↓	Best fitted order value (p-value)	AIC Value
Bengaluru	6	3.4469
Dongsi	8	5.3808

Figure 8. VAR lags order selection using AIC method for the city of Bengaluru

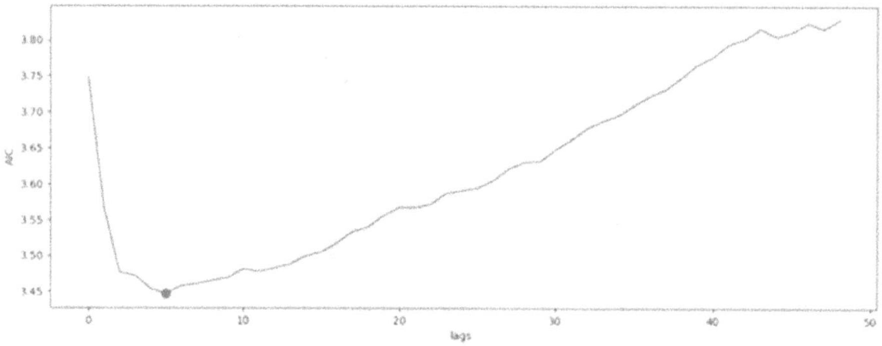

Figure 9. VAR lags order selection using AIC method for the city of Dongsi

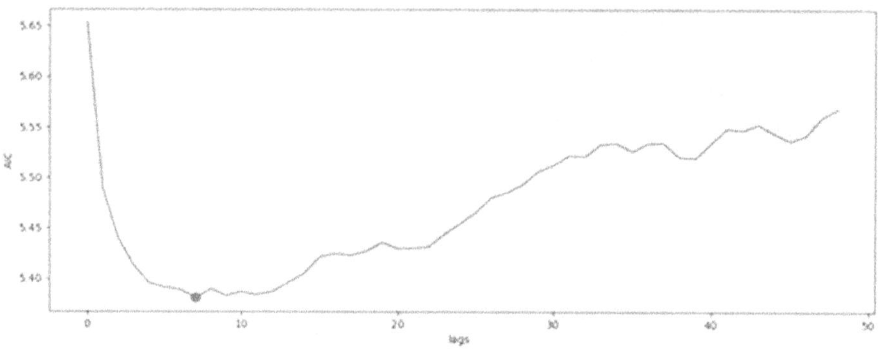

Selection of Hyperparameter and Simulation Analysis

The hyperparameter setting is essential while designing a deep learning model. The hyperparameter value setting for the suggested hybrid deep learning model is shown in the following Table. 4. The fitted VAR value helps to improve the training of the neural network. These values are fed to the GRU model for multi-step ahead forecasts. The model is developed using Keras-tensorflow framework, and the Keras-hypertune library is used to optimize the hyperparameter during training. All the models are executed on the Google Collab online platform. Fig. 10 and Fig. 11 depict the outcomes of the VAR-GRU model for different parameters of the cities of Bengaluru and Dongsi, respectively.

Table 4. Hyperparameter setting

Parameters	Values	Parameters	Values
Training Data	80%	GRU units	[128, **96**, 64, 32]
Testing Data	20%	Learning rate	[[7e-4, **1e-3**, 2e-3, 3e-3]
Activation function	tanh	Epoch	200
Optimizer	Adam	Batch Size	[60, 120, **156**, 240]
Loss Function	MSE		

Figure 10. Prediction analysis of (a) temperature (b) pressure (c) humidity (d) Wind speed for the city of Bengaluru using VAR-GRU model

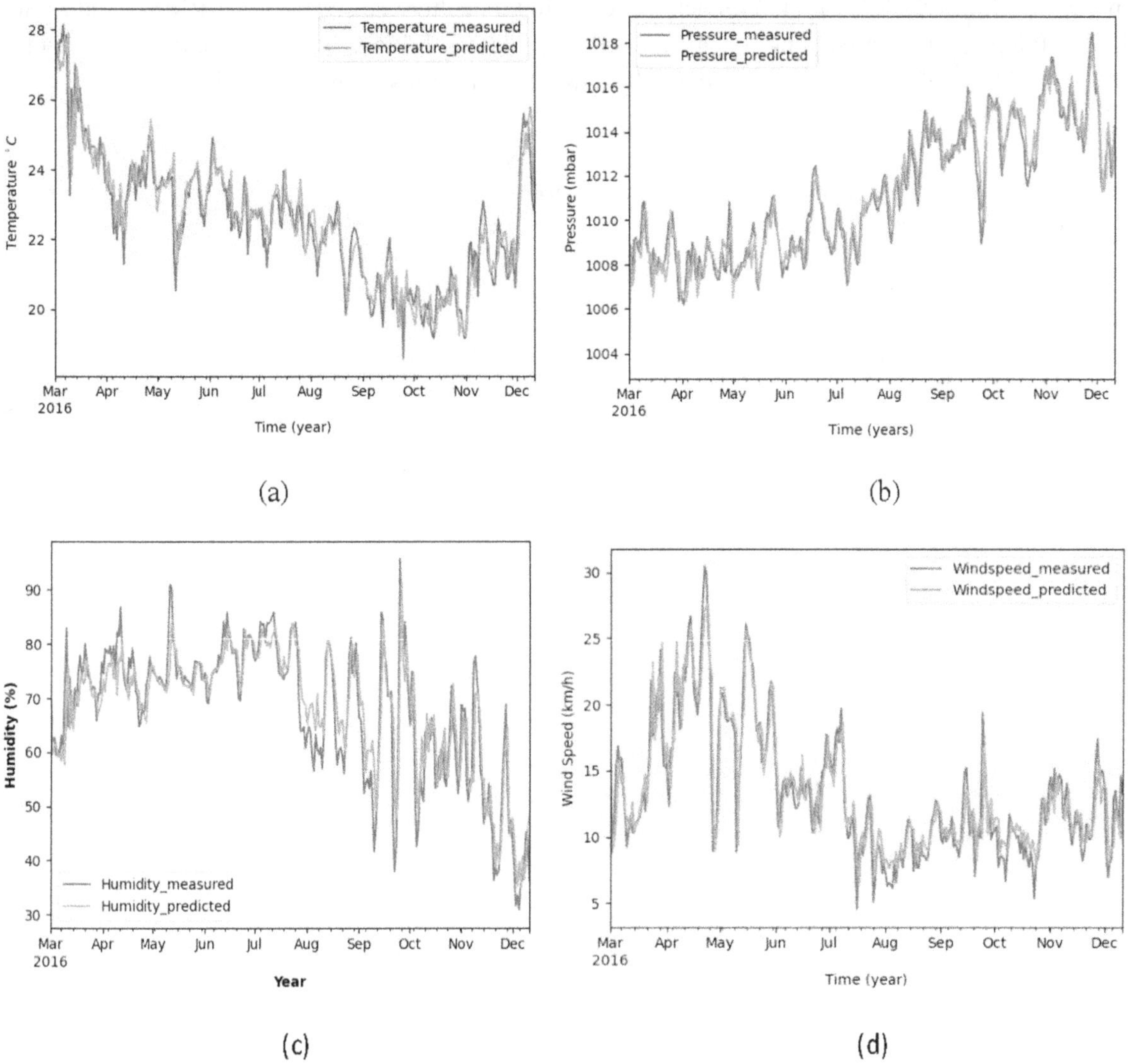

Comparative Analysis

The proposed model is investigated to compare with a few baseline models such as VAR, LSTM, GRU, and a hybrid model VAR-LSTM. All the models are evaluated with their MAE, MSE, RMSE, and R^2 Score. Table 5 – Table 8 represents the performance metrics value of different weather parameters for different studied cities. The lower value of RMSE and higher value of R^2 Score is considered as the best-predicted model for the time series climatology data. From Table 5 – Table 8, it is evident that the proposed VAR-GRU model exhibits superior for most of the weather parameters in terms of the majority of the performance metrics. In other words, the combination of the VAR and GRU models is efficient in capturing the long-term temporal dependency while doing regression for the time series forecasting of the multivariate weather parameters. However, a few baseline models, such as GRU and VAR-LSTM, perform slightly better in terms of MAE for the temperature and humidity of Bengaluru, as represented in Table 5. Similarly, VAR-LSTM models perform slightly better in terms of MSE, RMSE, and R^2 Score for the temperature of Bengaluru, as mentioned in Table 6 – Table 8. This may be due to the fact that VAR-LSTM is efficient in capturing the hidden trend in the temperature data of Bengaluru.

Figure 11. Prediction analysis of (a) temperature (b) pressure (c) dew point (d) Wind speed for the city of Dongsi using VAR-GRU model

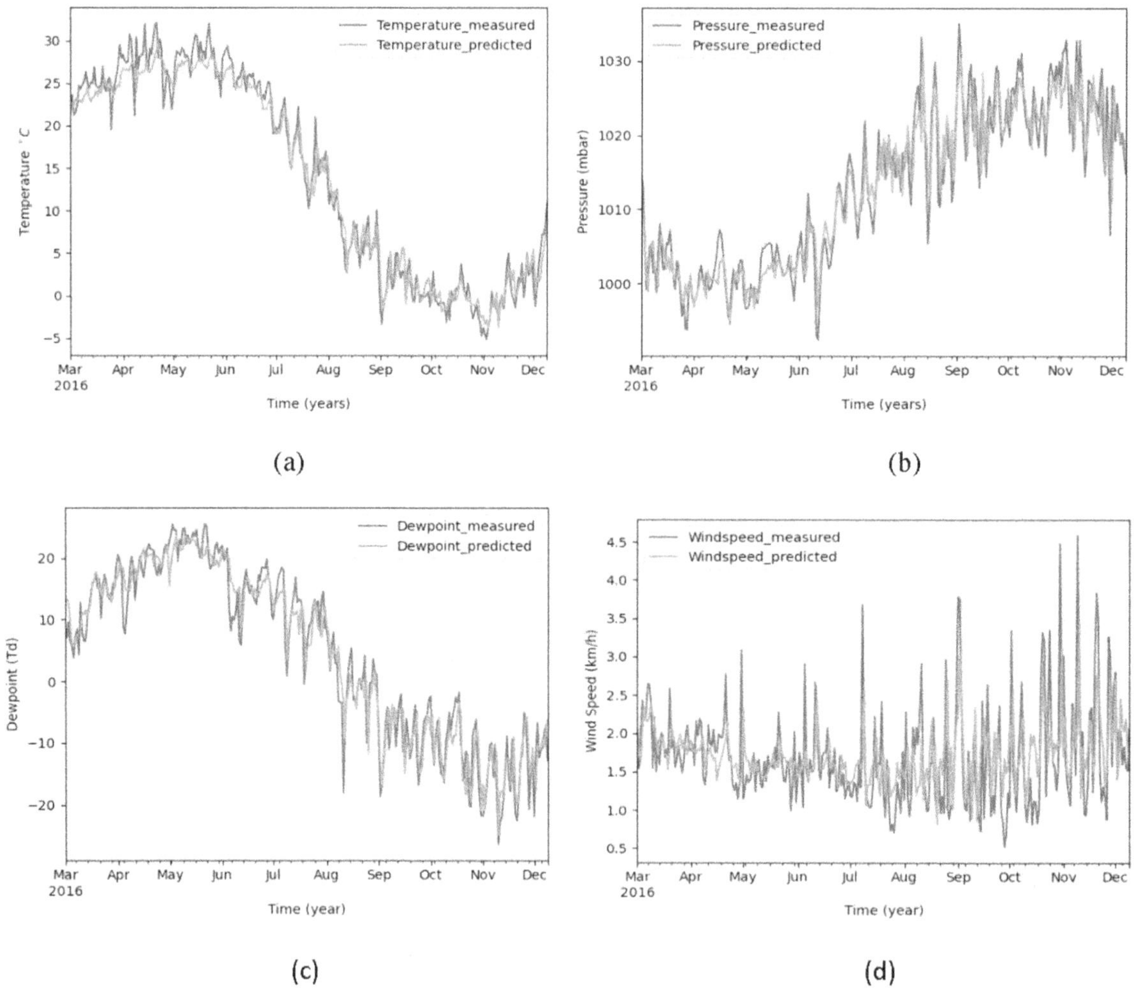

Table 5. MAE value of climatology parameter for the city of Bengaluru

Model	Benguluru				Dongsi			
	Temperature	**Pressure**	**Humidity**	**Wind Speed**	**Temperature**	**Pressure**	**Dew Point**	**Wind Speed**
LSTM	0.500	0.710	4.298	1.579	1.766	3.670	3.132	0.471
VAR	0.501	0.664	4.169	1.655	1.625	3.374	3.093	0.516
GRU	***0.478***	0.697	4.261	1.561	1.609	3.396	3.082	0.467
VAR-LSTM	0.501	0.667	***4.089***	1.551	1.741	2.965	2.831	0.429
VAR-GRU	0.494	**0.634**	4.125	**1.518**	**1.564**	**2.844**	**2.692**	**0.418**

Table 6. MSE value of climatology parameter for the city of Bengaluru

Model	Benguluru				Dongsi			
	Temperature	**Pressure**	**Humidity**	**Wind Speed**	**Temperature**	**Pressure**	**Dew-Point**	**Wind Speed**
LSTM	0.454	0.844	35.369	4.645	5.065	21.667	16.351	0.409
VAR	0.479	0.751	36.405	5.045	4.508	19.311	17.111	0.571
GRU	0.447	0.815	34.968	4.563	4.253	18.679	14.821	0.397
VAR-LSTM	***0.442***	0.735	32.517	4.545	4.791	14.663	14.181	0.349
VAR-GRU	0.466	**0.695**	**32.429**	**4.528**	**3.928**	**14.141**	**12.629**	**0.331**

Table 7. RMSE value of climatology parameter for the city of Bengaluru

Model	Benguluru				Dongsi			
	Temperature	**Pressure**	**Humidity**	**Wind Speed**	**Temperature**	**Pressure**	**Dew Point**	**Wind Speed**
LSTM	0.673	0.919	5.947	2.155	2.251	4.654	4.043	0.641
VAR	0.692	0.866	6.033	2.246	2.123	4.394	4.136	0.755
GRU	0.683	0.903	5.913	2.136	2.062	4.322	3.849	0.631
VAR-LSTM	***0.665***	0.857	5.702	2.131	2.188	3.829	3.765	0.591
VAR-GRU	0.675	**0.831**	**5.667**	**2.095**	**1.982**	**3.761**	**3.553**	**0.575**

Table 8. R^2-score value of climatology parameter for the city of Bengaluru

Model	Benguluru				Dongsi			
	Temperature	Pressure	Humidity	Wind Speed	Temperature	Pressure	Dew Point	Wind Speed
LSTM	0.861	0.899	0.781	0.808	0.964	0.811	0.915	0.062
VAR	0.852	0.908	0.776	0.792	0.968	0.831	0.911	-0.301
GRU	0.863	0.902	0.784	0.811	0.971	0.837	0.923	0.091
VAR-LSTM	***0.864***	0.912	0.799	0.812	0.966	0.871	0.926	0.201
VAR-GRU	0.857	**0.916**	**0.801**	**0.813**	**0.972**	**0.876**	**0.934**	**0.241**

CONCLUSION

Weather forecasting must be considered as an integral part of smart farming to achieve the best yields in quantity and quality. This chapter proposes a novel deep learning-based hybrid model (VAR-GRU) model for weather forecasting the four important weather parameters such as temperature, pressure, humidity, and wind speed for the cities of Bengaluru and temperature, pressure, dew point, and wind speed for the cities of Dongsi. The effectiveness of the proposed VAR-GRU model is proven by comparing its performance metrics with that of other baseline models such as LSTM, VAR, GRU, and another hybrid VAR-LSTM model. In the future, the model can be deployed in a web hosting platform to disseminate information with the help of interactive results digitally. However, the proposed model performance is sensitive to hyperparameter tuning and may sometimes fail in mapping nonlinear relationships among time series variables. Hence, further scope exists to improve the model performance using other DL networks, such as Bi-LSTM, BiGRU, CNN-GRU, and Temporal Convolutional Networks (TCN).

REFERENCES

Akaike, H. (1974). A new look at the statistical model identification. *IEEE Transactions on Automatic Control*, *19*(6), 716–723. doi:10.1109/TAC.1974.1100705

Al Sadeque, Z., & Bui, F. M. (2020). A deep learning approach to predict weather data using cascaded LSTM network. *2020 IEEE Canadian Conference on Electrical and Computer Engineering (CCECE)*, (pp. 1–5). IEEE. 10.1109/CCECE47787.2020.9255716

Anusha, N., Chaithanya, M. S., & Reddy, G. J. (2019). Weather prediction using multi linear regression algorithm. *IOP Conference Series. Materials Science and Engineering*, *590*(1), 012034. doi:10.1088/1757-899X/590/1/012034

Box, G. E., Jenkins, G. M., & Reinsel, G. (1970). Time series analysis: forecasting and control Holden-day San Francisco. *BoxTime Series Analysis: Forecasting and Control Holden Day1970.*

Chen, P., Niu, A., Liu, D., Jiang, W., & Ma, B. (2018). Time series forecasting of temperatures using SARIMA: An example from Nanjing. *IOP Conference Series. Materials Science and Engineering*, *394*, 052024. doi:10.1088/1757-899X/394/5/052024

Cheung, Y.-W., & Lai, K. S. (1995). Lag order and critical values of the augmented Dickey–Fuller test. *Journal of Business & Economic Statistics*, *13*(3), 277–280. doi:10.1080/07350015.1995.10524601

Cogato, A., Meggio, F., De Antoni Migliorati, M., & Marinello, F. (2019). Extreme weather events in agriculture: A systematic review. *Sustainability (Basel)*, *11*(9), 2547. doi:10.3390u11092547

Darmawan, H., Yuliana, M., & Hadi, M. Z. (2022). Realtime Weather Prediction System Using GRU with Daily Surface Observation Data from IoT Sensors. *2022 International Electronics Symposium (IES)*, (pp. 221–226). IEEE. 10.1109/IES55876.2022.9888468

Han, Y., Mi, L., Shen, L., Cai, C. S., Liu, Y., Li, K., & Xu, G. (2022). A short-term wind speed prediction method utilizing novel hybrid deep learning algorithms to correct numerical weather forecasting. *Applied Energy*, *312*, 118777. doi:10.1016/j.apenergy.2022.118777

Haq, K. R., & Harigovindan, V. P. (2022). Water quality prediction for smart aquaculture using hybrid deep learning models. *IEEE Access : Practical Innovations, Open Solutions*, *10*, 60078–60098. doi:10.1109/ACCESS.2022.3180482

Harvey, C. A., Saborio-Rodríguez, M., Martinez-Rodríguez, M. R., Viguera, B., Chain-Guadarrama, A., Vignola, R., & Alpizar, F. (2018). Climate change impacts and adaptation among smallholder farmers in Central America. *Agriculture & Food Security*, *7*(1), 1–20. doi:10.118640066-018-0209-x

Hasan, N., Nath, N. C., & Rasel, R. I. (2015). A support vector regression model for forecasting rainfall. *2015 2nd international conference on electrical information and communication technologies (EICT)*, (pp. 554–559). IEEE. 10.1109/EICT.2015.7392014

Kadiyala, A., & Kumar, A. (2014). Vector time series models for prediction of air quality inside a public transportation bus using available software. *Environmental Progress & Sustainable Energy*, *33*(2), 1069–1073. doi:10.1002/ep.11959

Karevan, Z., & Suykens, J. A. (2018). Spatio-temporal stacked LSTM for temperature prediction in weather forecasting. *arXiv preprint arXiv:1811.06341*. doi:/arXiv.1811.06341 doi:10.48550

Kogekar, A. P., Nayak, R., & Pati, U. C. (2021). A CNN-GRU-SVR based Deep Hybrid Model for Water Quality Forecasting of the River Ganga. *2021 International Conference on Artificial Intelligence and Machine Vision (AIMV)* (pp. 1-6). IEEE. 10.1109/AIMV53313.2021.9670916

Kwiatkowski, D., Phillips, P. C., Schmidt, P., & Shin, Y. (1992). Testing the null hypothesis of stationarity against the alternative of a unit root: How sure are we that economic time series have a unit root? *Journal of Econometrics*, *54*(1-3), 159–178. doi:10.1016/0304-4076(92)90104-Y

Lahouar, A., & Slama, J. B. (2017). Hour-ahead wind power forecast based on random forests. *Renewable Energy*, *109*, 529–541. doi:10.1016/j.renene.2017.03.064

Lawal, A., Rehman, S., Alhems, L. M., & Alam, M. M. (2021). Wind speed prediction using hybrid 1D CNN and BLSTM network. *IEEE Access : Practical Innovations, Open Solutions*, *9*, 156672–156679. doi:10.1109/ACCESS.2021.3129883

Liu, B., Gan, H., Chen, D., & Shu, Z. (2023). Research on Fault Early Warning of Marine Diesel Engine Based on CNN-BiGRU. *Journal of Marine Science and Engineering*, *11*(1), 56. doi:10.3390/jmse11010056

Mondal, A., Kundu, S., Mukhopadhyay, A., & ... (2012). Rainfall trend analysis by Mann-Kendall test: A case study of north-eastern part of Cuttack district, Orissa. *International Journal of Geology. Earth and Environmental Sciences*, *2*, 70–78.

Narasimha Murthy, K. V., Saravana, R., & Vijaya Kumar, K. (2018). Modeling and forecasting rainfall patterns of southwest monsoons in North–East India as a SARIMA process. *Meteorology and Atmospheric Physics*, *130*(1), 99–106. doi:10.100700703-017-0504-2

Oo, Z. Z., & Sabai, P. H. (2020). Time Series Prediction Based on Facebook Prophet: A Case Study, Temperature Forecasting in Myintkyina. *International Journal of Applied Mathematics Electronics and Computers*, *8*(4), 263–267. doi:10.18100/ijamec.816894

Ouhame, S., & Hadi, Y. (2019). Multivariate workload prediction using Vector Autoregressive and Stacked LSTM models. *Proceedings of the New Challenges in Data Sciences: Acts of the Second Conference of the Moroccan Classification Society*, (pp. 1–7). ACM. 10.1145/3314074.3314084

Peng, T., Zhang, C., Zhou, J., & Nazir, M. S. (2021). An integrated framework of Bi-directional long-short term memory (BiLSTM) based on sine cosine algorithm for hourly solar radiation forecasting. *Energy*, *221*, 119887. doi:10.1016/j.energy.2021.119887

Qing, X., & Niu, Y. (2018). Hourly day-ahead solar irradiance prediction using weather forecasts by LSTM. *Energy*, *148*, 461–468. doi:10.1016/j.energy.2018.01.177

Radhika, Y., & Shashi, M. (2009). Atmospheric temperature prediction using support vector machines. *International journal of computer theory and engineering*, *1*, 55. doi:10.7763/IJCTE.2009.V1.9

Ren, X., Li, X., Ren, K., Song, J., Xu, Z., Deng, K., & Wang, X. (2021). Deep learning-based weather prediction: A survey. *Big Data Research*, *23*, 100178. doi:10.1016/j.bdr.2020.100178

Sabat, N. K., Nayak, R., Pati, U. C., & Das, S. K. (2022). *A Comparative Analysis of Univariate Deep Learning-based Time-series Models for Temperature Forecasting of the Bhubaneshwar*. IEEE. doi:10.1109/iSSSC56467.2022.10051494

Salman, A. G., & Kanigoro, B. (2021). Visibility forecasting using autoregressive integrated moving average (ARIMA) models. *Procedia Computer Science*, *179*, 252–259. doi:10.1016/j.procs.2021.01.004

Shank, D. B., Hoogenboom, G., & McClendon, R. W. (2008). Dewpoint temperature prediction using artificial neural networks. *Journal of Applied Meteorology and Climatology*, *47*(6), 1757–1769. doi:10.1175/2007JAMC1693.1

Shivhare, N., Rahul, A. K., Dwivedi, S. B., & Dikshit, P. K. (2019). ARIMA based daily weather forecasting tool: A case study for Varanasi. *Mausam (New Delhi)*, *70*(1), 133–140. doi:10.54302/mausam.v70i1.179

Soneji, H. (2020). Historical Weather Data for Indian Cities. *Historical Weather Data for Indian Cities. Kaggle*. doi:10.34740/KAGGLE/DSV/1129180

Tejoyadav, M., Nayak, R., & Pati, U. C. (2022). A Comparative Analysis of Multivariate Statistical Time Series Models for Water Quality Forecasting of the River Ganga. In *Ambient Intelligence in Health Care: Proceedings of ICAIHC 2022* (pp. 429–441). Springer. 10.1007/978-981-19-6068-0_41

Toharudin, T., Pontoh, R. S., Caraka, R. E., Zahroh, S., Lee, Y., & Chen, R. C. (2020). Employing long short-term memory and Facebook prophet model in air temperature forecasting. *Communications in Statistics. Simulation and Computation*, 1–24. doi:10.1080/03610918.2020.1854302

Vining, K. C. (1990). Effects of weather on agricultural crops and livestock: An overview. *The International Journal of Environmental Studies*, *36*(1-2), 27–39. doi:10.1080/00207239008710581

Wang, J., Li, J., Wang, X., Wang, T., & Sun, Q. (2022). An quality prediction model based on CNN-BiNLSTM-attention. *Environment, Development and Sustainability*, 1–16. doi:10.100710668-021-02102-8

Wu, K., Wu, J., Feng, L., Yang, B., Liang, R., Yang, S., & Zhao, R. (2021). An attention-based CNN-LSTM-BiLSTM model for short-term electric load forecasting in integrated energy system. *International Transactions on Electrical Energy Systems*, *31*(1), e12637. doi:10.1002/2050-7038.12637

Zamani Joharestani, M., Cao, C., Ni, X., Bashir, B., & Talebiesfandarani, S. (2019). PM2. 5 prediction based on random forest, XGBoost, and deep learning using multisource remote sensing data. *Atmosphere (Basel)*, *10*(7), 373. doi:10.3390/atmos10070373

Zendehboudi, A., Baseer, M. A., & Saidur, R. (2018). Application of support vector machine models for forecasting solar and wind energy resources: A review. *Journal of Cleaner Production*, *199*, 272–285. doi:10.1016/j.jclepro.2018.07.164

Zhang, S., Guo, B., Dong, A., He, J., Xu, Z., & Chen, S. X. (2017). Cautionary tales on air-quality improvement in Beijing. *Proceedings - Royal Society. Mathematical, Physical and Engineering Sciences*, *473*(2205), 20170457. doi:10.1098/rspa.2017.0457 PMID:28989318

Zheng, H., & Wu, Y. (2019). A xgboost model with weather similarity analysis and feature engineering for short-term wind power forecasting. *Applied Sciences (Basel, Switzerland)*, *9*(15), 3019. doi:10.3390/app9153019

Zhou, X., Xu, J., Zeng, P., & Meng, X. (2019). Air pollutant concentration prediction based on GRU method. *Journal of Physics: Conference Series*, *1168*, 032058. doi:10.1088/1742-6596/1168/3/032058

Chapter 10
AI-Enabled Crop Recommendation System Based on Soil and Weather Patterns

Priyanka Sharma
https://orcid.org/0000-0002-9503-1170
Swami Keshvanand Institute of Technology, Management, and Gramothan, India

Pankaj Dadheech
https://orcid.org/0000-0001-5783-1989
Swami Keshvanand Institute of Technology, Management, and Gramothan, India

A. V. Senthil Kumar Senthil
https://orcid.org/0000-0002-8587-7017
Hindusthan College of Arts and Science, India

ABSTRACT

Agriculture is the foremost factor which is important for the survival of human beings. Farming contributes to a very big part of GDP; still, several areas exist where improvements are required. One of those is crop recommendation. Crop productivity is boosted as a result of accurate crop prediction. As crop production has already started to suffer from climate change, improving crop output is consequently desirable because agronomists are impotent to select the appropriate crop(s) depending on environmental and soil parameters, and the mechanism of forecasting the selection of the appropriate crops manually has failed. Factors like soil characteristics, soil types, climate characteristics, temperature, rainfall, area, humidity, geographic location etc. affect crop forecast. This chapter focuses mainly on building a recommendation system, i.e., suggesting the kind of the crop by applying various machine learning and deep learning techniques depending upon several parameters. The system would help the farmers for the appropriate decision to be taken regarding the crop type.

DOI: 10.4018/978-1-6684-8516-3.ch010

INTRODUCTION

Agriculture is the cornerstone of the nation's development. Due to its extraordinary agricultural areas and other resources, India is known as an agricultural country. Elements like temperature and soil moisture have an impact on how productive, contagious, and yield-producing agriculture has grown recently (Edwin, 2017). Agriculture-related issues have impeded the country's progress. It is vital to modernize the recognized farming practices nowadays. New agricultural trends are required to manage crops in a regulated atmosphere.

The choice of a crop is one of the most crucial elements that directly affects the final output; as a result, a farmer should always choose the best crop choice while taking environmental factors into account. Choosing the optimum crop for a specific farm can be difficult because there are so many different variables that could influence the yield. Although crop suggestions from specialists are frequently requested, many farms cannot afford or have access to this option due to its time and expense requirements. It has seen a huge increase due to their immense benefits in supporting users' needs by locating the most appropriate items based on information extracted from a collection of data.

The use of recommender systems in agricultural management has recently produced some intriguing and promising results. Similar to the Amazon shop, these technologies are essential in guiding users' actions by helping them increase profits or reduce risks. Recommender Systems are used for a variety of purposes in many modern digital enterprises like Yahoo, Google, Netflix etc. including consumer segmentation, fraud detection, financial banking, healthcare systems, and education (Iniyan et al., 2023).

Recommendation Systems are utilized to assist farmers in making wiser decisions. The crop recommendation discipline does not, however, have a comprehensive classification strategy for its algorithms and features. The lack of a systematic literature review that is explicitly focused on this issue and the diversity of techniques presented in the surveys are the main reasons behind this, as the majority of studies have only looked at the application of machine learning to agricultural yield prediction (Yamparla, 2022). As a result, choosing an algorithm and input parameters that meet one's needs can be difficult and confusing while developing a crop recommender system. Also, it would be challenging for academics to keep track of agricultural advancements and algorithm usage.

The motivation behind this research is to create a system that can advise which crop should be grown in a specific circumstance. Based on the levels of nitrogen, phosphorus, and other elements in the atmosphere, it makes forecasts (Sharma and Rathi, 2016). It chooses which crop to grow by measuring the amount of rainfall and other weather factors. Also, it offers predictions depending on the soil's current state. The initiative teaches farmers and the general public which crops may be produced in a specific area using a range of machine learning techniques. Both the agriculture business and general use can benefit from this effort.

The structure of this chapter is as follows: A comprehensive overview of crop recommendation systems is provided in Section 1. Review of the literature is presented in Section 2. Section 3 discusses the suggested work and dataset aspects. An examination of the outcomes is covered in Section 4. We wrap up the work in section 5 with a few potential future developments.

LITERATURE REVIEW

(Van et al., 2020) conducted a systematic literature assessment. The analysis shows that soil type, temperature and rainfall are the most frequently utilized features, while ANN`s are the most frequently used algorithm in this model. (Rashid et al., 2021) reviewed various ML algorithms for predicting the agricultural yield with extra special importance on palm oil yields. (Kalimuthu et al., 2020) have used an approach where Naive Bayes algorithm is used. (Sharma et al., 2021) have given a comprehensive evaluation of ML applications in the realm of agriculture. By identifying and diagnosing eating disorders, reproductive patterns, or behaviour prediction, machine learning with computer vision can be utilised to increase livestock productivity.

In order to create a pre-season, (Cunha et al., 2018) did a forecast of soybean/maize production without using NDVI data, a system is built that integrate satellite-derived statistics on soil parameters, physical model-based seasonal climate forecasting data, precipitation and other sources. (Pande et al., 2021) developed a practical and user-friendly yield prediction system using ML algorithms to estimate the crop yield and make recommendations on fertilisers to increase yield. Authors presented a practical and user-friendly yield prediction system (Reddy and Kumar, 2021). The most profitable crop list can be picked using machine learning algorithms, and they can also forecast crop yields for user-selected crops. Selected Machine Learning techniques like SVM, ANN, RF, Multivariate Linear Regression, and k-NN are used to predict crop yield. The research examined CYP using ML techniques (Tahaseen and Moparthi, 2021) that were distinct from the features that were selected based on the crop attributes, scale and geological position these decisions were largely influenced by the availability of the data set. ML techniques were used to estimate crop yields while taking into account variables like temperature and weather.

Various machine learning techniques are covered for weed and pest detection, crop production prediction, and plant leaf disease detection. (Sharma et al., 2021) discussed the current situation of agricultural yield around the world is followed by a quick introduction to widely used characteristics and forecasting techniques. ML and AI can be used to increase agricultural production in India to meet public demand. (Ray et al., 2022) proposed 22 types of crops using distribution analysis, correlation analysis, majority voting and ensembling, with accuracy of 99.54% and 98.52%. (Vashisht et al., 2022) proposed an extreme learning machine to predict rice crop yield based on geography, season, and cultivation area. (Gupta et al., 2022) demonstrated that ML algorithms can be used to segment large amounts of data to make crop yield predictions.

(Seireg et al., 2022) used cascading regression and stacking regression to predict wild blueberry yield with accuracy. (Rasheed et al., 2021) tested a decision-aiding tool on historical data of cultivated land segments in Pakistan to estimate net profit and predicted production. The process of using machine learning techniques to train a model for seeing trends in data and then using that model to predict crops are defined by (Pant et al., 2021). Predictive analytics was used by (Chandraprabha et al., 2021) to forecast soil nutrients for various crops using Tamilnadu's soil-based dataset. (Raja et al., 2022) presented classification and feature selection techniques to estimate plant cultivation yields, with an ensemble technique outperforming existing classification techniques in prediction accuracy.

For the purpose of forecasting annual crop yields in West Africa, (Cedric et al., 2022) presented a machine learning-based prediction method based on machine learning to predict annual crop yields in West Africa, using Decision Tree, multivariate logistic regression, and k-NN. (Ali et al., 2022) evaluated crop production using remote sensing techniques and direct empirical statistical models. (Pantazi et al., 2016) used an unsupervised learning algorithm to predict wheat yield using satellite imagery and soil data. Using ML regression approaches for Time-Series Imagery of Landsat 8 OLI, (Aghighi et al., 2018) performed the prediction of silage maize. Modified recursive feature elimination (MRFE), a unique feature selection technique is defined to choose the most pertinent features from a crop prediction dataset (Mariammal et al., 2021). The performance analysis justifies why the MRFE technique outperforms other FS methods with a 95% accuracy.

Three important steps: pre-processing the dataset, EDA and detection module are included (Kumar et al., 2021) and plant disease prediction has been done. More than 98% of predictions for each disease turned out to be accurate on average. This research establishes the viability of employing this method for less expensively and more quickly identifying plant diseases. APSIM crop model and Daily high-resolution CubeSat photos are combined by (Ziliani et al., 2022). APSIM and CubeSat photos combined to produce high-resolution yield maps, with $R^2 = 0.73$ and RMSE = 12%.

(Vlachopoulos et al., 2022) found that the best algorithm for GAI prediction was the random forests algorithm, with an MAE of 0.67 and RMSE of 10.86%. Yield maps created using various Kriging techniques and other mapping approaches (Birrell et al., 1996) showed the same broad patterns, but localised yield features were represented differently.

(Goel and Mishra, 2022) reported 95.64% accuracy using phenological data and deep learning algorithms. On the basis of input dimensions, the deep recurrent Q-learning network creates an environment for agricultural yield prediction, as mentioned by (Elavarasan and Vincent, 2020). Q-Network is a better option to forecast crop yield, with 93.7% accuracy. ANN method tested by (Haque et al., 2020) to show how different factors affect crop yield, with MSE and standard deviation used to illustrate error rate. (Cunha and Silva, 2020) created a productive crop yield model using crop calendars, weather forecast information, and remote sensing data. (Mondal and Banerjee, 2021) developed a model that incorporates rainfall, temperature, and chemical elements to calculate agricultural productivity. The results show that the proposed model performs fairly well. (Bose et al., 2016) developed Spiking Neural Networks for distant sensing spatiotemporal analysis of picture time series to estimate crop yield, with an average accuracy of 95.64% and an average error of 0.236 t/ha.

(Saeed and Lizhi et al., 2019) developed a DNN approach to improve prediction accuracy, reducing average yield to 11% and 46%. A multilayer deep learning model combines RNN and CNN to extract spatial and temporal features from soil property data, time-series remote sensing data, and model outputs (Sun et al., 2020). SSTNN is a novel deep learning architecture, developed by (Qiao et al., 2021) that combines recurrent neural networks and 3D convolutional neural networks to extract temporal relationships from multi-spectral images. Multi-parametric Multiple Kernel DNN (MMKDNN) by (Kalaiarasi and Anbarasi, 2022) enhanced learning capacity of MDNN to improve agricultural yield prediction for medium-scale data.

(Abbaszadeh et al., 2022) integrate the outputs of many deep neural networks, such as the 3DCNN and ConvLSTM, to produce more reliable and accurate soybean crop production estimates. The result is a probabilistic estimate of soybean crop yield. CNNs and hyperspectral imaging were used to model the spectrum data set and compare the impacts of PCA and multidimensional scattering correction (Pang et al., 2020). (Alebele et al., 2021) represented rice yield estimation based on Gaussian kernel regression that outperforms Bayesian linear inference and probabilistic Gaussian regression with higher prediction accuracy.

(Martinez et al., 2021) proposed Gaussian processes (GPs) that enable the identification of climate extremes, anomalies, and their associated causes that affect crop productivity. (Qiao et al., 2021) developed a 3-D convolutional neural multi-kernel network to capture hierarchical features for agricultural yield prediction. (Sivanantham et al., 2022) found that using orthogonal basis functions and quantile regression improved prediction accuracy and precision by 32% and 9%, respectively.

(Li et al., 2022) concentrated on the integration of a variety of solar-induced fluorescence (SIF), satellite data, land surface temperature (LST), and microwave vegetation optical depth (VOD), including optical-based vegetation indices with environmental data (soil, climate, topography and geography). Machine learning techniques outperformed linear regression techniques to predict maize, rice, and soybean yields when combined with satellite data and environmental data. (Gupta et al., 2021) hope to assist farmers in increasing the yield of their crops by gathering and analysing data on soil, temperature, wind speed, rainfall, crop productivity, seed and humidity (in some places). The MapReduce architecture is used to process and analyse data, and K-means clustering is used to predict crops.

(Liu et al., 2022) used MLR to predict the occurrence of plant diseases using IoT-based environmental parameters, with up to 91% accuracy. (Udutalapally et al., 2021) trained CNN to create an illness prediction system with 99.24% accuracy. (Makkithaya and G., 2022) trained deep residual network-based regression models ResNet-16 and ResNet-28 for soybean production prediction using federated averaging. (Mehta et al., 2021) evaluated the effectiveness of various crop yield prediction algorithms, including CNN, LSTM, and the CNN-LSTM framework. (Agarwal and Tarar, 2021) improved their model by using deep learning and machine learning to forecast an appropriate yield, with 97% accuracy. According to the analysis done by (Bodapati et al., 2022), data was first sorted using an artificial neural network technique, and harvest production was then predictably calculated using machine learning. During the testing phase, the yield value is roughly estimated based on specific restrictions.

(Mopidevi et al., 2022) have used machine learning and deep learning to predict Ficus stem benjamina stem in controlled areas. Comparative studies that evaluate crop development using ML techniques like SVR and RFR are described (Swarnakantha et al., 2022). This article introduces a platform where growers may connect with other growers and specialists through an agricultural forum, as well as sell their crops to domestic and worldwide buyers. (Bhansali et al., 2022) created a recommendation model using N-K-P, Ph.D. value, and rainfall as training parameters to identify type of disease and suggest how to treat or prevent it. (Nancy et al., 2022) have performed an Image based plant disease detection along with classification using ML & DL.

PROPOSED WORK

Building an intelligent machine learning model in Python is the task done here. It suggests the optimum crop to plant depending on the soil's composition, pH level, rainfall, and location. By examining variables such as district, state, season, and crop type using various supervised machine learning approaches, the model focuses on anticipating the agricultural production. The main goal of this method is to give farmers crop recommendations that will assist them choose a crop that will yield more profitably by letting them know the crop yield in advance.

The proposed system stores the information relevant for processing through the ML Algorithms. This module mainly focuses on the testing & training the models with the dataset which is prepared in the staring module and also to make those algorithms ready for prediction. All models are created, then their accuracy and their validation scores are checked and all the models are saved in pkl files for further use for predictions. Also, different outputs are predicted and its accuracy is checked through different - different types of inputs.

Dataset Preparation

We were able to develop a predictive model using the dataset collected from kaggle.com, which will recommend the best crops to plant on a particular farm based on a number of variables. The dataset dimension is (2200, 8). The ingredients present in soil are: Nitrogen, Phosphorous, Potassium.

Data fields:

N - Nitrogen ratio in soil
P - Phosphorous ratio in soil
K - Potassium ratio in soil
temperature - temperature in degree Celsius
humidity - relative humidity in %
pH - soil's pH value
rainfall –rainfall in mm

Figure 1. Dataset snapshot

```
In [5]: crop_df = pd.read_csv('Dataset.csv')
        crop_df.head(10)

Out[5]:
```

	N	P	K	temperature	humidity	ph	rainfall	label
0	90	42	43	20.879744	82.002744	6.502985	202.935536	rice
1	85	58	41	21.770462	80.319644	7.038096	226.655537	rice
2	60	55	44	23.004459	82.320763	7.840207	263.964248	rice
3	74	35	40	26.491096	80.158363	6.980401	242.864034	rice
4	78	42	42	20.130175	81.604873	7.628473	262.717340	rice
5	69	37	42	23.058049	83.370118	7.073454	251.055000	rice
6	69	55	38	22.708838	82.639414	5.700806	271.324860	rice
7	94	53	40	20.277744	82.894086	5.718627	241.974195	rice
8	89	54	38	24.515881	83.535216	6.685346	230.446236	rice
9	68	58	38	23.223974	83.033227	6.336254	221.209196	rice

Figure 2 shows the illustration of various features of the dataset: (a) Nitrogen (b) Phosphorus (c) Potassium (d) PH (e) Humidity (f) Temperature (g) Rainfall

Figure 2. Features of dataset

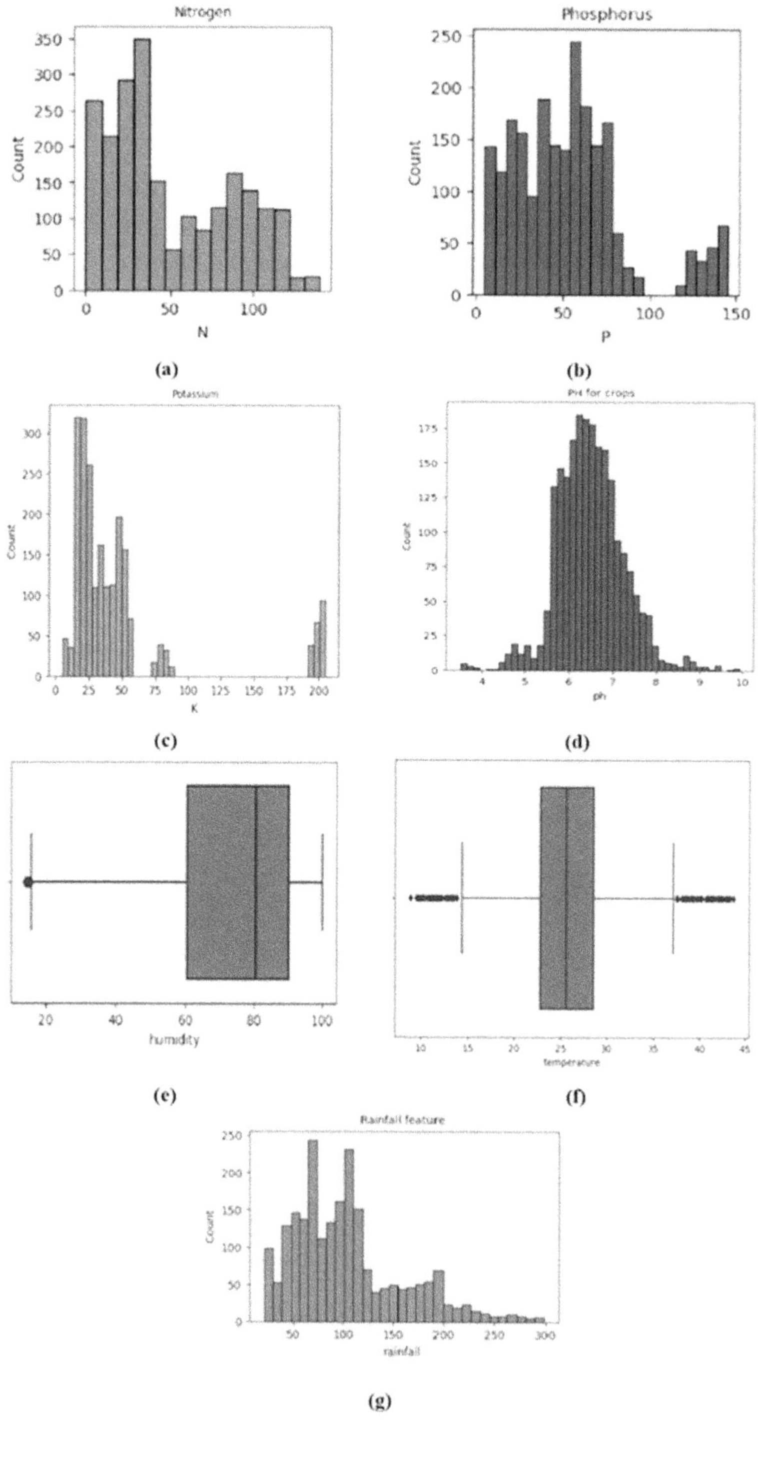

Technologies to be Used

Various technologies are used in making of this project. It includes Numpy, Pandas, Matplotlib etc. Also, pytorch and various python libraries are used. In this project, python files, .pynb files and .pkl files are used for model preparation. Also, various ML technologies are used. To run a ML Algorithm on a prepared dataset, various techniques are used to do this.

ML Algorithm Preparation

The algorithms are chosen appropriately and trained and tested with the dataset and many different models are created. The system is an intranet-based system so users can predict crops based on their input values that provides detail general information about the which crop to grow and in which condition. It enhances the agriculture management in adding, viewing and updating the crops that can be grown in a particular area according to the particular condition.

Users may theoretically have access to a large range of products within a specific site. One or more Web pages may serve as tangible representations of these objects. Conceptually, each of these objects represents a distinct kind of semantic entity. During a session on this website, a user may access several of these things at once. Figure 3 shows the approach used in the work done.

The algorithms applied on the dataset mentioned are: Decision Tree, Naïve Bayes Classifier, Random Forest, k-NN. SVM, XGBoost and Logistic Regression. The behaviour and performance of each model is shown in the result section.

Figure 4 represents confusion matrices for various models: (a) Decision Tree (b) Naïve Bayes Classifier (c) Logistic Regression (d) Random Forest (e) Support Vector Machine (f) XGBoost (g) k-NN

RESULTS

All the algorithms are tested and trained on that dataset and different models are created by taking predictions through this model. The algorithm which has to be chosen should work properly with the prepared datasets. So, all the algorithms are trained and tested with the dataset and many different models are created. After checking all those algorithms, the ML Algorithms has to be coded and then models are created so that the proper predictions can be made. Out of the 7 algorithms applied, Naïve Bayes, XGBoost gives the maximum and equal accuracy i.e. 99.55%. Next, Random Forest and SVM give an accuracy percentage of 99.31. k-NN and Logistics regression results into the accuracy values of 98.86 and 98.63 respectively. Decision tree performed the worst in given scenario by providing the accuracy of 85.91% only.

CONCLUSION

The process for proposing crops before cultivation is described in the article. The input data is first pre-processed in order to identify missing values, eliminate redundant information, standardize the data, and convert target attributes into factor attributes. The dataset is divided into training and testing phases before ML techniques are used to the enhanced attributes. To determine the type of crop that is most suitable for growing in a specific plot of land depending on the soil properties (i.e. ratio of nitrogen, phosphorus, and potassium), weather conditions, temperature, soil PH, rainfall, etc., various machine learning algorithms are trained using unknown samples from the training dataset. The testing dataset is used to use the learned classifier to forecast the crop that will be grown. A suitable crop is chosen, and the results are accurately evaluated. The study concludes that XGBoost and Naive Bayes classifiers have the highest accuracy and are the most efficient and appropriate approaches.

Figure 3. Flow chart of the proposed approach

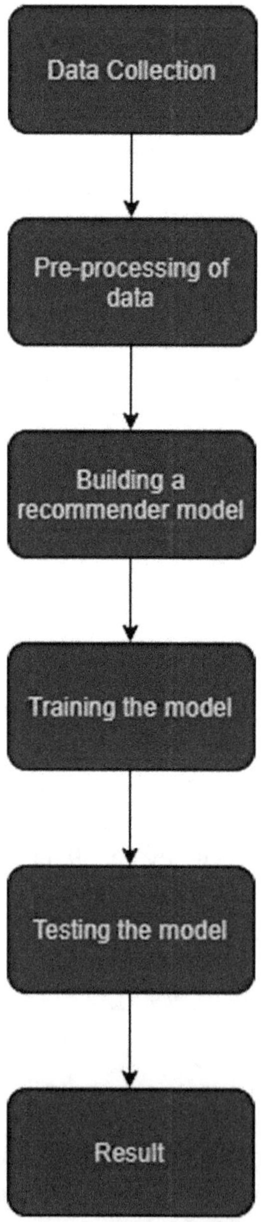

Figure 4. Confusion matrices of various models

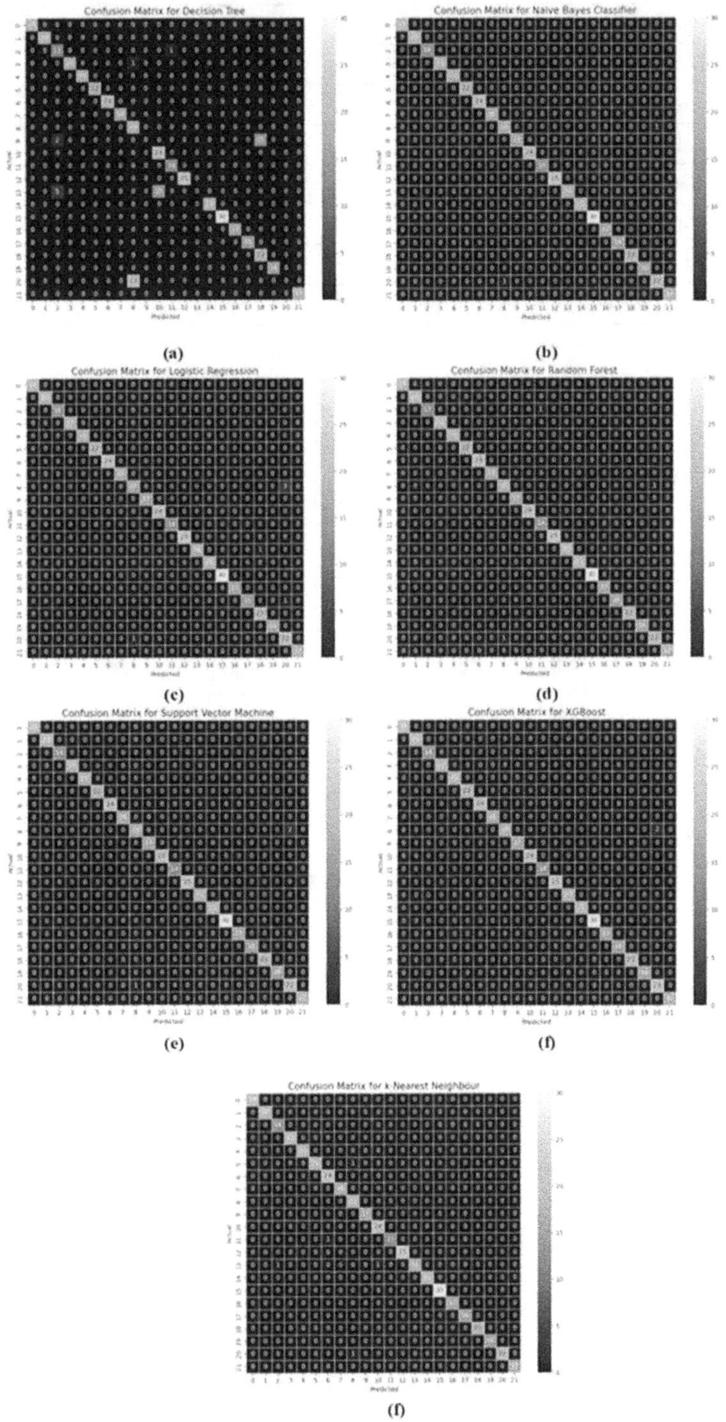

Table 1. Accuracy of various models applied

S. No.	Model	Accuracy (in Percentage)
1	Decision Tree	85.91
2	Naive Bayes Classifier	99.55
3	Random Forest	99.31
4	k-NN	98.86
5	SVM	99.31
6	XGBoost	99.55
7	Logistic Regression	98.63

Figure 5. Accuracy comparison

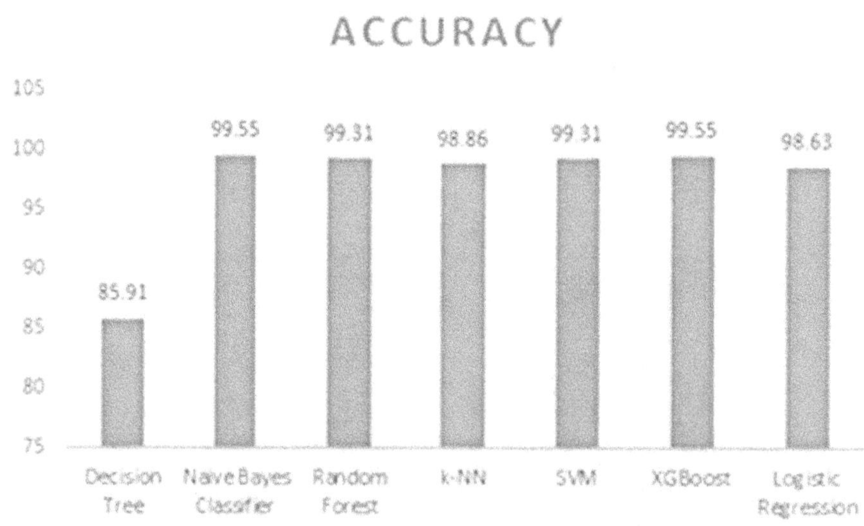

REFERENCES

Abbaszadeh, P., Gavahi, K., Alipour, A., Deb, P., & Moradkhani, H. (2022). Bayesian multi-modeling of deep neural nets for probabilistic crop yield prediction. *Agricultural and Forest Meteorology, 314*, 108773. doi:10.1016/j.agrformet.2021.108773

Agarwal, S., & Tarar, S. (2021). A hybrid approach for crop yield prediction using machine learning and deep learning algorithms. *Journal of Physics: Conference Series, 1714*(1), 012012. doi:10.1088/1742-6596/1714/1/012012

Aghighi, H., Azadbakht, M., Ashourloo, D., Shahrabi, H. S., & Radiom, S. (2018, December). Machine learning regression techniques for the silage maize yield prediction using time-series images of landsat 8 OLI. *IEEE Journal of Selected Topics in Applied Earth Observations and Remote Sensing*, *11*(12), 4563–4577. doi:10.1109/JSTARS.2018.2823361

Alebele, Y., Wang, W., Yu, W., Zhang, X., Yao, X., Tian, Y., Zhu, Y., Cao, W., & Cheng, T. (2021). Estimation of crop yield from combined optical and SAR imagery using Gaussian kernel regression. *IEEE Journal of Selected Topics in Applied Earth Observations and Remote Sensing*, *14*, 10520–10534. doi:10.1109/JSTARS.2021.3118707

Ali, A. M., Abouelghar, M., Belal, A. A., Saleh, N., Yones, M., Selim, A. I., Amin, M. E. S., Elwesemy, A., Kucher, D. E., Maginan, S., & Savin, I. (2022). Crop yield prediction using multi sensors remote sensing (Review Article). *Egyptian Journal of Remote Sensing and Space Science, 25*(3), 711–716. doi:10.1016/j.ejrs.2022.04.006

Bhansali, S., Shah, P., Shah, J., Vyas, P., & Thakre, P. (2022). Healthy harvest: Crop prediction and disease detection system. *7th International conference for Convergence in Technology (I2CT),* (pp. 1–5). IEEE Publications. 10.1109/I2CT54291.2022.9825446

Birrell, S. J., Sudduth, K. A., & Borgelt, S. C. (1996). Comparison of sensors and techniques for crop yield mapping. *Computers and Electronics in Agriculture*, *14*(2–3), 215–233. doi:10.1016/0168-1699(95)00049-6

Bodapati, N., Himavaishnavi, J., Rohitha, V., Jagadeeswari, D. L., & Bhavana, P. (2022). Analyzing crop yield using machine learning. *IEEE International Conference on Electronics and Renewable Systems (ICEARS),* (pp. 1–8). IEEE. 10.1109/ICEARS53579.2022.9752242

Bose, P., Kasabov, N. K., Bruzzone, L., & Hartono, R. N. (2016, November). Spiking neural networks for crop yield estimation based on spatiotemporal analysis of image time series. *IEEE Transactions on Geoscience and Remote Sensing*, *54*(11), 6563–6573. doi:10.1109/TGRS.2016.2586602

Cedric, L. S., Adoni, W. Y. H., Aworka, R., Zoueu, J. T., Mutombo, F. K., Krichen, M., & Kimpolo, C. L. M. (2022). *Crops yield prediction based on machine learning models: Case of West African countries.* Smart Agricultural Technology.

Chandraprabha, M., & Dhanaraj, R. K. (2021). Soil based prediction for crop yield using predictive analytics. *3rd IEEE International Conference on Advances in Computing, Communication Control. and Networking (ICAC3N),* (pp. 265–270). IEEE. 10.1109/ICAC3N53548.2021.9725758

Cunha, R. L. F., & Silva, B. (2020). Estimating crop yields with remote sensing and deep learning. IEEE Latin American GRSS y ISPRS Remote Sensing Conference (LAGIRS), (pp. 273–278). IEEE.

Cunha, R. L. F., Silva, B., & Netto, M. A. S. (2018). A scalable machine learning system for pre-season agriculture yield forecast. *14th International Conference on e-Science (e-Science),* (pp. 423–430). IEEE Publications. 10.1109/eScience.2018.00131

Edwin, T. (2017) [Online]. Onion, tomato price spike: Season not the only reason. *The Hindu Business Online.* https://www.thehindubusinessline.com/economy/agri-business/onion-tomato-price-spike-season-not-the-only-reason/article9957255.ece#.

Elavarasan, D., & Vincent, P. M. D. (2020). Crop yield prediction using deep reinforcement learning model for sustainable agrarian applications. *IEEE Access : Practical Innovations, Open Solutions*, *8*, 86886–86901. doi:10.1109/ACCESS.2020.2992480

Goel, L., & Mishra, A. (2022). A survey of recent deep learning algorithms used in Smart farming. *IEEE Region 10 Symposium (TENSYMP),* (pp. 1–6). IEEE. 10.1109/TENSYMP54529.2022.9864477

Gupta, M. V, S. K. B., B, K., Narapureddy, H. R., Surapaneni, N., & Varma, K. (2022). Various crop yield prediction techniques using machine learning algorithms. *IEEE Second International Conference on Artificial Intelligence and Smart Energy (ICAIS),* (pp. 273–279). IEEE. 10.1109/ICAIS53314.2022.9742903

Gupta, R., Sharma, A. K., Garg, O., Modi, K., Kasim, S., Baharum, Z., Mahdin, H., & Mostafa, S. A. (2021). WB-CPI: Weather based crop prediction in India using big data analytics. *IEEE Access : Practical Innovations, Open Solutions*, *9*, 137869–137885. doi:10.1109/ACCESS.2021.3117247

Haque, F. F., Abdelgawad, A., Yanambaka, V. P., & Yelamarthi, K. (2020). Crop yield prediction using deep neural network. *6th World Forum on Internet of Things (WF-IoT),* (pp. 1–4). IEEE Publications. 10.1109/WF-IoT48130.2020.9221298

Iniyan, S., Varma, V. A., & Naidu, C. T. (2023). Crop yield prediction using machine learning techniques. *Advances in Engineering Software*, *175*, 103326. doi:10.1016/j.advengsoft.2022.103326

Kalaiarasi, E., & Anbarasi, A. (2022). Multi-parametric multiple kernel deep neural network for crop yield prediction. *Materials Today: Proceedings, 62*(7), 4635–4642. doi:10.1016/j.matpr.2022.03.115

Kalimuthu, M., Vaishnavi, P., & Kishore, M. (2020). Crop prediction using machine learning. *Third IEEE International Conference on Smart Systems and Inventive Technology (ICSSIT),* (pp. 926–932). IEEE. 10.1109/ICSSIT48917.2020.9214190

Kumar, M., Kumar, A., & Palaparthy, V. S. (2021). Soil sensors-based prediction system for plant diseases using exploratory data analysis and machine learning. In IEEE Sensors Journal, 21(16). doi:10.1109/JSEN.2020.3046295

Li, Z. Ding, L., & Xu, D. (2022). Exploring the potential role of environmental and multi-source satellite data in crop yield prediction across Northeast China. *Science of the Total Environment, 815*. doi:10.1016/j.scitotenv.2021.152880

Liu, Z., Bashir, R. N., Iqbal, S., Shahid, M. M. A., Tausif, M., & Umer, Q. (2022). Internet of things (IoT) and machine learning model of plant disease prediction–blister blight for tea plant. *IEEE Access : Practical Innovations, Open Solutions*, *10*, 44934–44944. doi:10.1109/ACCESS.2022.3169147

Makkithaya, K., & N. V. G. (2022). A federated learning-based crop yield prediction for agricultural production risk management. IEEE Delhi Section Conference (DELCON), 2022 (pp. 1–7). IEEE.

Mariammal, G., Suruliandi, A., Raja, S. P., & Poongothai, E. (2021, October). Prediction of land suitability for crop cultivation based on soil and environmental characteristics using modified recursive feature elimination technique with various classifiers. *IEEE Transactions on Computational Social Systems*, *8*(5), 1132–1142. doi:10.1109/TCSS.2021.3074534

Martínez-Ferrer, L., Piles, M., & Camps-Valls, G. (2021, December). Crop yield estimation and interpretability with Gaussian processes. *IEEE Geoscience and Remote Sensing Letters*, *18*(12), 2043–2047. doi:10.1109/LGRS.2020.3016140

Mehta, T., Manolkar, O., Multaikar, S., Patil, S., & Gawande, N. (2021). Intensification of agriculture using deep learning and machine learning: A survey. *10th IEEE International Conference on Internet of Everything, Microwave Engineering,* (pp. 1–6). IEEE. 10.1109/IEMECON53809.2021.9689196

Mondal, A., & Banerjee, S. (2021). Effective crop prediction using deep learning. *IEEE International Conference on Smart Generation Computing, Communication and Networking (SMART GENCON).* IEEE. 10.1109/SMARTGENCON51891.2021.9645872

Mopidevi, S., Singitham, V., Thippani, B., Shamanthula, R., & Satya Phanindra Vallabhaneni, N. S. (2022). Plant growth and yield prediction using ML and DL algorithms. *IEEE International Conference on Electronics and Renewable Systems (ICEARS),* (pp. 1470–1477). IEEE. 10.1109/ICEARS53579.2022.9751906

Nancy, P., Pallathadka, H., Naved, M., Kaliyaperumal, K., Arumugam, K., & Garchar, V. (2022). Deep learning and machine learning based efficient framework for image based plant disease classification and detection. *IEEE International Conference on Advanced Computing Technologies and Applications (ICACTA),* (pp. 1–6). IEEE. 10.1109/ICACTA54488.2022.9753623

Pande, S. M., Ramesh, P. K., Anmol, A., Aishwarya, B. R., Rohilla, K., & Shaurya, K. (2021). Crop recommender system using machine learning approach. *5th IEEE International Conference on Computing Methodologies and Communication (ICCMC).* (pp. 1066–1071). IEEE. 10.1109/ICCMC51019.2021.9418351

Pang, L., Men, S., Yan, L., & Xiao, J. (2020). Rapid vitality estimation and prediction of corn seeds based on spectra and images using deep learning and hyperspectral imaging techniques. *IEEE Access : Practical Innovations, Open Solutions, 8,* 123026–123036. doi:10.1109/ACCESS.2020.3006495

Pant, J., Pant, R. P., Kumar Singh, M., Pratap Singh, D., & Pant, H. (2021). Analysis of agricultural crop yield prediction using statistical techniques of machine learning. *Materials Today, 46*(20), 10922–10926. doi:10.1016/j.matpr.2021.01.948

Pantazi, X. E., Moshou, D., Alexandridis, T., Whetton, R. L., & Mouazen, A. M. (2016). Wheat yield prediction using machine learning and advanced sensing techniques. *Computers and Electronics in Agriculture, 121,* 57–65. doi:10.1016/j.compag.2015.11.018

Qiao, M. He, X., Cheng, Xijie, Li, Panle, Luo, Haotian, Zhang, Lehan, & Tian, Z. (2021). Crop yield prediction from multi-spectral, multi-temporal remotely sensed imagery using recurrent 3D convolutional neural networks. *International Journal of Applied Earth Observation and Geoinformation, 102,* 102436. doi:10.1016/j.jag.2021.102436

Qiao, M., He, X., Cheng, X., Li, P., Luo, H., Tian, Z., & Guo, H. (2021). Exploiting hierarchical features for crop yield prediction based on 3-D convolutional neural networks and multikernel Gaussian process. *IEEE Journal of Selected Topics in Applied Earth Observations and Remote Sensing, 14,* 4476–4489. doi:10.1109/JSTARS.2021.3073149

Raja, S. P., Sawicka, B., Stamenkovic, Z., & Mariammal, G. (2022). Crop prediction based on characteristics of the agricultural environment using various feature selection techniques and classifiers. *IEEE Access : Practical Innovations, Open Solutions, 10,* 23625–23641. doi:10.1109/ACCESS.2022.3154350

Rasheed, N., Khan, S. A., Hassan, A., & Safdar, S. (2021). A decision support framework for national crop production planning. *IEEE Access : Practical Innovations, Open Solutions*, *9*, 133402–133415. doi:10.1109/ACCESS.2021.3115801

Rashid, M., Bari, B. S., Yusup, Y., Kamaruddin, M. A., & Khan, N. (2021). A comprehensive review of crop yield prediction using machine learning approaches with special emphasis on palm oil yield prediction. *IEEE Access : Practical Innovations, Open Solutions*, *9*, 63406–63439. doi:10.1109/ACCESS.2021.3075159

Ray, R. K., & S. K. (2022). Das and S. Chakravarty. Smart Crop Recommender System-à Machine Learning Approach. *12th IEEE International Conference on Cloud Computing, Data Science & Engineering (Confluence),* (pp. 494–499). IEEE.

Reddy, D. J., & Kumar, M. R. (2021). Crop yield prediction using machine learning algorithm. *5th IEEE International Conference on Intelligent Computing and Control. Systems (ICICCS),* (pp. 1466–1470). IEEE. 10.1109/ICICCS51141.2021.9432236

Saeed, Khaki, & Lizhi, W. (2019). ISSN=1664-462X. Crop yield prediction using deep neural networks. *Frontiers in Plant Science*, 10.

Seireg, H. R., Omar, Y. M. K., El-Samie, F. E. A., El-Fishawy, A. S., & Elmahalawy, A. (2022). Ensemble machine learning techniques using computer simulation data for wild blueberry yield prediction. *IEEE Access : Practical Innovations, Open Solutions*, *10*, 64671–64687. doi:10.1109/ACCESS.2022.3181970

Sharma, A., Jain, A., Gupta, P., & Chowdary, V. (2021). Machine learning applications for precision agriculture: A comprehensive review. *IEEE Access : Practical Innovations, Open Solutions*, *9*, 4843–4873. doi:10.1109/ACCESS.2020.3048415

Sharma, P., & Rathi, Y. (2016). Efficient densit- based clustering using automatic parameter detection. *Proceedings of the International Congress on Information and Communication Technology*. Springer.

Sharma, S. K., Sharma, D. P., & Verma, J. K. (2021). Study on Machine-Learning Algorithms in Crop Yield Predictions specific to Indian Agricultural Contexts. *IEEE International Conference on Computational Performance Evaluation (ComPE),* (pp. 155–166). IEEE. 10.1109/ComPE53109.2021.9752260

Sivanantham, V., Sangeetha, V., & Alnuaim, A. A. Hatamleh, Wesam A., Anilkumar, C., Hatamleh, A. A., & Sweidan, D. (2022). Quantile correlative deep feedforward multilayer Perceptron for crop yield prediction. *Computers and Electrical Engineering, 98*. doi:10.1016/j.compeleceng.2022.107696

Sun, J., Lai, Z., Di, L., Sun, Z., Tao, J., & Shen, Y. (2020). Multilevel deep learning network for county-level corn yield estimation in the U. S. Corn Belt. *IEEE Journal of Selected Topics in Applied Earth Observations and Remote Sensing*, *13*, 5048–5060. doi:10.1109/JSTARS.2020.3019046

Swarnakantha, S., Chathurika, B., Weragoda, K. V., Bowatte, W. M. I. K., Thalawala, E. V., & Bandara, M. M. U. L. (2022). Decision-making platform for SMART plantation agriculture using machine learning and image processing. *7th International conference for Convergence in Technology (I2CT),* (pp. 1–6). IEEE Publications. 10.1109/I2CT54291.2022.9825063

Tahaseen, M., & Moparthi, N. R. (2021). An assessment of the machine learning algorithms used in agriculture. *5th International Conference on Electronics, Communication and Aerospace Technology (ICECA),* (pp. 1579–1584). IEEE Publications. 10.1109/ICECA52323.2021.9675918

Udutalapally, V., Mohanty, S. P., Pallagani, V., & Khandelwal, V. (2021). sCrop: A novel device for sustainable automatic disease prediction, crop selection, and irrigation in Internet-of-agro-things for Smart agriculture. In IEEE Sensors Journal, 21(16), 17525–17538. doi:10.1109/JSEN.2020.3032438

van Klompenburg, T., Kassahun, A., & Catal, C. (2020). Crop yield prediction using machine learning: A systematic literature review. *Computers and Electronics in Agriculture, 177.* doi:10.1016/j.compag.2020.105709

Vashisht, S., Kumar, P., & Trivedi, M. C. (2022). Improvised extreme learning Machine for crop yield prediction. *3rd IEEE International Conference on Intelligent Engineering and Management (ICIEM),* (pp. 754–757). IEEE. 10.1109/ICIEM54221.2022.9853054

Vlachopoulos, O., Leblon, B., Wang, J., Haddadi, A., LaRocque, A., & Patterson, G. (2022). Evaluation of crop health status with UAS multispectral imagery. *IEEE Journal of Selected Topics in Applied Earth Observations and Remote Sensing, 15,* 297–308. doi:10.1109/JSTARS.2021.3132228

Yamparla, R., Shaik, H. S., Guntaka, N. S. P., Marri, P., & Nallamothu, S. (2022). Crop yield prediction using random forest algorithm. *7th International Conference on Communication and Electronics Systems (ICCES),* (pp. 1538–1543). IEEE. 10.1109/ICCES54183.2022.9835756

Ziliani, M. G., Altaf, M. U., Aragon, B., Houborg, R., Franz, T. E., Lu, Y., Sheffield, J., Hoteit, I., & McCabe, M. F. (2022). Early season prediction of within-field crop yield variability by assimilating CubeSat data into a crop model. *Agricultural and Forest Meteorology, 313.* doi:10.1016/j.agrformet.2021.108736

Chapter 11
AI-Enabled IoT and WSN-Integrated Smart Agriculture System

Ashok Kumar Koshariya
https://orcid.org/0000-0002-7427-4835
Department of Plant Pathology, School of Agriculture, Lovely Professional University, India

D. Kalaiyarasi
Department of Electronics and Communication Engineering, Panimalar Engineering College, India

A. Arokiaraj Jovith
College of Engineering and Technology, SRM Institute of Science and Technology, India

T. Sivakami
Bharath Institute of Science and Technology, India

Dler Salih Hasan
Department of Computer Science and Information Technology, University of Salahaddin, Erbil, Iraq

Sampath Boopathi
https://orcid.org/0000-0002-2065-6539
Muthayammal Engineering College, India

ABSTRACT

Agriculture and farming have gotten smarter as a result of the use of current technology such as Wireless Sensor Networks (WSN) and the Internet of Things (IoT). Smart farming is an enhanced agriculture system that offers data such as temperature, soil moisture, and so on, to assist in the growth of plants and cattle. It integrates wireless sensors and the internet to collect and communicate information with farmers. The priority event-based energy efficient algorithm developed in this study is utilized for accurate and efficient information transmission regarding power consumption and node priority. The major goal of the IoT-sensor network in this chapter is to increase farm productivity and extend its lifespan by applying intelligent algorithms such as Artificial Neural Network (ANN) to recognize environmental conditions and improve total production. Priority event-based energy efficient method reduces energy usage and increases the lifetime using Dijkstra's algorithm.

DOI: 10.4018/978-1-6684-8516-3.ch011

INTRODUCTION

The internet of things (IoT) is a promising technology that provides answers to problems in a variety of sectors. Kevin Ashton of MIT's Auto-ID lab invented the term in 1999. The IoT is a network of billions of connected devices that can sense, collect, and transmit data without human intervention, affecting a wide range of industries including health care monitoring, building automation, logistics, connected vehicles, smart city infrastructure, smart grid, smart home, smart retail, smart agriculture, and smart farming. This chapter describes the use of IoT in Smart Agriculture, including an introduction to IoT, Wireless Sensor Networks, Smart Agriculture Using Wireless Sensor Networks, Motivation for Research, Challenges, Research Objectives, and Research Contributions (Haseeb et al., 2020).

Agriculture has experienced multiple revolutions, including plant and animal domestication, crop rotations, and the "green revolution." ICT is viewed as the catalyst for a fourth agricultural revolution. Smart farming is a management approach that employs cutting-edge technology to measure, monitor, automate, and evaluate operations. It is managed by sensors and controlled by software. Because of population expansion, growing use of technology, and climate-smart agriculture, smart farming is becoming increasingly vital. It is a cyber-physical system that controls and manages the whole farm system using smart devices connected to the Internet. Traditional tools are improved by smart gadgets, which provide autonomous context awareness, built-in intelligence, and the capacity to undertake autonomous or remote operations. Humans are still involved in the process, but at a higher cognitive level, with robots performing the majority of operational activities(Alghazzawi et al., 2021).

The Internet of Things (IoT) enables wireless hardware to share data via a network, resulting in a significant increase in the number of electrical, connecting devices in the 1980s and 1990s. M2M and IoT communication have facilitated the growth of linked devices, with Cisco forecasting that there would be 50 billion connected devices by 2020. The Internet of Things is envisioned as the foundation of a networked, safe, intelligent, and inventive civilization of the future. The Internet of Things (IoT) is an ubiquitous computer system that enables devices to connect directly with one another and exchange related information, allowing humans to concentrate on choices and actions rather than filtering and integrating data.

Wireless Sensor Network (WSN) technology has improved greatly, allowing the use of motes and sensor nodes to monitor ecological occurrences across a vast geographic area. By engaging with a gateway, sensor nodes may communicate wirelessly and relay data to a base station or coordinator node. WSNs can monitor a broad range of surroundings and acquire exact information since the communication is based on numerous sensors. The detecting, storage, processing, and transmission capabilities of sensor nodes have increased. WSNs have found applications in the military, agriculture, sports, medicine, and industry. Precision agriculture (PA) aims to enhance field management by avoiding the same management routine no matter what the site conditions are. PA lowers pesticide waste while also ensuring crops receive the nutrients they require, leading in efficient, ecologically responsible agriculture. PA is a management strategy that employs information technology to improve agricultural quality and output. It consists of five steps: data collection, diagnosis, data analysis, precision field operation, and evaluation. WSNs are used to boost agricultural production and anticipate crop health and product quality by predicting irrigation plans based on weather and soil moisture. Additional sensor nodes can be added to the present WSN to improve the monitoring characteristics of the smart farming system and make the network scalable. WSN deployment techniques, measurement times, routing protocols, energy efficiency, cost, communication range, scalability, and fault tolerance have all been challenges. Although dispersed sensor node deployment might assist increase network lifetime, selecting a distribution zone can be problematic. WSNs are battery-powered, which eliminates the requirement for connections to the main power source. It is critical to decrease power depletion and extend battery life in order to reduce power depletion and extend battery life(Rao et al., 2022).

IoTs give important data and learning opportunities, and a WSN created a system to examine plant-related sensor data for social events. This review employed WSN data to discover learning through data mining. Leaf spot disease was monitored using remote sensors and field level identification. As an Internet-based observation approach for IoTs, distributed computing was proposed. Knowledge mining was utilized to extract useful data and learning; however this activity might jeopardize the information provided by IoT. Data mining uses IoTs to extract useful and actionable knowledge from enormous amounts of data, hence addressing Information Creation Mechanisms. The electronic framework assists an executive in tracking the water requirements of yields and may be enhanced to estimate harvest water requirements. The suggested system is divided into three layers: the natural layer information, the information and correspondence layer, and the application layer. The ecological information collecting layer gathers data from sensors and device control on natural elements, and the framework layer utilizes the gathered data to track and monitor harvesting and deliver information to the board(Burugari et al., 2021).

ZIGBEE PROTOCOL

The IEEE 802.15.4 standard for control and control systems for wireless personal area networks is central to Zigbee connectivity (WPANs). These WPANs operate on wavelengths of 868 MHz, 902-928 MHz, and 2.4 GHz, with a data rate of 250 kbps for occasional two-way information transfer. Zigbee is a low-power communication technology that can manage and track applications with a range of 10-100 meters. It is more dependable and straightforward than other short-run remote sensors, and it may be enlarged to need the use of several hubs to interface with one another.

Zigbee Architecture

A Zigbee network is composed of three components: an organizer Zigbee, a modem, and an end computer, with one facilitator serving as the system's root and extension. Zigbee switches are used to transfer data between devices, hence minimizing battery use. The number of switches, facilitators, and end machines is determined by the system type. The Zigbee module for wireless channel data transfer is seen(Nandal & Dahiya, 2021).

Physical Layer: Signals are modulated and demodulated by the layer, disclosing their frequency, date rate, and channel size.

MAC Layer: Secure data transmission is accomplished by accessing the carrier background and issuing connection coordination beacon packets.

Network Layer: All system-specific tasks are handled by the system layer.

Access management sub-Layer: The layer enables network layers to connect Zigbee device and client artifact resources to data management systems.

Application Framework: ZDO connects program artifacts to Zigbee systems by detecting, activating, and binding the connection to other devices. It has two primary values, two types of data sources, and a default messaging application.

RESEARCH METHODOLOGY

Due to rising demand, the need for safe drinking water and clean water for agriculture is increasing. In locations where water is scarce, water management is crucial, and water adaptation approaches are being studied to provide an appropriate supply of water. Commercial sensor-based systems for monitoring agricultural fields and managing irrigation systems are costly, but manufacturers are now developing low-cost sensors that can be linked to nodes to construct cost-effective irrigation management and farm monitoring systems. WSNs have emerged as a viable technology for smart farming and precision agriculture, enhancing productivity and crop output. Precision agriculture using autonomous irrigation systems collects real-time data from the field using a sensor system to monitor soil conditions correctly and dependably. Farmers are not required to check their fields. Smart farming necessitates the use of the internet as well as sensitive technologies to gather data on soil conditions, fertility, moisture, temperature, and food storage. Data may be saved on the cloud, allowing farmers to access it at the appropriate moment. The systematic approach for establishing wireless sensor nodes in smart farming is covered in this chapter(Al-Tarawneh et al., 2022).

Region-Based Clustering Method

Precision agriculture necessitates the collection of data regarding soil conditions from every area of the field, therefore the routing protocol must assure full coverage with sensors distributed across the land. To partition the field into fixed regions, the region-based static clustering approach is applied. The agricultural area was split into nine fixed-size zones, and information from the region was collected using two types of sensors. These sensors were temperature and humidity sensors, and they were in charge of monitoring environmental aspects such as weather, dryness, and flooding. This paper presents a hybrid routing system for precision agriculture that is sensitive to periodic thresholds. It uses region-based static clustering algorithms to distribute sensor nodes, ensuring efficient coverage of the whole agricultural area. When the user-defined timeframe or detected attribute, value exceeds the necessary threshold, data is delivered to the base station. This network employs three types of dissimilar packet sensor nodes: Type 1, Type 2, and Type 3. Type 1 sends data straight to the BS, but types 2 and 3 use fuzzy-based grouping to reduce energy consumption and increase network longevity. To split the agricultural area into equal parts, region segmentation is done based on the availability of different types of heterogeneous nodes in the network and their sensing range. To split the agricultural area into equal parts, region segmentation is done based on the availability of different types of heterogeneous nodes in the network and their sensing range.

Sensitive Threshold Procedure

Irrigation in an agricultural area does not necessitate continuous data delivery to the BS, hence a threshold-sensitive protocol has been devised to send data only when the measured value falls under the user-defined range. The threshold sensitivity technique has a limit, which means the farmer is oblivious of the field conditions if the detected value does not exceed the thresholds. The threshold-sensitive process constantly observes the environmental parameter, but data is only provided to the BS when it is within the user-specified range. Routing protocols are used to transfer data after a set amount of time has passed. The farmer/user collects temperature and moisture content data using a PT and sends it to the BS through hybrid routing. Type 1 nodes are put in R-9, while Type 3 and Type 2 nodes are installed in the center and outermost zones, respectively(Indrapandi & Rizwana, 2021).

Existing System

The current precision agricultural data transfer system employs a region division-based clustering protocol, in which each sensor node observes an environmental parameter and transmits data to the BS when it crosses user-defined data-transmission thresholds. For data transmission, the cluster head (CH) is employed, with one CH from each area chosen to convey information. Eight CHs are chosen each round based on the availability of Type 2 and Type 3 nodes. In the event of detecting a large field, more CH nodes or additional Type-2 and Type-3 nodes planted in an outermost or intermediate zone can be used to balance data-aggregation energy. When the CH is picked, data is transferred to the base station.

PROPOSED METHODOLOGY

Hybrid Sensor, Region Division, Node Deployment, Periodic Threshold Checking, Event Checker, Routing Protocol, and Performance Evaluation were the seven modules presented for the proposed system. During critical situations, the Region division module is utilized to find priority-based sensor nodes for effective data transfer to the Base Station packet, resulting in optimal energy utilization. Hybrid Sensor, Region Division, Node Deployment, Periodic Threshold Checking, Event Checker, Routing Protocol, and Performance Evaluation are the seven modules presented in the proposed system.

Hybrid Sensor Module

The Hybrid sensor module collects data from the region using two sensors: temperature and humidity. The network employs three types of dissimilar packet sensor nodes: Type-1, Type-2, and Type-3. Type-1 sends data straight to the Base Station, but Types-2 and 3 use the best way to the Base Station.

Region Division Module

For effective data transmission, the proposed system splits the field into six regions, with the outermost regions R-1, R-2, R-3, R-4, and R-5 placed far from the BS node.

Node Deployment Module

Type-2 and Type-3 nodes in the outermost zone use more energy to transfer data to the Base Station than Type-1 nodes in the innermost region. This is owing to their proximity to the BS.

Periodic Threshold Checking Module

Irrigation does not necessitate constant data transfer to the BS, therefore nodes in the Periodic Threshold Checking Module monitor environmental factors continually. Data transfer to the BS happens only when the detected value falls within the user-specified range. However, threshold sensitivity has a limit, therefore after a given duration, a different routing method is used to convey the data. A Periodic Threshold Checking Module is used by the farmer/user to gather temperature and moisture content data at each set time period. To determine when to start the water system in the field, a periodic dynamic threshold system calculates two state variables, surface temperature and moisture level.

Event Checker Module

The Event Checker Module validates multi-node events and sends them to the base station in priority order, using a trust value to determine the optimum routing.

Routing Protocol Module

Each node's trust value is assigned by the periodic table, which also routes data packets.

Performance Evaluation Module

The suggested technique assumes that after organization, both sensor networks and the BS remain static, and that the BS is situated in the detector sector's middle condition. This route distance is utilized to compute energy usage and do performance evaluations.

HYBRID REGION DIVISION AND NODE DEPLOYMENT MODULE

Farmers can communicate data via wireless sensor networks (WSNs). Precision agriculture (PA) is a management approach that employs data innovation to increase quality and yield. Farm management will increase PA by synchronizing appropriate seed nutrients and wasting chemicals used to control plants, pests, and infections. This research examines the recent use of wireless networks (WSNs) in agricultural analysis and classes. It investigates different wireless communication systems, vitality-compliant and vitality-collection approaches, as well as the relationship between early research and farming-based WSNs. It also examines the problems and obstructions of WSNs in the horticulture field, as well as how these approaches might be employed to handle IoT data. Sensors are utilized to collect inputs for processing, including temperature and humidity sensors detecting and collecting data from external conditions. Weather conditions are important in smart farming, with dry settings requiring more water and wet situations requiring less water from canals(Kadam et al., n.d.).

Assumptions

We've picked a grass field with specific soil types. It is critical to explain the 9 step margins of soil moisture tests in order to create an accurate soil moisture chart at varied depths. It is 30 cm long and separated by 10 cm by sensors, allowing for near-constant forecasts. To get an IoT stage structure, this study organized a soil moisture test using low-power equipment. Each neighbouring SN monitors the battery speed and detects and saves the solar vitality acquired in order to process a vitality spending profile. The most significant aspects are that SNs can range from tiny instruments to enormous implanted levels, and that limited capability is required for sign planning, device identification, a smartphone, a low memory level, and onboard access. These devices have the capability of wireless exchange, making them suitable for a variety of applications.

Region Division and Node Deployment

To transport data to the Base Station packet, this work employs three types of packet sensor nodes: Type-1, Type-2, and Type-3 (BS). All sensors are distributed in a specified land by packet (A*A) unit square for real-time sensing and data transfer. For efficient data transmission, the field is split into six regions, with R-1, 2, 3, and 4 nodes in the outermost regions and 6 nodes in the centre of the innermost area. Type-3 nodes use more resources than Type-1 nodes, whereas Type-2 nodes use less energy but are more powerful than Type-1 nodes. Because of their reduced energy use, type-1 nodes occupy the innermost region. The network setup includes three types of homogeneous sensor nodes that provide data to the BS and employ fuzzy-based grouping algorithms to increase efficiency and network life span. A periodic dynamic threshold system can use the suggested model to determine two state variables: surface temperature and moisture level. According to the unit's hypotheses, both sensor networks and the BS are static after organization, and the BS is located in the detector sector's middle situation.

To detect and relay data, the agricultural zone is separated into six fixed sections, with three types of sensor nodes delivered to an A/A unit. Field estimates are used to divide land into fields. Depending on the efficacy of various devices and their identification range, region segregation should be possible. Type-3 nodes with greater vitality than Type-1 and Type-2 nodes should be placed in secondary locations because they have greater vitality than Type-1 and Type-2 nodes. Type-2 nodes with high vitality should be put in strategic areas, whereas Type-1 nodes with low vitality should be distributed across the deepest sector. Type three nodes in the primary regions, two nodes in the supplemental areas, and one node in the deeper area in the suggested template.

Constant data is not necessary for submitting a water conservation plan to the BS since the water network only functions when there is a critical demand for water. When perceived meaning falls beyond the user-defined range, knowledge transitions to the BS, letting farmers to make decisions about water conservation measures. Because the capability limitation restricts data transmission to the BS, we propose a guiding approach that conveys data after a defined intermediate period. The transition, however, is dependent not only on the hard or soft sides, but also on the PT. Using a PT, the farmer/user collects data associated with humidity and temperature information. Half-and-half steering is employed to provide data to the BS. Type-3 and type-2 nodes delivered to the core and peripheral areas exchange information via the CH node. Type-3 clusters with the lowest output are grouped in core areas, receiving extra feedback from the CH node at the outer edge region(Islam & Dey, 2019).

Type-2 nodes in the central locations place the box in front of Type 3 nodes, resulting in increased data accumulation obligations and a gap in the central locations. Because of the inclusion gap in the core locations, the member networks on the outside edge areas are unable to communicate with the BS. To remove the incorporation disconfirmation model, the suggested definition employs type-3 nodes in the core locales. Only one CH from each outer edge and centerfield is picked, and a minimum of eight CHs are chosen for knowledge transmission. To guarantee that the strain is correctly modified at each location outside/center, the number of CH nodes must match the available nodes of type-2 and type-3. By picking only one CH from each exterior and center field, the suggested technique preserves the life of a CH node's knowledge creation. Separate CH nodes may be selected to change the data's agglomeration potency if additional type-2 and type-3 nodes are supplied in an outermost or center location. By allowing other community element nodes to communicate data to their CH nodes, which then pass it to the BS, the suggested technique aids in determining the proper number of CHs. The stiff threshold (T1) is the edge of an opportunity for monitoring prior values, while the mellow edge (T2) is an incentive for tiny observed accuracy changes. The two edges (T1 and T2) are employed for information transmission and can be chosen for yield dependent on air temperature and water levels. Consumers can determine these boundaries based on climate circumstances and agricultural water.

ADAPTIVE PRIORITY AND EVENT-BASED ENERGY EFFICIENT CLUSTER SELECTION ALGORITHM

The sink node's duty scheduling and packet planning are critical for controlling the utilization of Wireless Sensor Networks (WSNs). Current scheduling strategies employ inefficient First-Come-First-Served (FCFS) algorithms. Healthy nodes necessitate a shift in information transmission requirements, taking into account needs and conveying cutoff time. WSN business schedule formulae frequently do not adhere to traffic components, such as an ongoing need scheduler, which presents a need plot for three groups of packet planning. This is due to the fact that information is classed into knowledge categories(Janardhana et al., 2023; Samikannu et al., 2023; Trojovský et al., 2023)

- The most crucial concept is that priorities should be set.
- Priority is given to non-continuous wireless packets arriving from sensor nodes.
- At the current sensor node, non-continuous neighboring parcels are seen.
- Sensor nodes employ non-continuous information boxes to decrease the start-to-finish postponement of information packets generated in various locations. Constant information may be acquired at several axes, with the lowest priority3 being prioritized.
- Nodes are built in a leveled structure, with BS packet separation and Time-Division Multiplexing Access (TDMA) given at different speeds. Time slots 1 and 2 can be assigned to nodes at the lowest and second-lowest levels, respectively. Nodes are built in a leveled structure, with BS packet separation and Time-Division Multiplexing Access (TDMA) given at different speeds. Time slots 1 and 2 can be assigned to nodes at the lowest and second-lowest levels, respectively.

The paper provides a need-based scheduler for large-scale sensor systems, in which knowledge packets with the greatest separation from a source node to a BS have the highest demand and a short cutoff length. RACE is a packet booking approach that guides existing large-scale sensor systems by using EDF preparation estimate and MAC to reduce overhead traffic and increase handle time. Prioritizing the neighborhood at a single RACE node is inadequate since packets from different senders would compete over a shared radio channel. The suggested assignment booking approach includes a scheduling component to emphasize "unexpected problem detection" and allocate high-need tasks (organizing errands) and low-need tasks (identification). Neighborhood undertakings are not required for system assignments at any point in the obligation process. During the obligation procedure, the need for neighborhood assignments is greater than the demand for system undertakings. The suggested phased line scheduler plot employs equal line quantities, with level-1 nodes having one thread and level Lk-2 nodes having two threads. When a node receives a packet, it determines the parcel's requirement and routes it to the appropriate line(Reddy et al., 2023).

The sink nodes' wireless network is in charge of sensing instances and existing packets, turning them into packets, and transmitting them to a destination. This ensures the revitalization of target nodes and their inexhaustible resilience. Dijkstra's approach uses DS-RSS to effectively utilise the wireless sensor's energy by combining Route Routing and Rest Wake Scheduling techniques. Data packets are conveyed using hand-off and CDMA methods, which improves the lifespan architecture while decreasing viability. The most important notion is that the most important idea is that the most important idea is that the most important idea. The check node declares Tb Period guides and waits for their ACK response for an intermediate period Ts. If the node does not get an ACK response or if there are no nodes in the receiving box, it returns to the current state and restores the reference points. When there is no data for a sensor node to submit, it enters rest mode. The rest cycle intermediate time is determined arbitrarily for the network with wireless sensor nodes, and it may be raised for large devices. The sensor node wakes up for a brief period of time before returning to the time-rest phase of T sleep(S. et al., 2022).

Because sensor node tickers are updated and extra information is transmitted, wireless sensor networks are secure and simple to run. The wake-up event can be used in a Poisson approach with a fixed rate, but the intermediate time estimate of T sleep is unknown. Upside down Poisson reserve is a reduced storage product that assures the entire number of data packets provided per unit time is known. Dijkstra's calculation is the smallest operation of the lone source, determining the shortest path from the source node to each exterior node in the detector coordinate separation. Dijkstra's estimate works by determining the shortest path from the source node to its near-end neighbors. It is the sole method for resolving edge trust in distant sensing since it has the least difference and keeps the positive paths from the source node to the assessed goal node. By keeping the current sets of nodes in a line of precedence, Dijkstra's computation determines the appropriate node. The lowest difference between the root and itself is zero, and the prize for various nodes is set to infinity. Following each node's processing and application of the computation, the smallest good paths from topic to query node are obtained.

DIJKSTRA'S ALGORITHM

- Individual confidence is outperformed by the Dijkstra algorithm in distinguishing between primary and source nodes, resulting in a more efficient system.
- Network isolation promotes adoration for each node, zero separation for the basic network, and vastness for individual nodes.
- Uncertified nodes are separated until the start node is con d as the current node, resulting in the formation of a cluster.
- An evaluated node is reserved for evaluating each current node neighbor and removing them from the uninspected community.
- An evaluated node is reserved for evaluating each current node neighbor and removing them from the uninspected community.
- Dijkstra's computation is performed by picking a current node on each circle and indicating the displacement from the basic or important node to the full other node on the constant partnership. The schematic representation is selected as the schematic representation, and the difference between them is zero.

If the entire disparity between the estimate and the forecast is not the same, the current node is re-labeled as full estimation. After restoring the separations of each nearby node, the node with the least degree of divergence from the current or existing node is chosen as the inspected node. The uninspected node is the current node with the shortest distance to the essential node. The technique is carried out until the surrounding nodes having the least discrepancy are obtained. The current node is marked as being tested and is transferred to the nearest uninspected node until the target is marked as being investigated(Boopathi, 2022b, 2022c, 2022a; Domakonda et al., 2023; Kannan et al., 2022; Vennila et al., 2023; Yupapin et al., 2023).

PROPOSED ALGORITHM

To decrease energy-intensive routing, the Priority Event-based Energy Efficient algorithm is applied. The cluster routing protocol is divided into three phases: cluster head generation, creation, and communication. Cluster setups and secure information transfer are included in the Priority Event-based Energy Efficient method. The Priority Event-based Energy Efficient method selects cluster heads at random and distributes energy evenly to each node, resulting in lower energy usage and increased network longevity. The management technique of the Adaptive Priority Event-based Energy Efficient algorithm is intermittent. To save resources, we employ clusters and data transmission, however the time of reliable data transfer is substantially longer than the outcomes.

Nodes can generate a number between 0 and 1 at random; if it is smaller than the T(n) threshold, it is a cluster member. Nodes designated as cluster heads will notify all nearby nodes that they have been chosen as cluster leaders. The proportions of cluster nodes p, r, N, and G account for the total number of nodes. Cluster heads create a TDMA, or a timing gap, for each node in the cluster and broadcast it to the rest of the cluster. Each node will send data during its time gap, while the other will save resources by going to sleep. During safe data transmission, non-cluster nodes must transmit monitored data to the cluster head, which can be divided into numerous frames with their own times. Each node's data is part of a frame, each having its own period. At the conclusion of each cycle, cluster chiefs and clusters are re-elected and require resources. In-round stable stage length is longer than cluster establishment. The connectivity state is saved so that data may be collected from nodes at any time. To decrease duplicated data, the cluster head performs data processing such as data fusion. The Priority Event-based Energy Efficient method does not take remaining node energy into account when selecting cluster heads, resulting in cluster heads with low energy and early cluster death, lowering the network's lifespan.

PRIORITY EVENT BASED ENERGY EFFICIENT ALGORITHM

Algorithms 1 and 2 illustrate the primary process of Dijkstra's algorithm, as well as the overall process shown in Algorithm 1. Area A of the network is split into four sections (Xi, Yi) where nodes Ni are placed. Nodes are divided into subgroups based on their behaviour and usage, such as temperature and moisture, and all other nodes are similarly divided into subsets. Each node is placed at an identical distance apart. During deployment, nodes are initialized with the required settings, such as energy, trust, and distance thresholds. All attributes are assigned threshold values in order to identify the appropriate node for route design and data transfer. Nodes are investigated on a regular basis depending on their criteria, and the best nodes are chosen for routing. To decrease energy usage, this suggested solution makes advantage of sophisticated characteristics of TDMA and CDMA systems. Nodes are allocated modes and priorities, which reduces energy use while sleeping and listening. All nodes are re-clustered in their position and behavior after successful data transfer. Priority is assigned to nodes in the network, such as high, normal, and low, and their parameters are compared to threshold levels before and after route formation. If a node's parameter value is less than the threshold value, its values are reallocated. The network's nodes are adept at monitoring, recording, and sending data(Reddy et al., 2023; Samikannu et al., 2023; Selvakumar et al., 2023).

To identify the shortest path between two nodes, the Dijkstra method is utilized. It begins with the source node and calculates the distance between the nodes, but only considers the node with the shortest distance for further shortest tree building. The Queue model is used to hold the distance data of each neighbor node. Finally, depending on the least distance value, the method finds the shortest tree from the source node to the destination node and removes extraneous nodes.

SMART AGRICULTURAL MANAGEMENT INFORMATION SYSTEMS

Amis is an IoT-Sensor-based agricultural information management system with a visual description and testing data to help you comprehend its effectiveness. It gives aberrant data information to help farmers enhance their yields.

Figure 1. Smart agricultural management information systems

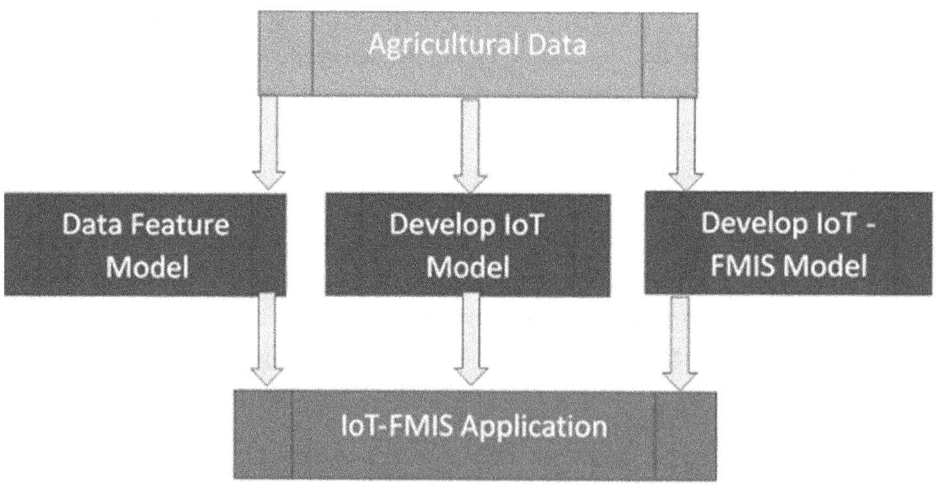

Agriculture Management Information System (AMIS)

The AIMS model for the smart agriculture is illustrated in Figure 1. This research study includes an intelligent algorithm for an intelligent information management system (AMIS). It operates on the product-line engineering process method, which includes two activities: Domain Engineering and FMIS Development. Domain Engineering focuses on planning and building AMIS design and code, whereas

FMIS Development focuses on a specific IoT-based FMIS. The entire AMIS architecture is comprised of three steps: establishing the IoT AMIS family feature model, developing the reference architecture for IoT-based AMIS, and generating reusable components required to construct the AMIS-based reference architecture. The initial phase in AMIS development is to create an IoT-FMIS application feature model, then the particular AMIS application architecture is created using the IoT-based AMIS reference architecture, and finally the AMIS is implemented. The initial stage in Domain engineering is the creation of a family feature model for an IoT-based AMIS application. A feature diagram, often known as a tree, represents similarities and contrasts within a domain by using root nodes to represent characteristics and sub-features to indicate a hierarchical tree(Boopathi, Siva Kumar, et al., 2023; Harikaran et al., 2023; Kumara et al., 2023; Senthil et al., 2023; Vanitha et al., 2023).

Data Engineering creates a family feature model, reference architecture, and IoT FMIS components. The main aim is to create an IoT FMIS application that includes a feature model, architecture, and application architecture. The communication protocol member is represented by the collection of features in the feature configuration, which allows users to pick features to construct configurations that are controlled by feature restrictions.

Feature Model For IOT

The session layer, which initiates connection and communication protocols, is required in the top-level feature diagram of an IoT architecture. The protocol type is required for the top-level feature diagram, which can be MQTT, XAMPP, AMCP, DDS, or CoAP. Protocols used to link devices, servers, and other systems include MQTT, XMPP, AMQP, DDS, and CoAP. MQTT gathers data and talks with servers, XMPP connects devices to servers, AMQP links servers to other systems, DDS is the quickest data bus, and CoAP is used for restricted nodes and networks(Boopathi, Siva Kumar, et al., 2023; Palaniappan et al., 2023; S. et al., 2022; Sampath et al., 2022).

Feature Model For FMIS

Domain analysis is used to find common and variation elements in smart farming, which is an essential domain for IoT-based smart farming. FMIS software, with a high-level feature diagram and function, is a key component of smart farming. The process of gathering data from sensors and other agricultural devices is known as data acquisition. The administration of AMIS processing, external system connectivity, visualization, and data collecting is referred to as system management. Scalability, security, dependability, and extensibility are among its benefits. The communication of the external system is the ultimate layer of communication with the external system.

Feature Model for IOT Based FMIS

The various features of the IoT and WSN Model is illustrated in Figure 2. The main decomposition is AMIS, and IoT is integrated with multiple AMIS feature trees. Data collection comprises both traditional and IoT data, which is supported by five different session layer protocols. The ANN architecture is used in the IoT-based farm management information system (FMIS) to pick protocols for data gathering, data processing, image and video processing, data logging, decision-support, and data mining. Depending on the application type, different characteristics are employed for different processing.

Figure 2. Various feature of IoT

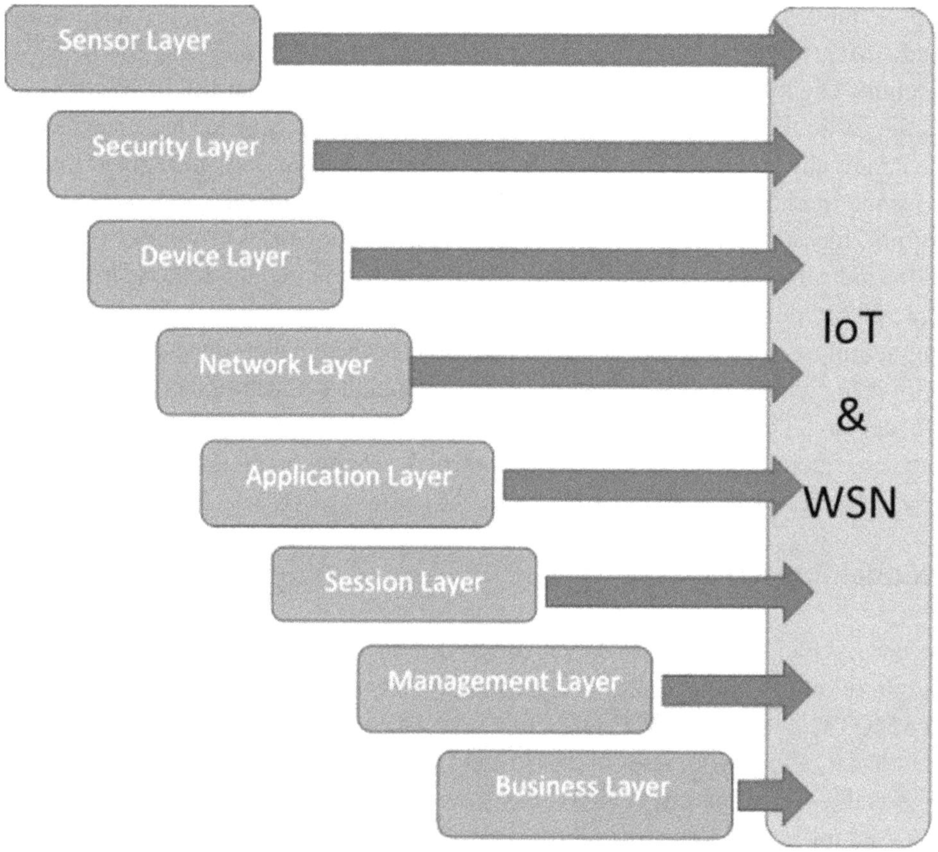

INTELLIGENT IOT BASED AGRICULTURE MANAGEMENT

The study presented an ANN (Artificial Neural Network) to lessen negative affects on the agricultural environment by randomly and evenly distributing a N number of IoT sensor devices in the environment. With data collecting, pre-processing, analyzing, and decision making, IoT devices provide security, safety, and greater output. For intelligent data processing and categorization, the data analytics model employs ANN. IOT and WSN Model for Agricultural Management System is depicted in Figure 3.

IoT sensors transfer monitored data to nearby data storage using TCP and UDP-based data transmission protocols. However, certain sensors continually create environmental data and send it to storage, which is referred to as time-series data. The ANN algorithm is used in this study to increase the efficiency of processing IoT data, particularly time series data. ANN is a viable prediction model for agricultural forecasting systems that can cope with both short and long time series data(Babu et al., 2023; Boopathi, Arigela, et al., 2023; Jeevanantham et al., 2023).

Figure 3. IOT and WSN Model for agricultural management system

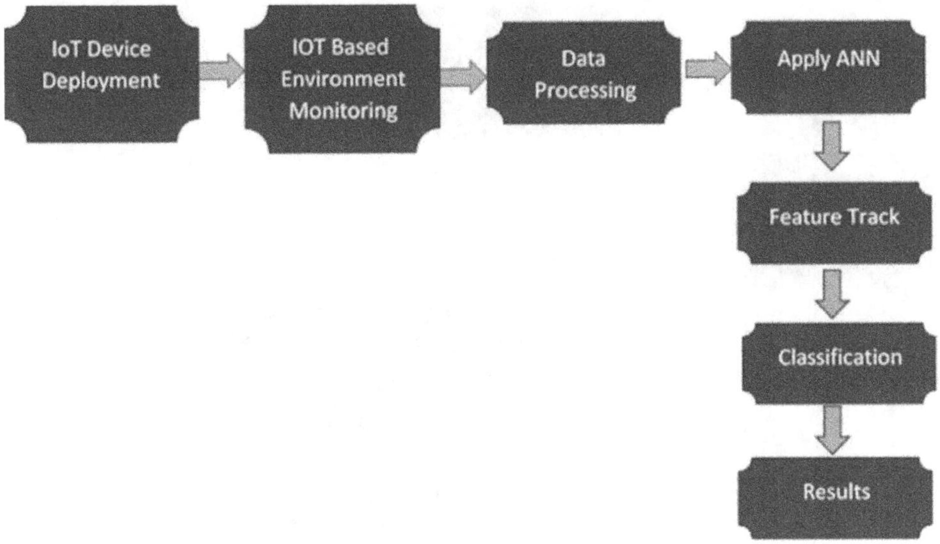

Artificial Neural Network (ANN)

An artificial neural network mimics the operation of the human brain by learning and gaining information using mathematical expressions. It is believed that the output value will be between -1 and 1. To increase output estimation and minimize error values, the data processing approach employs two steps. It trains the ANN model and adjusts hyperparameters like bias and weights. Back Propagation (BP) and Feed Forward (FF) are two built-in algorithms of the ANN. It is used to anticipate the quantity of soil moisture for stable agricultural uses. Because the link between soil moisture and its characteristics is nonlinear, it is obtained using a hybrid modified flower pollination method directed by the ANN algorithm(Boopathi, Lewise, et al., 2021; Boopathi, Thillaivanan, et al., 2021; Haribalaji et al., 2021; Sampath et al., 2021).

Data Set

Various Modules of AIMS is illustrated in Figure 4. The suggested ANN method was tested in a Python programming environment using an FAO dataset. The suggested technique was developed, trained, and validated using a neural network of 10 neurons and 6 delay components. External Input-NARX was discovered to be the most exact and effective Nonlinear Autoregressive Network. The NARX framework trains a neural network using historical input datasets, with 80 percent of the data utilized for training and 20 percent for testing. The proportionate is described using K-cross-validation, and quick supervised algorithms are implemented using the Levenberg-Marquardt paradigm. Backpropagation is used to predict time series. The yield, yield, area, production, and production input numbers are all the same. Python-driven algorithms are utilized to forecast closed-loop networks, and the dataset is subjected to time series data analysis.

Figure 4. Various modules of AIMS

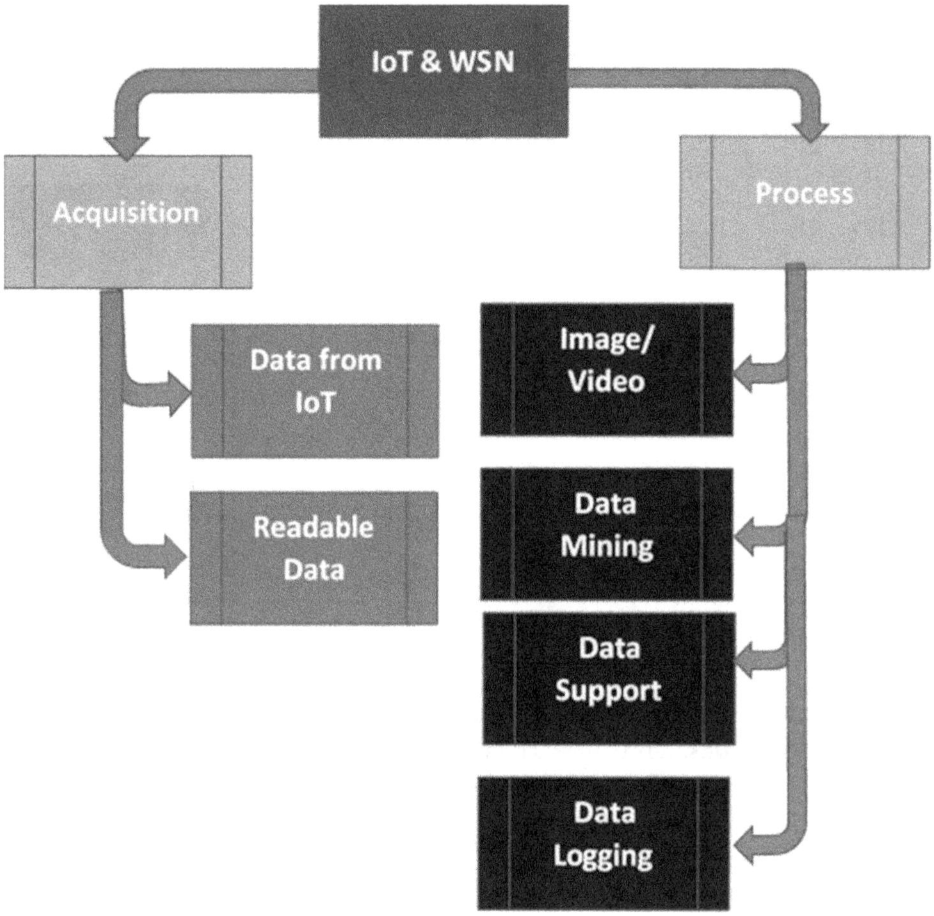

Experimental Results

The dataset was taken from FAO and experimented with the ANN algorithm in Python. It estimated the amount of product obtained yearly and verified it, which changes depending on year and season and environmental conditions. The data was aggregation of official, semi-official, estimated, or calculated data. The average amount of food production is predicted from data based on a constant time interval. ANN is a better method for data analytics in agriculture, as it accurately predicts food crop production based on harvesting area, time, and crop time. It is a better method for data analytics in agriculture.

Results and Discussions

Performance metrics are used to assess proposed algorithms, implementation parameters, and results. For a non-homogeneous wireless sensor network model, the Priority Event-based Energy Efficient Method was compared to Dijkstra's current method. It was suggested that 40% of nodes be deployed as type-2 nodes, 60% as type-3 nodes, and the remainder as type-1 nodes. To reduce energy usage, a hybrid routing strategy was adopted.

Energy Consumption: The LEACH protocol, like the Priority Event-based Energy Efficient algorithm, is divided into two parts: transmit and receive. When nodes transport data, they demand energy, and their amplifiers use energy when they process data.

Test Case Vs. Distance: For all sample test situations, the suggested Priority Event-based Energy-Efficient Method resulted in shorter routing lengths than the classic Dijkstra's algorithm.

Test Case Vs. Energy Consumption: In terms of improved energy usage, the suggested Priority Event-based Energy-Efficient Algorithm surpassed the classic Dijkstra's algorithm for the majority of sample test cases.

Throughput: In all sample test situations, the suggested Priority Event-based Energy-Efficient Algorithm surpassed the classic Dijkstra's algorithm in terms of throughput.

CONCLUSION

The suggested smart farming system divides the land into six equal-sized areas and deploys three types of nodes: type-1, type-2, and type-3. The creation of an adaptive priority and event-based cluster selection method for successful sensor node communication are addressed, when compared to Dijkstra's algorithm, the suggested Priority Event-based Energy Efficient method reduces energy usage and increases the lifetime of a smart farming system.

REFERENCES

Al-Tarawneh, L., Mehyar, A., Alasasaf, S. E., & Al-Mariat, M. (2022). Environmental Tracking System using IoT Based WSN: Smart Agriculture. *2022 4th IEEE Middle East and North Africa COMMunications Conference, MENACOMM 2022*, (pp. 147–152). IEEE. 10.1109/MENACOMM57252.2022.9998269

Alghazzawi, D., Bamasaq, O., Bhatia, S., Kumar, A., Dadheech, P., & Albeshri, A. (2021). Congestion Control in Cognitive IoT-Based WSN Network for Smart Agriculture. *IEEE Access : Practical Innovations, Open Solutions*, *9*, 151401–151420. doi:10.1109/ACCESS.2021.3124791

Babu, B. S., Kamalakannan, J., Meenatchi, N., M, S. K. S., S, K., & Boopathi, S. (2023). Economic impacts and reliability evaluation of battery by adopting Electric Vehicle. *IEEE Explore*, (pp. 1–6). IEEE. doi:10.1109/ICPECTS56089.2022.10046786

Boopathi, S. (2022a). An Extensive Review on Sustainable Developments of Dry and Near-Dry Electrical Discharge Machining Processes. *Journal of Manufacturing Science and Engineering*, *144*(5), 50801. doi:10.1115/1.4052527

Boopathi, S. (2022b). Cryogenically treated and untreated stainless steel grade 317 in sustainable wire electrical discharge machining process: A comparative study. *Environmental Science and Pollution Research International*, 1–10. doi:10.100711356-022-22843-x PMID:36057706

Boopathi, S. (2022c). Experimental investigation and multi-objective optimization of cryogenic Friction-stir-welding of AA2014 and AZ31B alloys using MOORA technique. *Materials Today. Communications*, *33*, 104937. doi:10.1016/j.mtcomm.2022.104937

Boopathi, S., Arigela, S. H., Raman, R., Indhumathi, C., Kavitha, V., & Bhatt, B. C. (2023). Prominent Rule Control-based Internet of Things: Poultry Farm Management System. *IEEE Explore*, (pp. 1–6). IEEE. doi:10.1109/ICPECTS56089.2022.10047039

Boopathi, S., Lewise, K. A. S., Subbiah, R., & Sivaraman, G. (2021). Near-dry wire-cut electrical discharge machining process using water-air-mist dielectric fluid: An experimental study. *Materials Today: Proceedings*, *49*(5), 1885–1890. doi:10.1016/j.matpr.2021.08.077

Boopathi, S., Siva Kumar, P. K., & Meena, R. S. J., S. I., P., S. K., & Sudhakar, M. (2023). Sustainable Developments of Modern Soil-Less Agro-Cultivation Systems. In Human Agro-Energy Optimization for Business and Industry (pp. 69–87). IGI Global. doi:10.4018/978-1-6684-4118-3.ch004

Boopathi, S., Thillaivanan, A., Pandian, M., Subbiah, R., & Shanmugam, P. (2021). Friction stir processing of boron carbide reinforced aluminium surface (Al-B4C) composite: Mechanical characteristics analysis. *Materials Today: Proceedings*, *50*(5), 2430–2435. doi:10.1016/j.matpr.2021.10.261

Burugari, V. K., Selvaraj, P., & Palaniappan, K. (2021). Smart agriculture using WSN and IoT. In *Artificial Intelligence and IoT-Based Technologies for Sustainable Farming and Smart Agriculture* (pp. 192–212). IGI Global. doi:10.4018/978-1-7998-1722-2.ch012

Domakonda, V. K., Farooq, S., Chinthamreddy, S., Puviarasi, R., Sudhakar, M., & Boopathi, S. (2023). Sustainable Developments of Hybrid Floating Solar Power Plants. In *Human Agro-Energy Optimization for Business and Industry* (pp. 148–167). IGI Global. doi:10.4018/978-1-6684-4118-3.ch008

Haribalaji, V., Boopathi, S., & Asif, M. M. (2021). Optimization of friction stir welding process to join dissimilar AA2014 and AA7075 aluminum alloys. *Materials Today: Proceedings*, *50*, 2227–2234. doi:10.1016/j.matpr.2021.09.499

Harikaran, M., Boopathi, S., Gokulakannan, S., & Poonguzhali, M. (2023). Study on the Source of E-Waste Management and Disposal Methods. In *Sustainable Approaches and Strategies for E-Waste Management and Utilization* (pp. 39–60). IGI Global. doi:10.4018/978-1-6684-7573-7.ch003

Haseeb, K., Din, I. U., Almogren, A., & Islam, N. (2020). An energy efficient and secure IoT-based WSN framework: An application to smart agriculture. *Sensors (Basel)*, *20*(7), 2081. doi:10.339020072081 PMID:32272801

Indrapandi, A., & Rizwana, S. (2021). Smart Agriculture Monitoring System based Data Aggregation Process in IoT and WSN using Multiple Sensor Nodes. *Turkish Online Journal of Qualitative Inquiry*, *12*(7).

Islam, M. S., & Dey, G. K. (2019). Precision agriculture: Renewable energy based smart crop field monitoring and management system using WSN via IoT. *2019 International Conference on Sustainable Technologies for Industry 4.0, STI 2019*, (pp. 1–6). IEEE. 10.1109/STI47673.2019.9068017

Janardhana, K., Anushkannan, N. K., Dinakaran, K. P., Puse, R. K., & Boopathi, S. (2023). *Experimental Investigation on Microhardness, Surface Roughness, and White Layer Thickness of Dry EDM*. Engineering Research Express. doi:10.1088/2631-8695/acce8f

Jeevanantham, Y. A., A, S., V, V., J, S. I., Boopathi, S., & Kumar, D. P. (2023). Implementation of Internet-of Things (IoT) in Soil Irrigation System. *IEEE Explore*, 1–5. doi:10.1109/ICPECTS56089.2022.10047185

Kadam, A. A., Peddi, P., & Gujar, S. N. (n.d.). *Design Of Deep Cnn Approach For Optimal Water Control Using Iot Based Wsn For Smart Agriculture Applications*.

Kannan, E., Trabelsi, Y., Boopathi, S., & Alagesan, S. (2022). Influences of cryogenically treated work material on near-dry wire-cut electrical discharge machining process. *Surface Topography : Metrology and Properties*, *10*(1), 15027. doi:10.1088/2051-672X/ac53e1

Kumara, V., Mohanaprakash, T. A., Fairooz, S., Jamal, K., Babu, T., & B., S. (2023). Experimental Study on a Reliable Smart Hydroponics System. In *Human Agro-Energy Optimization for Business and Industry* (pp. 27–45). IGI Global. doi:10.4018/978-1-6684-4118-3.ch002

Nandal, V., & Dahiya, S. (2021). Energy Efficient Data Aggregation Protocol Based Iot-WSN Framework For Smart Agriculture. *Webology, 18*(6), 4032–4045.

Palaniappan, M., Tirlangi, S., Mohamed, M. J. S., Moorthy, R. M. S., Valeti, S. V., & Boopathi, S. (2023). Fused deposition modelling of polylactic acid (PLA)-based polymer composites: A case study. In Development, Properties, and Industrial Applications of 3D Printed Polymer Composites (pp. 66–85). IGI Global. doi:10.4018/978-1-6684-6009-2.ch005

Rao, A. K., Nagwanshi, K. K., & Pathak, S. (2022). Empirical Study on Energy-Efficient IoT-Based WSN Routing Protocols for Smart Agriculture System. In *Lecture Notes in Networks and Systems* (Vol. 392, pp. 259–271). Springer. doi:10.1007/978-981-19-0619-0_23

Reddy, M. A., Reddy, B. M., Mukund, C. S., Venneti, K., Preethi, D. M. D., & Boopathi, S. (2023). Social Health Protection During the COVID-Pandemic Using IoT. In *The COVID-19 Pandemic and the Digitalization of Diplomacy* (pp. 204–235). IGI Global. doi:10.4018/978-1-7998-8394-4.ch009

S., P. K., Sampath, B., R., S. K., Babu, B. H., & N., A. (2022). Hydroponics, Aeroponics, and Aquaponics Technologies in Modern Agricultural Cultivation. In *Trends, Paradigms, and Advances in Mechatronics Engineering* (pp. 223–241). IGI Global. doi:10.4018/978-1-6684-5887-7.ch012

Samikannu, R., Koshariya, A. K., Poornima, E., Ramesh, S., Kumar, A., & Boopathi, S. (2023). Sustainable Development in Modern Aquaponics Cultivation Systems Using IoT Technologies. In *Human Agro-Energy Optimization for Business and Industry* (pp. 105–127). IGI Global. doi:10.4018/978-1-6684-4118-3.ch006

Sampath, B., Naveenkumar, N., Sampathkumar, P., Silambarasan, P., Venkadesh, A., & Sakthivel, M. (2021). Experimental comparative study of banana fiber composite with glass fiber composite material using Taguchi method. *Materials Today: Proceedings*, *49*(5), 1475–1480. doi:10.1016/j.matpr.2021.07.232

Sampath, B. C. S., & Myilsamy, S. (2022). Application of TOPSIS Optimization Technique in the Micro-Machining Process. In Trends, Paradigms, and Advances in Mechatronics Engineering (pp. 162–187). IGI Global. doi:10.4018/978-1-6684-5887-7.ch009

Selvakumar, S., Adithe, S., Isaac, J. S., Pradhan, R., Venkatesh, V., & Sampath, B. (2023). A Study of the Printed Circuit Board (PCB) E-Waste Recycling Process. In Sustainable Approaches and Strategies for E-Waste Management and Utilization (pp. 159–184). IGI Global.

Senthil, T. S. R. Ohmsakthi vel, Puviyarasan, M., Babu, S. R., Surakasi, R., & Sampath, B. (2023). Industrial Robot-Integrated Fused Deposition Modelling for the 3D Printing Process. In Development, Properties, and Industrial Applications of 3D Printed Polymer Composites (pp. 188–210). IGI Global. doi:10.4018/978-1-6684-6009-2.ch011

Trojovský, P., Dhasarathan, V., & Boopathi, S. (2023). Experimental investigations on cryogenic friction-stir welding of similar ZE42 magnesium alloys. *Alexandria Engineering Journal*, *66*(1), 1–14. doi:10.1016/j.aej.2022.12.007

Vanitha, S. K. R., & Boopathi, S. (2023). Artificial Intelligence Techniques in Water Purification and Utilization. In *Human Agro-Energy Optimization for Business and Industry* (pp. 202–218). IGI Global., doi:10.4018/978-1-6684-4118-3.ch010

Vennila, T., Karuna, M. S., Srivastava, B. K., Venugopal, J., Surakasi, R., & B., S. (2023). New Strategies in Treatment and Enzymatic Processes. In *Human Agro-Energy Optimization for Business and Industry* (pp. 219–240). IGI Global. doi:10.4018/978-1-6684-4118-3.ch011

Yupapin, P., Trabelsi, Y., Nattappan, A., & Boopathi, S. (2023). Performance Improvement of Wire-Cut Electrical Discharge Machining Process Using Cryogenically Treated Super-Conductive State of Monel-K500 Alloy. *Iranian Journal of Science and Technology. Transaction of Mechanical Engineering*, *47*(1), 267–283. doi:10.100740997-022-00513-0

Chapter 12
AI-Based Plant Disease Detection and Classification Using Pretrained Models

Honey Mehta
Pandit Deendayal Energy University, India

Rajeev Kumar Gupta
Pandit Deendayal Energy University, India

Arti Jain
https://orcid.org/0000-0002-3764-8834
Jaypee Institute of Information Technology, Noida, India

Santosh Kumar Bharti
https://orcid.org/0000-0002-0627-6433
Pandit Deendayal Energy University, India

Nilesh Kunhare
Vellore Institute of Technology, Bhopal, India

ABSTRACT

Plants growth is crucial to an agricultural industry, and to the economy of a nation. Consequently, taking care of plants is essential. Like humans, plants are susceptible to several bacterial, fungal, and viral diseases. To avoid the plants from being destroyed, prompt disease detection and treatment are crucial. The goal of this chapter is to introduce plant disease detection and classification using deep learning and transfer learning based pre-trained models. Using images of leaves, the models can identify several plant ailments. The models, namely- Convolutional Neural Network (CNN), MobileNet, and VGG16 are applied for plant disease identification. To train them, the PlantVillage dataset is used, and to enhance the sample size, the dataset is augmented. For experimentation purposes, images of both healthy and damaged plants are taken. Experiment results reveal that VGG16 has outperformed CNN and MobileNet models for the detection of tomato, potato, and apple plant diseases. The accuracy of the VGG16 model is 0.89, 0.92, and 0.95 for the tomato, potato, and apple plant diseases respectively.

DOI: 10.4018/978-1-6684-8516-3.ch012

INTRODUCTION

Being an integral component of our ecosystem, plants are crucial to maintain life on the planet earth. The plants support overall health, well-being of our planet, give us food and air. But much like people and animals, plants can get sick from different pathogens like bacteria, viruses, fungi, and nematodes (Saleem et al., 2019; Kulkarni et al., 2021). These plant diseases can have an adverse effect on the economy and environment, degrade crop quality, and cause large output losses. Effective disease management, which comprises prompt and focused treatments to stop the spread of infections and lessen their negative effects on crop yield, depends on early and accurate diagnosis of plant diseases. Traditional techniques of plant disease identification rely on visual inspection by experienced professionals (Alston et al., 2014) which can be time-consuming, labour-intensive, and subjective that frequently lead to misdiagnosis and delayed treatment. In order to enable prompt intervention and reduce crop losses, there is a rising need for quick, trustworthy, and non-destructive approaches for early plant disease identification (Barbedo, 2018; Eunice et al., 2022).

Recent technological developments have completely changed the way how plant diseases are found, especially in the areas of imaging, sensing, and data analysis. These technological developments have resulted in the creation of numerous cutting-edge methods and instruments for the identification of plant diseases, ranging from molecular techniques and Artificial Intelligence (AI) algorithms to remote sensing and spectroscopy (Narayana et al., 2018). These techniques have potential to provide real-time, precise, and high-throughput plant disease detection, enabling prompt management choices and targeted application of control measures (Talaviya et al., 2020). The objective of this chapter is to present a thorough analysis of the developments in the plant disease detection, with a special emphasis on the different methods and equipments that are emerging in the recent years. This research discusses about the fundamentals, benefits, drawbacks, and applications of several approaches that are used to identify plant diseases, with a focus on their potential for early and non-destructive disease detection (Muhammed, 2022; Javidan et al., 2023). The review also focuses on the difficulties and potential directions in the field of plant disease detection, such as requirements for integrating various strategies, standardizing methods, and creating user-friendly instruments for field/crop use.

There are several worthwhile utilities of plant disease detection and classification models that are stated herewith.

Early Detection and Intervention: The research findings can help to generate early warning systems for plant illnesses, giving farmers the ability to spot diseases in their early stages. This can assist farmers in making early and precise interventions, such spraying pesticides, establishing crop rotation, or modifying irrigation schedules, to stop the spread of illnesses and reduce crop losses (Eunice et al., 2022).

Increased Crop Yields and Food Security: Farmers can use effective disease control techniques and increase crop yields by correctly detecting plant diseases (Li et al., 2021). This can contribute to global food security by ensuring a sufficient and wholesome food supply, especially in areas where agriculture is a major source of income.

Sustainable Agriculture Practices: By limiting the indiscriminate use of pesticides, research can support sustainable agricultural practices (Khan et al., 2021). Farmers can maximize the use of pesticides by detecting diseases early, making focused interventions, lowering costs, and promoting ecologically friendly farming practices.

Cost-Effective Disease Management: Farmers can use cost-effective interventions to optimize their disease management methods by using the research findings. For farmers, this can lead to lower production costs, greater profitability, and higher economic sustainability (Dubey and Jalal, 2013).

Advancement of Agricultural Sciences: By investigating and creating novel strategies for plant disease detection, the research study can help to improve the field of agricultural sciences. The application of advanced technologies, including deep learning, computer vision, and sensor-based systems, can create new research opportunities and pave the way for future developments in the study of plant pathology (Saleem et al., 2019).

LITERATURE REVIEW

This section discusses on literature review related to the plant disease detection models. One of the most frequently used deep learning model is Convolutional Neural Network (CNN) which is quite effective at extracting simple low-level features from the plant images. However, due to high computational cost of training deep CNN layers, it is challenging (Ouhami et al., 2021). Several researchers have suggested transfer learning-based methods to address these problems i.e. VGG-16, ResNet, DenseNet, and Inception are a few well-known models. Also, rapidly grooming of sensor-based technologies such as mobile phones and internet of things enabled sensors, provide real-time and on-site monitoring of plant health data (Kamilaris and Prenafeta-Boldú, 2018). These sensors can detect changes in plant physiological and biochemical characteristics like electrical conductivity, gas exchange, and chlorophyll fluorescence, which can act as early warning signs of plant illnesses (Li et al., 2021). The principles, uses, and difficulties of these sensor-based technologies, as well as their promise for plant disease detection in various cropping systems are discussed. The research findings contribute to the development of automated and efficient plant disease detection systems, aid farmers in making informed decisions for disease control and crop protection (Prajapati et al., 2017). Multi-class classification for plant disease detection is a critical research area in agriculture and plant sciences (Muhammed, 2022; Javidan et al., 2023). Researchers have analyzed different approaches such as Convolutional Neural Networks (CNNs), Support Vector Machines (SVMs), and Ensemble methods for plant disease classification (Madhavan et al., 2021). The combination of ResNet50 and XGBoost classifier, termed as ConRXG is applied for plant leaf disease detection (Yadav et al., 2022). Table 1 summarizes the literature review related to the plant disease detection in the recent past.

PREREQUISITE

Hardware and Software Requirements

A GPU NVIDIA GeForce GTX workstation has served as the standard machine for evaluating the experiments. Windows 10 is used as the operating system, along with a Core i5 8th generation processor and 8 GB of RAM. The libraries Anaconda3, Kera's, OpenCV, NumPy CuDNN, and Theano are used for the software implementation. Simple libraries called CUDNN and CUMeM are created specifically to implement deep learning faster and with reduced memory usage. NVIDIA created both libraries to function in the Theano backend. Linux, Windows, Mac OS, iOS, Python, Java, and Android interfaces are all supported by the OpenCV platform, which aids in the development of this task. In this study, the training and testing accuracy for each trial are assessed. Also, the loss values for each model during training and testing phases are calculated. In order to speed up the learning rate of the CNN with transfer learning models, the models are trained using the PlantVillage dataset. The models- CNN, VGG-16, and MobileNet are the pretrained models that are selected for the investigation. These models have previously been trained over the ImageNet dataset, which contains 1.2 M images and 1,000 image categories.

Table 1. Literature review of plant disease detection

References	Methodology	Advantages	Disadvantages
Javidan et al., 2023	MobileNet and transfer learning	Lightweight model for resource constrained devices	May sacrifice some accuracy compared to larger models
Eunice et al., 2022	SVM with image processing	Robustness to noise and variability	Limited to handcrafted feature engineering
Ouhami et al., 2021	CNN with transfer learning	High accuracy and generalization	Requires large labeled datasets for training
Lu et al., 2020	SVM and Random Forest	Interpretable features for disease detection	Limited ability to capture complex patterns
Qiang et al., 2019	Inception-V3 neural network model transfer learning and fine-tuning	Leveraging pretrained models for improved performance	Potential overfitting due to limited plant disease specific data
Kamilaris and Prenafeta-Boldú, 2018	CNN with data augmentation	High accuracy in disease classification	Large amount of labeled data is required
Prajapati et al., 2017	K-means clustering, SVM	Image processing operations prepare images, more suitable for feature extraction	Longer training time and lesser accuracy on test images

Description of Dataset

The PlantVillage dataset is chosen which contains a collection of healthy and diseased plant leaves. It contains images of various plant species affected by different diseases. The dataset does not explicitly include annotated bounding box coordinates for disease detection, but can be used for plant disease classification task. The dataset includes varied plant leaf images which are divided into training, validation, and test sets. The images are in high-resolution format and have varying file formats. This dataset is being utilized for plant disease detection and classification using machine learning algorithms (Lu et al., 2020). Figure 1 shows sample images of 30 different types of leaf diseases.

Preprocessing and Data Augmentation

Preprocessing and data augmentation play crucial roles in enhancing the accuracy and reliability of plant disease detection system. This chapter focuses on investigating the impact of pre-processing techniques and data augmentation methods on plant disease detection using image-based data (Li et al., 2012; Dubey and Jalal 2013). The preprocessing techniques such as image normalization, resizing, and filtering are applied, added feature extraction and the classification performance is evaluated. Additionally, different data augmentation methods such as rotation, flipping, and zooming are applied to assess their effectiveness in improving the ability of model to generalize towards unseen data (Salleh et al., 2011). The findings of this research would provide insights into the importance of preprocessing and data augmentation in plant disease detection and offer guidelines for optimizing these techniques for improved disease detection accuracy in agricultural applications.

Figure 1. Sample images from PlantVillage dataset for 30 types of leaf diseases

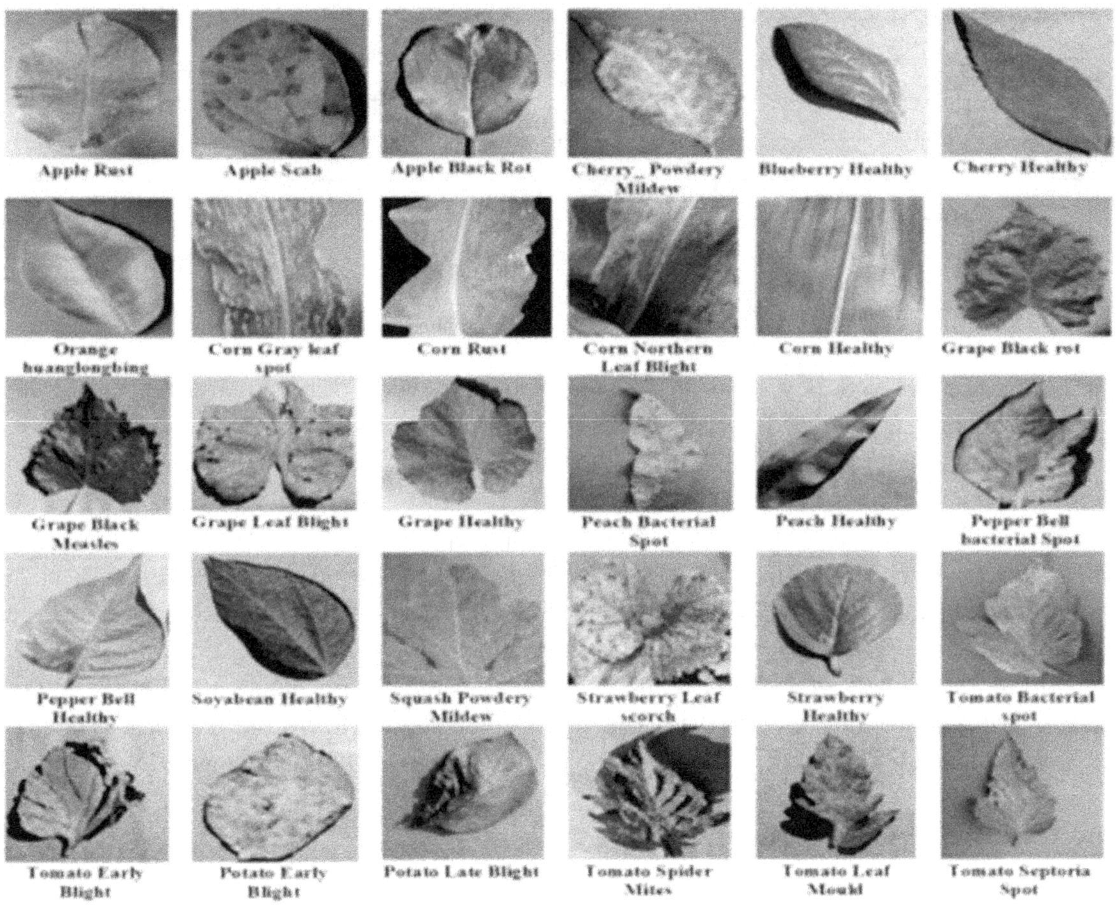

Fine-Tuning of Hyperparameters in Pretrained Models

Fine-tuning of hyperparameters in pretrained models is a critical research area in the plant disease detection using deep learning techniques. This chapter focuses on exploring the impact of hyperparameters tuning in pretrained models for plant disease detection (Subramanian et al., 2022). Table 2 investigates various hyperparameters such as learning rate, epochs, optimizer, etc., and evaluate their effects on the performance of pre-trained models. The interdependency between different hyperparameters and their effects on the accuracy and convergence of the models are analyzed. The findings of this research would provide insights into the optimal hyperparameters settings for fine-tuning pretrained models in the context of plant disease detection, helping to improve the accuracy and efficiency of plant disease detection system, and enabling more effective plant disease management in agriculture.

Table 2. Hyperparameters in pretrained models

Hyperparameter	Value
Dropout	0.5
Epochs	30
Activation	ReLu
Regularization	Batch Normalization
Optimizer	Stochastic Gradient Descent (SGD)
Learning Rate	0.001
Output Classes	38

PROPOSED METHODOLOGY

Figure 2 discusses the work flow of the proposed plant disease detection and classification task. Transfer learning models are considered to uplift the evaluation measures such as accuracy. The process consists of dataset collection, pre-processing, model training, post processing, visualization steps. First and foremost step of dataset collection is important as it explains the domain or use case that brings the impact over plant disease identification. After collection of the data, preprocessing step cleans the data with splitting of dataset into training, validation and testing phases. Then in the next step deep learning model is implemented with the usability of transfer learning such as MobileNet and VGG-16. The model accepts pre-trained weights of the parent model and does transfer learning to the edge layers for the given data. The performance analysis as post-processing step calculates the inference output and gives it in more human readable format of the given data. The visualization step performs the graphical operation of output of validation/test data and shows the comparative analysis of the overall methodology.

Convolutional Neural Networks

CNN (Siddiqui et al., 2020; Yadav et al., 2021) is one of the best deep learning models for image detection and classification tasks. It is a widely used model for plant disease detection by automatically extracting important features from the plant images, such as patterns and textures, that are indicative of specific diseases (Yadav et al., 2022). The model outperforms traditional methods of disease detection, such as visual inspection by experts or manual feature extraction. Furthermore, CNN model generalizes well to new images and detects previously unseen diseases, empowering as a powerful tool for reducing the crop losses and improving food security.

Figure 3 depicts the CNN architecture by applying a sequence of layers- convolutional, pooling, and dense layers to raw input data, such as images, in a hierarchical manner. The convolutional layer applies filters to an input data, extracts local features and preserves spatial information. The pooling layer reduces the spatial resolution of the output, while retaining the most important features (Li et al., 2021).

The dense layer performs the final classification task, mapping the learned features to the target labels. The model is trained using backpropagation, adjusting parameters of each layer to minimize differences between predicted and actual labels. This process continues until the model reaches a satisfactory level of accuracy .

Figure 2. Flow diagram of proposed work

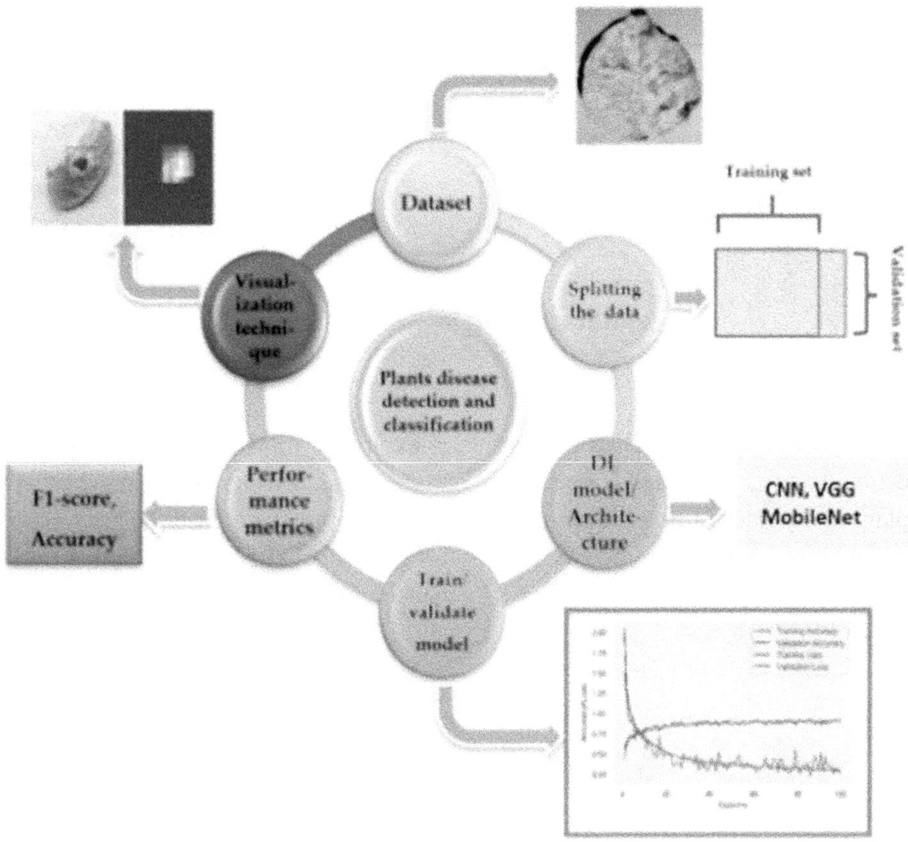

Despite the potential benefits of CNN, there are several drawbacks associated with this model, such as lengthy training process and pre-requisite for a larger dataset. Hence, Deep CNN models are necessary to extract the low-level and complicated characteristics from the plant images. In other words, difficulties can be solved using the Transfer Learning strategies. The pre-trained networks are utilized in transfer learning, which allow model parameters to be discovered on one dataset, utilize this training to be applied on different issues.

Transfer Learning Approach

The usage of transfer learning in plant disease detection research has gained increasing attention due to its potential for leveraging pretrained deep learning models and to improve the accuracy and efficiency

Figure 3. CNN architecture
(De Silva and Brown, 2022)

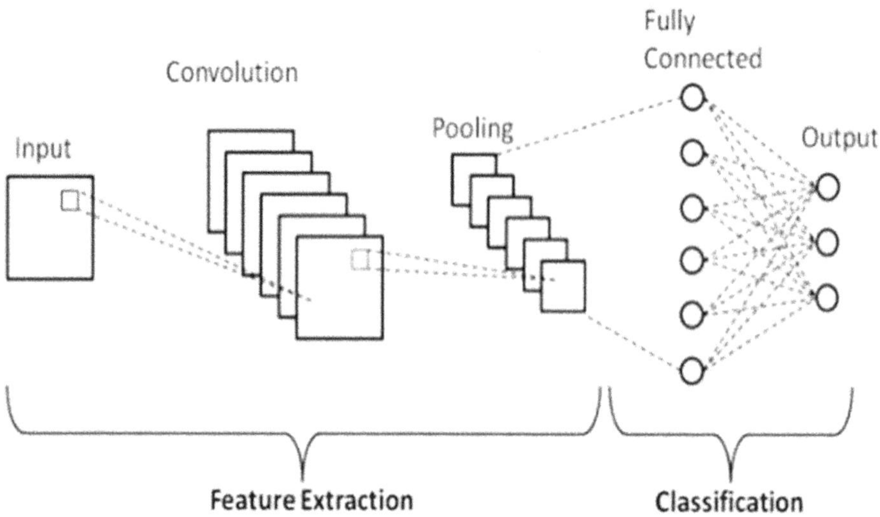

of the disease classification. This chapter focuses on exploring transfer learning approaches for plant disease detection, where knowledge learned from one domain, such as natural images, is transferred to the task of plant disease detection using limited or sparse plant image datasets. The chapter investigates different transfer learning concepts, including fine-tuning, feature extraction, and domain adaptation, and evaluate their effectiveness in improving the performance of plant disease detection models (Kamilaris and Prenafeta-Boldú, 2018).

VGG-16

VGG-16, on the other hand, is a deeper CNN model with a larger number of layers (Table 3) that is capable of capturing complex features from plant images. This model is being applied by researchers (Sujatha et al., 2021). They have explored the efficacy of VGG-16 in extracting relevant features from the plant images and its performance in multi-class classification task. The research findings contribute to the understanding of VGG-16 potential as a powerful tool for plant disease detection and provide insights for further improving the accuracy and efficiency of plant disease detection systems using deep learning approaches.

MobileNet

MobileNet (Ashwinkumar et al., 2022) is a lightweight deep learning architecture that is used for plant disease detection due to its efficiency and accuracy. The model utilizes depth wise separable convolutions to reduce the number of parameters and increases the speed of computation, while maintaining high accuracy. This architecture is particularly useful for resource-constrained environments, such as smartphones or edge devices. It has shown to achieve high accuracy in the plant disease detection tasks, while reducing the computational burden, making it a promising model for real-world applications.

Figure 4 shows the MobileNet architecture which consists of a series of depth wise separable convolutions, separates the spatial and depth dimensions of the input tensor. The depth wise convolution applies a single filter to each input channel, which extracts local features, while the point-wise convolution applies a 1x1 convolutional layer to combine the extracted features across channels (Yun et al., 2015). This process reduces the number of parameters while maintaining accuracy.

Table 3. Layers in VGG-16 model

Network Model	VGG-16
Total Layers	16
Max-pool Layers	5
Dense Layers	3
Drop-out Layers	2
Flatten Layers	1
Filter Size	3 x 3
Strides	2 x 2
Trainable Parameters	41.2 M

Figure 4. MobileNet Architecture

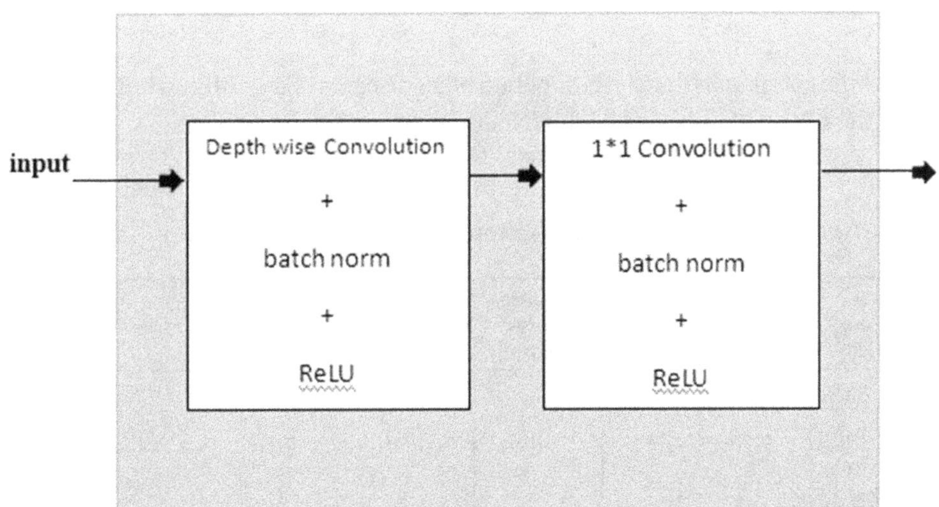

RESULTS AND DISCUSSION

This work undergoes three deep learning and transfer learning models- CNN, VGG16 and MobileNet for plant disease detection. Table 4-7 indicate that all three models are capable of accurately detecting plant diseases from leaf images. The tables include columns for the model type, plant disease, number of images, time taken, accuracy, precision, recall and F-score. They give information about how well one model is able to correctly identify the diseased plants over the other models.

Table 4 shows the comparative analysis of tomato plant diseases- Early Blight, Late Blight and Healthy Blight by using the CNN, VGG16 and MobileNet models.

Table 4. Comparison of CNN, VGG16, MobileNet models for tomato diseases

Model	Plant Disease (Tomato)	Numberof Images	Time Taken (hrs)	Training Accuracy (average)	Validation Accuracy (average)	Testing Accuracy (average)	Precision	Recall	F1-Score
CNN	Early Blight	3,500	4.5	0.92	0.90	0.89	0.91	0.88	0.89
	Late Blight						0.87	0.91	0.89
	Healthy Blight						0.93	0.95	0.94
VGG16	Early Blight	3,500	3	0.94	0.92	0.91	0.92	0.89	0.90
	Late Blight						0.86	0.93	0.89
	Healthy Blight						0.95	0.93	0.94
MobileNet	Early Blight	3,500	3.5	0.96	0.93	0.92	0.93	0.87	0.90
	Late Blight						0.85	0.94	0.89
	Healthy Blight						0.94	0.96	0.95

Table 5 shows the comparative analysis of potato plant diseases- Early Blight, Late Blight and Healthy Blight by using the CNN, VGG16 and MobileNet models.

Table 5. Comparison of CNN, VGG16, MobileNet models for potato diseases

Model	Plant Disease (Potato)	Numberof Images	Time Taken (hrs)	Training Accuracy (average)	Validation Accuracy (average)	Testing Accuracy (average)	Precision	Recall	F1-Score
CNN	Early Blight	7,000	3	0.91	0.90	0.90	0.92	0.89	0.90
	Late Blight						0.86	0.93	0.89
	Healthy Blight						0.94	0.91	0.93
VGG16	Early Blight	7,000	5	0.93	0.91	0.91	0.90	0.88	0.89
	Late Blight						0.86	0.91	0.89
	Healthy Blight						0.93	0.94	0.93
MobileNet	Early Blight	7,000	4	0.95	0.92	0.93	0.89	0.87	0.88
	Late Blight						0.86	0.91	0.88
	Healthy Blight						0.94	0.93	0.93

Table 6 shows the comparative analysis of apple plant diseases- Early Blight, Late Blight and Healthy Blight by using the CNN, VGG16 and MobileNet models.

Table 6. Comparison of CNN, VGG16, MobileNet models for apple diseases

Model	Plant Disease (Apple)	Numberof Images	Time Taken (hrs)	Training Accuracy (average)	Validation Accuracy (average)	Testing Accuracy (average)	Precision	Recall	F1-Score
CNN	Early Blight	5,500	4	0.92	0.92	0.91	0.92	0.88	0.90
	Late Blight						0.85	0.92	0.86
	Healthy Blight						0.94	0.96	0.96
VGG16	Early Blight	5,500	4.5	0.94	0.94	0.92	0.91	0.89	0.90
	Late Blight						0.88	0.92	0.90
	Healthy Blight						0.94	0.95	0.94
MobileNet	Early Blight	5,500	4	0.96	0.93	0.94	0.92	0.90	0.91
	Late Blight						0.91	0.93	0.92
	Healthy Blight						0.95	0.96	0.96

Table 7 shows the comparative analysis of tomato, potato and apple plant diseases by using the CNN, VGG16 and MobileNet models.

Table 7. Comparative Analysis of CNN, VGG16, and mobilenet models for plant diseases

Model	Plant Species	Plant Disease	Number of Images	Time Taken	Accuracy	Precision	Recall	F1-Score
CNN	Tomato	Early Blight	2,000	3 hours	0.92	0.89	0.95	0.92
VGG16	Tomato	Early Blight	2,000	4 hours	0.95	0.93	0.96	0.94
MobileNet	Tomato	Early Blight	2,000	2 hours	0.94	0.92	0.95	0.93
CNN	Tomato	Late Blight	1,500	2.5 hours	0.88	0.85	0.89	0.86
VGG16	Tomato	Late Blight	1,500	3 hours	0.91	0.89	0.92	0.90
MobileNet	Tomato	Late Blight	1,500	1.5 hours	0.89	0.87	0.90	0.88
CNN	Potato	Late Blight	1,800	3 hours	0.91	0.88	0.93	0.91
VGG16	Potato	Late Blight	1,800	4 hours	0.94	0.92	0.95	0.93
MobileNet	Potato	Late Blight	1,800	2 hours	0.92	0.89	0.93	0.91
CNN	Apple	Apple Scab	2,500	4 hours	0.95	0.93	0.96	0.94
VGG16	Apple	Apple Scab	2,500	3.5 hours	0.96	0.95	0.97	0.96
MobileNet	Apple	Apple Scab	2,500	2.5 hours	0.94	0.94	0.96	0.95

The following observations are noted from the Table 7.

VGG!6: VGG16 outperformed CNN and MobileNet models for the detection of tomato, potato and apple plant diseases. The accuracy of the VGG16 model is 0.89, 0.92 and 0.95 for the tomato, potato and apple plants respectively.

CNN: CNN is a baseline model that provides a simple and fast solution for plant disease detection but with slightly lower accuracy in comparison with VGG16 and MobileNet models. On the test dataset, it achieves an accuracy of 0.88, precision of 0.85, recall of 0.89 and F1-Score of 0.86 respectively for tomato plant late light disease, and so on.

MobileNet: MobileNet is a lightweight CNN model that is designed for mobile and embedded devices with limited computational resources. It is observed that the test dataset acheives an accuracy of 0.94, precision of 0.92, recall of 0.95, and F1-Score of 0.93 respectively for the tomato plant, early blight disease, and so on. MobileNet has relatively faster inference time which makes it suitable for real-time plant disease detection applications on mobile devices.

CONCLUSION

This chapter presents a comprehensive exploration of the application of AI techniques in the field of plant diseases. By leveraging the power of machine learning and computer vision algorithms, this research develops an automated and efficient system for identification and classification of plant diseases. The results presented in this chapter are based on a specific dataset, pre-processing techniques, and hyperparameters chosen. Therefore, it's important to interpret the results in the context of the specific use cases and to carefully evaluate the performance of the deep learning and transfer learning models.

The chapter focuses on investigating the impact of CNN, VGG16 and MobileNet models on the accuracy and performance of plant disease detection system. It analyzes the pros and cons of different models and their suitability for plant disease detection task. MobileNet offers a good trade-off between accuracy and computational efficiency, making it suitable for mobile and embedded devices. VGG16 provides higher accuracy but with slightly longer inference time due to its deeper architecture. CNN offers a simple and fast solution for plant disease detection, albeit with slightly lower accuracy.

The experiments could be extended to evaluate performance of the three models on larger datasets, different plant species, and under varying environmental conditions to assess their generalization capabilities. The work would inspire further studies and innovations in this field to mitigate crop losses, improve food security, and promote sustainable agricultural applications.

REFERENCES

Alston, J. M., & Pardey, P. G. (2014). Agriculture in the global economy. *The Journal of Economic Perspectives*, *28*(1), 121–146. doi:10.1257/jep.28.1.121

Ashwinkumar, S., Rajagopal, S., Manimaran, V., & Jegajothi, B. (2022). Automated plant leaf disease detection and classification using optimal MobileNet based convolutional neural networks. *Materials Today: Proceedings*, *51*, 480–487. doi:10.1016/j.matpr.2021.05.584

Barbedo, J. G. (2018). Factors influencing the use of deep learning for plant disease recognition. *Biosystems Engineering*, *172*, 84–91. doi:10.1016/j.biosystemseng.2018.05.013

De Silva, M., & Brown, D. (2022, August). Plant Disease Detection using Deep Learning on Natural Environment Images. In *2022 International Conference on Artificial Intelligence, Big Data, Computing and Data Communication Systems (icABCD)* (pp. 1-5). IEEE. 10.1109/icABCD54961.2022.9855925

Dubey, S. R., & Jalal, A. S. (2013). Adapted approach for fruit disease identification using images. In *Image processing: Concepts, methodologies, tools, and applications* (pp. 1395–1409). IGI Global. doi:10.4018/978-1-4666-3994-2.ch069

Eunice, J., Popescu, D. E., Chowdary, M. K., & Hemanth, J. (2022). Deep Learning-Based Leaf Disease Detection in Crops Using Images for Agricultural Applications. *Agronomy (Basel)*, *12*(10), 2395. doi:10.3390/agronomy12102395

Javidan, S. M., Banakar, A., Vakilian, K. A., & Ampatzidis, Y. (2023). Diagnosis of grape leaf diseases using automatic K-means clustering and machine learning. *Smart Agricultural Technology*, *3*, 100081. doi:10.1016/j.atech.2022.100081

Kamilaris, A., & Prenafeta-Boldú, F. X. (2018). A review of the use of convolutional neural networks in agriculture. *Journal of Agricultural Science*, *156*(3), 312–322. doi:10.1017/S0021859618000436

Khan, M. A., Akram, T., Sharif, M., Alhaisoni, M., Saba, T., & Nawaz, N. (2021). A probabilistic segmentation and entropy-rank correlation-based feature selection approach for the recognition of fruit diseases. *EURASIP Journal on Image and Video Processing*, *2021*(1), 1–28. doi:10.118613640-021-00558-2

Kulkarni, P., Karwande, A., Kolhe, T., Kamble, S., Joshi, A., & Wyawahare, M. (2021). Plant disease detection using image processing and machine learning. *arXiv preprint arXiv:2106.10698.*

Li, G., Ma, Z., & Wang, H. (2012). Image recognition of grape downy mildew and grape powdery mildew based on support vector machine. In *Computer and Computing Technologies in Agriculture V: 5th IFIP TC 5/SIG 5.1 Conference*. Springer.

Li, L., Zhang, S., & Wang, B. (2021). Plant disease detection and classification by deep learning—A review. *IEEE Access : Practical Innovations, Open Solutions*, *9*, 56683–56698. doi:10.1109/ACCESS.2021.3069646

Lu, B., Dao, P. D., Liu, J., He, Y., & Shang, J. (2020). Recent advances of hyperspectral imaging technology and applications in agriculture. *Remote Sensing (Basel)*, *12*(16), 2659. doi:10.3390/rs12162659

Madhavan, M. V., Thanh, D. N. H., Khamparia, A., Pande, S., Malik, R., & Gupta, D. (2021). Recognition and classification of pomegranate leaves diseases by image processing and machine learning techniques. *Computers, Materials & Continua*, *66*(3), 2939–2955. doi:10.32604/cmc.2021.012466

Muhammed, F. (2022). Computer vision-based plant disease identification system: A review. *Computer*, *1*(1), 59–78.

Narayana, B. V. V. S., Ravi, K. S., & Ramesh, N. V. K. (2018). A review on advanced crop field monitoring system in agriculture field through top notch sensors. *Journal of Advanced Research in Dynamical and Control Systems*, *10*(6), 1572–1578.

Ouhami, M., Hafiane, A., Es-Saady, Y., El Hajji, M., & Canals, R. (2021). Computer vision, IoT and data fusion for crop disease detection using machine learning: A survey and ongoing research. *Remote Sensing (Basel)*, *13*(13), 2486. doi:10.3390/rs13132486

Prajapati, H. B., Shah, J. P., & Dabhi, V. K. (2017). Detection and classification of rice plant diseases. *Intelligent Decision Technologies*, *11*(3), 357–373. doi:10.3233/IDT-170301

Qiang, Z., He, L., & Dai, F. (2019). Identification of plant leaf diseases based on inception V3 transfer learning and fine-tuning. In *Smart City and Informatization: 7th International Conference, iSCI 2019*. Springer.

Saleem, M. H., Potgieter, J., & Arif, K. M. (2019). Plant disease detection and classification by deep learning. *Plants*, *8*(11), 468. doi:10.3390/plants8110468 PMID:31683734

Salleh, M. A. M., Mahmoud, D. K., Karim, W. A. W. A., & Idris, A. (2011). Cationic and anionic dye adsorption by agricultural solid wastes: A comprehensive review. *Desalination*, *280*(1-3), 1–13. doi:10.1016/j.desal.2011.07.019

Siddiqui, F., Gupta, S., Dubey, S., Murtuza, S., & Jain, A. (2020, January). Classification and diagnosis of invasive ductal carcinoma using deep learning. In *2020 10th International Conference on Cloud Computing, Data Science & Engineering (Confluence)* (pp. 242-247). IEEE. 10.1109/Confluence47617.2020.9058077

Subramanian, M., Shanmugavadivel, K., & Nandhini, P. S. (2022). On fine-tuning deep learning models using transfer learning and hyper-parameters optimization for disease identification in maize leaves. *Neural Computing & Applications*, *34*(16), 13951–13968. doi:10.100700521-022-07246-w

Sujatha, R., Chatterjee, J. M., Jhanjhi, N. Z., & Brohi, S. N. (2021). Performance of deep learning vs machine learning in plant leaf disease detection. *Microprocessors and Microsystems*, *80*, 103615. doi:10.1016/j.micpro.2020.103615

Talaviya, T., Shah, D., Patel, N., Yagnik, H., & Shah, M. (2020). Implementation of artificial intelligence in agriculture for optimisation of irrigation and application of pesticides and herbicides. *Artificial Intelligence in Agriculture*, *4*, 58–73. doi:10.1016/j.aiia.2020.04.002

Yadav, A. K., Jain, A., Morato, J. L., & Yadav, D. (2021). Retinal Blood Vessel Segmentation using Convolutional Neural Networks. In KDIR (pp. 292-298). doi:10.5220/0010719500003064

Yadav, D., Gupta, A., Jain, A., & Yadav, A. K. (2022). Plant leaf disease detection using CNN with transfer learning and XGBoost. *International Journal of Data Analysis Techniques and Strategies*, *14*(3), 244–265. doi:10.1504/IJDATS.2022.128273

Yun, S., Xianfeng, W., Shanwen, Z., & Chuanlei, Z. (2015). PNN based crop disease recognition with leaf image features and meteorological data. *International Journal of Agricultural and Biological Engineering*, *8*(4), 60–68.

Chapter 13
Comparative Analysis of AI Techniques for Plant Disease Detection and Classification on PlantDoc Dataset

Macha Sarada
https://orcid.org/0000-0003-2910-4623
Jawaharlal Nehru Technological University, Hyderabad, India

ABSTRACT

Any developing economy can be thought of as having its foundation in the agricultural sector. One industry where automation and the internet of things can have a significant impact is agriculture and modern farming. Due to the country's expanding population and rising food demands, agriculture is important in India. As a result, agricultural yield needs to be increased. Diseases brought on by bacteria, fungus, and viruses are one of these significant factors in decreased crop yields. By using methods for plant disease detection, this can be avoided and managed. Farmers must have access to the latest technologies and practices in order to get the most yield possible from their crops. A maximum agricultural output must be maintained, which means keeping healthy plants and keeping an eye on their surroundings to spot or detect diseases. Artificial intelligence (AI), machine learning, and deep learning have all been proven to be incredibly useful in the field of modern agriculture as a means of sophisticated image analysis.

INTRODUCTION

The agriculture sector has long relied on contemporary science to supply food for 7 billion people. The food security of human society is, however, threatened by a number of dangers that those in the agriculture sector must deal with. As far as we are aware, some of the threats are changes in climate, grazing of animals, diseases in plants, etc One of the biggest risks is plant disease, which not only results in large losses of crops meant for human consumption but also has an adverse effect on farmers whose main source of revenue is from the production of healthy crops (Somowiyarjo, 2011; Aadi-A1,2017). Human

DOI: 10.4018/978-1-6684-8516-3.ch013

experts laboriously inspect mature plants during the plant harvesting process to ensure they are free of disease and fit for human consumption before removing them. Finding them costs a lot of effort and money the illness that affects a particular plant using the usual visual method, especially if the farmhouse is big and there are lots of plants. (Gawande and Gavhale, 2014). Moreover, given the seeming daily expansion in populace around the world, it only makes sense that procedure be mechanised in order to satisfy the expanding demands of the public. Compared to traditional visual identification of illnesses of plant, initial discovery of plant ailments is now significantly simpler, faster, and less expensive thanks to the introduction of ML models. ML models are gradually replacing traditional plant disease identification in the market as a result of recent research in this field(Ngugi, Abelwahab and Abo Zahhad, 2021).

LITERATURE

D.A. Bashish et al. (2010) divided the leaf picture into four clusters using k-means segmentation and squared Euclidean distances. The Colour Co-occurrence approach was used to feature extraction for both colour and texture characteristics. Using a neural network detection method built on the Back Propagation methodology, classification is finally completed. The whole system's accuracy in classifying and detecting illnesses was estimated to be 93%.

M. Banghe and associates (2015) A web-based tool for diagnosing fruit ailments has been created as a result of user contributions of fruit photos to the system. The characteristics were extracted using CCV (colour coherence vector) metrics, morphology, and colour. The clustering process employed the k-means method. SVM is employed to ascertain a person's infection status. The accuracy of this study's diagnosis of pomegranate disease is 82%.

To detect fungal infections on plant leaves, J.D. Pujari et al. (2015) employed a range of crop types, including commercial crops, cereal crops, and fruit and vegetable crops. There have been several crop management tactics employed. • K-means clustering was used to segment fruit crops, and texture features were focused on and identified using ANN and closest neighbour algorithms, attaining an overall average accuracy of 90.723%.

- The chan-vase strategy is used for segmenting vegetable crops, local binary patterns are used to extract texture features, and SVM and the k-nearest neighbour algorithm are used to extract texture features. Overall average accuracy of 87.825% in classification
- Commercial crops were segmented using the grab-cut algorithm. With an overall average accuracy of 84.825%, wavelet-based feature extraction was used with Mahalnobis distance and PNN as classifiers.
- K-means clustering and the canny edge detector were used to segment the cereal crops. Colour, shape, texture, texture colour, and random transform features were retrieved. SVM and closest neighbour classifiers were applied, yielding a total accuracy of 83.72%.

In order to automate the detection and categorization of plant diseases, V. Singh et al. (2016) employed an image segmentation approach called a genetic algorithm. Using a limited number of images, four plant leaves—banana, rose, and lemon—were taught and evaluated.. beans, The colour co-occurrence strategy, which considered both texture and colour data, was used for feature extraction. The diseases were classified with 86.54% and 95.71% accuracy using the Minimum Distance Criterion with k-mean clustering and the SVM classifier, respectively. The accuracy increases to 93.63% when the genetic algorithm is paired with the Minimum Distance Criterion classifier.

E. Kiani et al. (2017) employed a fuzzy decision maker to spot diseased leaves in an outdoor strawberry crop. The detection time was 1.2 seconds, and the total accuracy of detection and segmentation of plant illnesses was 97%.

H. Ali and associates (2017) Their study aims to use the E colour difference approach, which has a 99.9% overall accuracy, to separate the diseased region and detect illnesses using colour histogram and textural clues. Classifiers such fine KNN, Cubic SVM, Boosted tree, and Bagged tree have all been used. The bagged tree classifier outperforms the competition by obtaining 99.5%, 100%, and 100% accuracy on the RGB, HSV, and LBP features, respectively. With accuracy rates of 88.9%, 90.1%, and 50.90%, respectively, Fine KNN, Cubic SVM, and Boosted tree classifiers did well.

An approach for constructing an automated plant disease detection system was developed by G. Saradhambal et al. (2018). The contaminated region of the leaves was predicted using the Otsu's classifier and the k-means clustering technique. Both shape and texture characteristics were retrieved in the recommended study. While texture-oriented properties included contrast, correlation, energy, homogeneity, and mean, shape-oriented parameters recovered in this study included area, colour axis length, eccentricity, solidity, and perimeter.

STATE OF THE ART

Monoculture's Effects on the Economy and Environment

The establishment of monoculture was made in an effort to meet the rising demand for food. Planting a single type of crop on the same land is known as this type of agriculture. The agronomist may focus only on producing one type of yields, and they only require the tools and machinery necessary to cultivate and harvest that one type of produce. This technique of farming is advantageous for farmers due to its benefits for the economy and the environment. (Helly, M. E.etal., 2003) The monoculture boosts productivity and efficiency. Additionally, compared to polyculture, this style of horticulture is easier to manage because it needs less efforts. It ensures farmers will earn more money. However, due to its detrimental effects on the soil, farmers are obligated to apply fertilisers to promote plant development and the soil. (Otani, Saria, et al 2019). Farmers also use a lot of pesticides to protect plants from diseases and pests. The factor that accounts for the negative environmental impact of monoculture due to its excessive use of chemical inputs. Numerous chemicals are hazardous and will harm the soil, even if farmers make every effort to use chemicals that are less harmful or at very low concentrations (Singh, Vijai et al. 2017). Additionally, there are fewer pests and insects in monocropping, This aids in stopping the illness from spreading among plants. It underlines once again the significance of using pesticides and other chemicals that could contaminate the ecosystem and have harmful effects on it all. Because monoculture depletes the soil and decreases the amount of nutrients, pesticide usage will deplete the soil. Farmers choose monocropping because it requires less equipment and workers, despite its drawbacks. Additionally, it ensures harvesting and planting efficiency, resulting in an increase in productivity and consequently a profit. Consequently, monoculture farming contributes to an increase in farm industry productivity and efficiency.

ARTIFICIAL INTELLIGENCE

Artificial intelligence refers to any technique that imitates human behaviour. In essence, it is a science that seeks to teach computers to solve challenging and complicated issuesIn other words, artificial intelligence is the engineering and science of building smart machines. AI need not be limited to methods that can be identified by medical tools, unlike using computers to investigate human intelligence (Jean, W. 2009). Artificial intelligence is the term used to describe the intelligence exhibited by robots or other devices. It is a component of computer science (Manu BN 2022). It has greatly enhanced human life, and it is now a prominent and practical area of study in computer science.

DEEP LEARNING

DL is a generous of ML. uses ANNs to familiarize, acquire from massive volumes of statistics. Additionally, high-level abstractions are learned from data via hierarchical architectures (Khirade, Sachin D., etal., 2015). DL, which combines machine learning and artificial intelligence, replicates how people learn certain types of information. Along with statistics and predictive modelling, deep learning is a crucial part of data science. Deep learning accelerates and simplifies the process of obtaining, analysing, and interpreting enormous volumes of data for data scientists. Profound learning is frequently viewed as a means to automate predictive testing at its most fundamental level. Instead of standard AI calculations, which are linear, deep learning calculations are piled inside an order of increasing intricacy and consideration. It is a novel method that has previously been applied in well-known AI applications. Three key causes, including the development of chip processing, the availability of inexpensive computer hardware, and machine learning techniques, have contributed to the current boom in deep learning (Kumar, Surender, etal., 2015) Numerous deep learning algorithms have been thoroughly examined and debated in recent years.

MACHINE LEARING

ML is a kind of AI that makes decisions without human input by using procedures and statistics to analyse situations . It talks about how computers carry out activities on their own depending on their past knowledge (Jean, W. 2009). We may contend that machine language experience is how artificial intelligence is created. Knowing that the human does not provide this system any programmes; rather, it is educated and taught via usage of the data that is acquired allows us to differentiate between conventional computer software and machine learning.

Supervised, Unsupervised, Semi-Supervised, Reinforcement are different types of learning techniques.

Supervised

Supervised learning refers to defined datasets that include both input and output bounds for the purposes of creating models. When developing a model, an 80:20 ratio is maintained between data collection and testing. Classification and regression are further supervised learning classifications. Classification is further subdivided into binary and multiclass.

The arrangement is classified as a supervised learning problem since the result is a discrete value. This distinct worth might be classified or parallelized. Relapse is a SL model having a broad and lasting impact. The relapse's purpose is to anticipate a worth that is more in accordance with production esteem.

Nearest Neighbor, Linear Regression, Support Vector Machine (SVM), Decision Trees, Random Forest, and Gaussian Naive Bayes are few examples of supervised learning models.

Unsupervised

Targets are not supplied for the model to produce displays for in unsupervised learning; instead, just input boundaries are provided. Unsupervised learning can be classified as either association and clustering . Information organised as assemblages created by several examples distinguished by the machine model is subject to clustering. While the term "Association" refers to a procedure based on standards for classifying relationships between the boundaries of a sizable informational collection.

Clustering Using Hierarchies, BIRCH - Balanced Iterative Reducing, and K-Means Clustering, are examples of unsupervised learning models.

Semi-Supervised

Semi-supervised learning functions similarly to the previously stated approaches. When working with information, both named and unnamed information are dealt with using this type of learning. The unsupervised technique is used to compute marks, and then supervised learning strategies are used to deal with the determined attributes. This process is more widely used in image datasets where a sizable fraction of the photos are unnamed.

Reinforcement

The model's application improves with disapproval in order to learn from examples and behaviour. Every time information is handled, something new about it is learnt and added to the knowledge that is being produced. As a result, as it acquires more, it gets more prepared and experienced. Q-based algorithm, Temporal Difference is used for reinforcement learning.

THE DISTINCTION BETWEEN ML & DL

Machine learning specifically refers to deep learning. In terms of technology, it works in a similar way to machine learning but employs distinct methods.It is based on how human neurons work and gave rise to the notion of artificial neural networks. While the ability to tell a cat from a dog in a picture may be used to show how machine learning works. In this case, the machine learning model inputs photos of both animals and extracts several features, such as shape, height, and eyes, from these two images. It next runs the classification algorithm and concludes by determining whether the image belongs to a dog or a cat.

THE AGRICULTURAL SECTOR AND THE FIELD OF ARTIFICIAL INTELLIGENCE

Farming is in upheaval as a result of AI. Artificial intelligence has increased yield output, which has also improved harvesting, real-time monitoring,, processing, and marketing. (Paras M et.al., 2019)

Detection of Disease

Images of plant leaves are categorised as unhealthy area, and healthy area of the leaf, according to image recognition and exploration. The diseased or contaminated region is then taken out and submitted to the lab for additional testing.

IOT (Internet of Things)-Driven Development

Health, manufacturing, energy, communications, and agriculture are just a few of the sectors and enterprises that will be impacted by the Internet of Things (IoT). In order to boost productivity, eminence, and turnover, IoT applications in agriculture are designed to give farmers decision-making tools and automation technologies that seamlessly integrate products, information, and services.

Expert System

It is possible to demonstrate the need for expert systems for the transmission of technical knowledge in agriculture by pointing out shortcomings in the current system of technology transfer and demonstrating how expert schemes can help to address those problems and are likely to be improved.

The Development of Insight Constructed on Photographs

Crop monitoring, field surveys, and other operations can benefit from drone imagery. Farmers may connect them to IOT and PC vision technology to ensure quick responses. These streams can provide farmers with constant weather notifications.

Field Management

Real-time estimations during the cultivation phase may be made utilising highquality images from the copter and drone systems by mapping the field and identifying places where the crop is most abundant. in which crops need water, fertiliser, and insecticides.

Monitoring of Yield Health

To develop crop metrics over thousands of acres of arable land, remote sensing (RS) techniques, hyperspectral photography, and 3D laser scanning are needed.

Robotics in Agriculture

Agribot, frequently referred to as Agbot, is a type of agricultural robot. By reducing the amount of physical effort needed, it helps farmers produce more yield with less work. In the future, it's possible that these agricultural robots may manage planting, tilling, harvesting, and other farm tasks independently. These farm robots will take care of weeding, disease management, and pest control.

AI TECHNIQUES ARE BEING APPLIED IN THE AGRICULTURAL SECTOR

Image Manipulation

A technique for measuring a disease's afflicted area and identifying colour changes in that area is image processing.

The surveys demonstrate the use of image processing for disease detection (*Nilay Ganatra et.al 2018)*. A robot for agriculture has been introduced that uses image processing to find leaf disease (Vijay Kumar et.al., 2018).

Machine Learning

Applications of AI for diagnosing illnesses have been successfully developed in the modern period. Machine learning algorithms can swiftly and reliably diagnose any illness. Shima Ramesh et al. (2018) used deep learning and machine learning algorithms to better diagnose plant diseases and increase the accuracy of their findings. According to Mukesh Kumar Tripathi et al., the Support Vector Machine (Machine Learning Algorithm) is the optimum method for diagnosing diseases.

Deep Learning

Deep learning assists in identifying a key association in the data and also records information about current clients that may be useful to patients who have similar symptoms or ailments.

Expert System

Agriculture proficient systems would consist of decision aids, Integrated Crop Management, fertilisation,irrigation, weed control, nutritional problems, herbicidal sprays and cultivation, (*Nilay Ganatra et.al. 2018)*.

CNN (Convolutional Neural Networks):

CNNs (Convolutional Neural Networks) are said to be state-of-the-art in picture recognition and are capable of making a quick and precise diagnosis.

AI For Detecting Plant Diseases

A computer programme in the field of agriculture was developed to solve various problems with plant diseases. Artificial intelligence is the most often used approach for detecting plant diseases due to its excellent performance and accuracy. Numerous issues in the realm of agriculture were identified and resolved with the aid of various AI approaches. Convolutional neural networks, deep learning, machine learning, and image processing are some of the techniques that have been used to detect disease in a variety of crops, including rice, tomato, banana, and others.

The most recent technology, including artificial intelligence, has transformed the agricultural industry as a whole. It significantly boosts output and real-time plant disease monitoring.

METHODOLOGY

Types Of Plant Diseases

"Anything that prevents a plant from performing to its maximum potential" is plant disease, is catergorized as Non-infectious (abiotic) and Infectuous (biotic) plant diseases (Figure 1) (Ravinderjit Kaur et.al 2019)

Figure 1. Plant disease classification

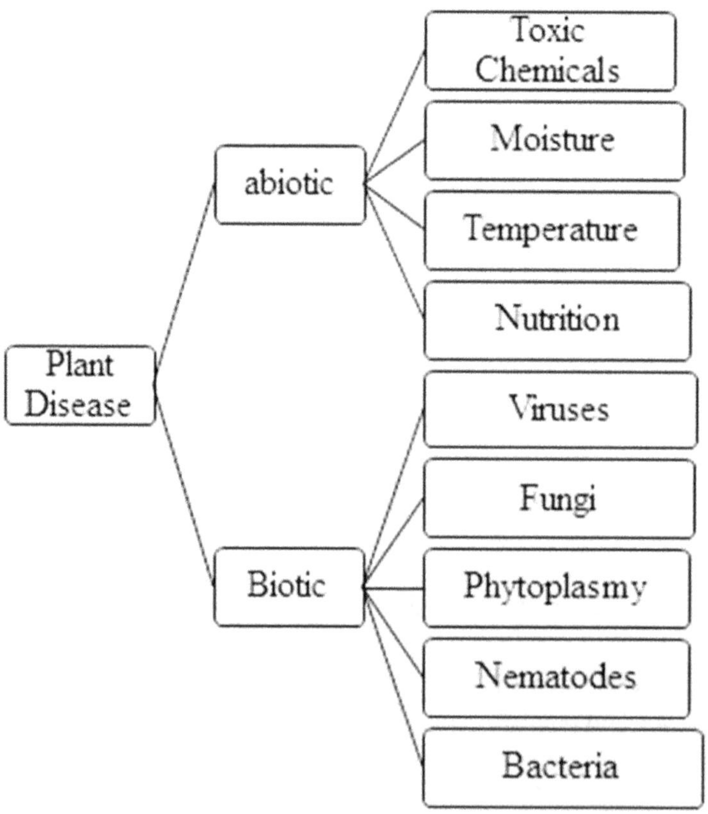

Non-Infectious or Abiotic Diseases

Instead of being brought on by living things, these illnesses are brought on by external environmental factors. Despite the fact that they cannot be passed from one plant to another, they are very typical and should be taken into account when determining the health of any plant. Examples of abiotic diseases include nutritional deficiencies, soil compaction, salt damage, ice, and sun burn. (Figure 2. a).

Aboiotic infections are caused by toxic chemicals, moisture, temperature and nutritions

Infectious or Biotic Disease

These diseases are the result of living creatures. They are known as plant pathogens when they infect plants. We will solely talk about plant disease pathogens when addressing plant pathology. Pathogens that may spread from one plant to another can infect many types of plant tissue, including leaves, shoots, stems, crowns, roots, tubers, fruit, seeds, and vascular tissues (Figure 2 b).

Boiotic infections are caused by viruses, fungi, phytoplasmy, nematodes and bacteria.

Figure 2. Plant disease classification

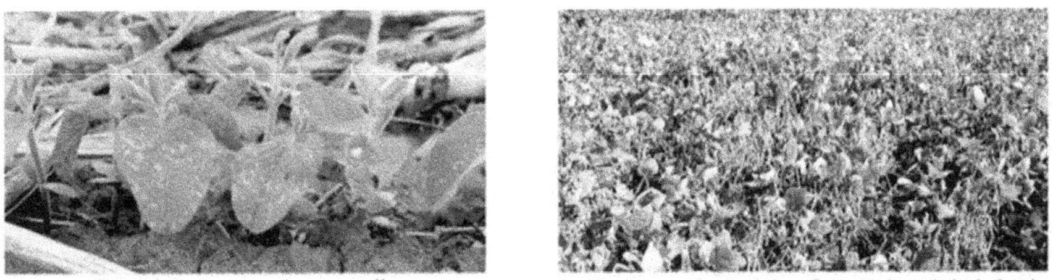

a. Frost injury on soybean seedings.

b. Soybean plants dying from Sclerotinia infection.

c. Soybean infected with Sclerotinia.

d. Soybean infected by bacterial blight.

e. Aster yellow phytoplasm infecting aster. (photo courtesy of Colorado State University's Whitney Cranshaw and Bugwood.org.)

f. Virus that causes the Bean Pod Mottle in Soybeans.

g. Adu Female adult soybean cyst nematodes arising from the roots of soybeans.

To distinguish between abiotic disorder and biotic disease concerns by looking for a few telltale signs.

- Many plant species frequently suffer from abiotic harm. Several different sorts of plants are probably going to suffer harm from chemical or drought stress. Problems with biotic illness, on the other hand, are more specie-specific. For instance, sweet corn is not harmed by the same fungus that cause tomato leaf blight.

- Over time, abiotic harm does not transfer from one plant to another. In addition to spreading within a single plant, biotic diseases can also spread to nearby plants of the same species. Disease pathogens frequently travel from plant to plant via wind-blown rain.
- Pathogen/s physical symptoms like bacterial slime, fungal growth, nematode cysts, or the existence of mites or insects, can occasionally be seen in biotic infections. Abiotic diseases do not exhibit disease symptoms.

DATASETS

Prior to being used for plant disease diagnosis, AI algorithms must be proficient on huge data-sets that can be produced from various leaf photos, such as fresh and diseased leaf images. Models will be able to distinguish among fresh and sick plants as a result. These data-sets frequently consist of photographs of several plant species and combinations, such as sick and healthy plants. The PlantVillage data-set contains more than 50,000 images of both healthy and sick leaves (Singh et.al. 2020;Hughes, D et.al., 2015) recently created a collection of 2,598 photos of 13 different crops in 17 different classifications (Figure 3). Image classifiers are a crucial element related to plant imagery. Datasets contain a variety of coloured, sized images.

Figure 3. Diverse leaf pictures from the PlantDoc data-set
("available at: https://github.com/ pratikkayal/PlantDoc-Dataset").

Datasets for Common Diseases

1) PlantVillage includes 54309 photos of plant leaf disease that are now available as an open data-set. Oranges, bell peppers, tomatoes, blueberries, raisins, raspberries, peaches, and potatoes are among the 14 different fruit and vegetable crops covered by the law. Corn has 26 infections, including 12

photos of a healthy crop of leaves, as well as 26 different viruses, fungi, mite and bacteria-caused diseases(*J. H. Kamdar et.al).*

2) "Plant Pathology Challenge" for CVPR 2020-FGVC7, includes 3,651 high-quality annotated RGB images of 865 healthy apple leaves, 1,399 with cedar apple rust signs, 187 complex disease patterns, 1,200 with apple scab, and 1,200 with both diseases.
3) Some are datasets of actual photos gathered by the authors for their research, while others are not (soybeans, tea, cucumbers, corn, grapes, apples).
4) The data gathering strategy is frequently used in apps that cultivate the plants and administer the virus, which utilise hyperspectral pictures to diagnose diseases.

In order to train any models utilising a variety of image classifiers, such as residual neural network (ResNet), DenseNets, and Visual Geometry Group (VGG) net, images are scaled or normalised (224x224 or 299x299 pixelsFor neural networks (NN), convolutional neural networks (CNN), and other types of networks, normalisation is used to balance pixel values in the range of 255. Categorical variables are also plotted in accordance with prototypes.

I. **ResNet**: The ResNet idea was put into practise by ResNet (He.k,Zhang, X et.al 2016) Its accuracy score of 96.43% placed it first. With final universal average merging, it is a network-in-network design. The web-like micro-NN is produced by the model using residual units (He, K et.al., 2016). This makes it possible to abstract the data from more intricate systems. ResNet is deeper than VGG network thanks to its shortcut connection. The vanishing gradient issue is taken care of by the model's depth. Versions of ResNetXX have a local gradient of 1. Identity mappings would be more dependable in 2016 with the ResNet upgrade (He et al. 2016).

II. **DenseNet** knows CNN's architecture, which links all layers in a forward direction to prevent data loss. Several fine-grained pictures made up the mapping between the layers.The mapping between the layers was composed of a series of fine-grained images. Because of the naturally existing continuing flow, these pictures' properties were retrieved and utilised as input in subsequent layers.. This configuration had benefits like eliminating gradient difficulties and requiring fewer parameters. As a result, DenseNet with 121 perfect layers was produced (Huang, G et.al., 2017).

III. **GoogleNet**: The "Inception" notion was first introduced in the GoogleNet architecture with Inception V4. Inception VN, where N is the version number, is regarded as the most basic implementation of the GoogleNet architecture. Later, (Szegedy et al. 2016) reported on an improved version of Inception V4. To hasten the training of Inception networks, this design is paired with residual connections. The conv layers in the Inception model are divided into four parallel levels: max pooling, a 5x 5 convolutional layer, a 3x 3 conv layer, and a 1x 1 conv layer. For depth reduction, the 1 x1 convolutional layer is employed.

IV. **VGG net model**: The VGG net model was created by Simonyan, Zisserman etal., which finished second in the competition with an error rate of 7.5%. The paradigm is often represented by its simplicity, consisting of simply 3x3 N linked arrays. Volume size minimization is handled through max pooling (downsampling). Two further layers with a total of 4,096 nodes each are fully connected.

V. **AlexNet**: Deep Convolutional Neural Network, or DCN-Network For the 2010 challenge, AlexNet was developed to divide around a million high-goal photos into 1000 different classes. (The ILSVR test). There are 650,000 neurons spread throughout 8 buried layers that make up virtually all of it. (Aggarwal, S et.al 2021).

DETECTION OF DISEASES IN AGRICULTURE BY USING AI TECHNIQUES

Figures 4 show a simple plant disease detection system with the following modules: capture, pre-processing, segmentation, extraction of features, and classification (or recognition). Its two stages are training and testing. The first step in the training process is to take a picture of a particular component, such as a leaf, a stem, a root, or a branch. The pre-processing of images to improve blur, reduce noise, and correct different geometric misrepresentations is optional. Segmentation detects unique or supplementary regions from an diseased training leaf picture and splits the areas of awareness from background. The classifier is then trained using feature vectors that were extracted from the areas of interest. During the testing process, a test leaf picture passes through pre-processing, segmentation, and feature extraction modules. The trained classifier identifies whether the test image is of a healthy or diseased sample.

Acquisition of Images

To create a reliable detection model, images of plants are acquired in a variety of biological settings. Numerous images with various resolutions can be found in the collection. Different tools, such as smartphones, tablets, and conventional RGB cameras, are used to take pictures of plants. Atmosphere, geographic location, and lighting conditions at the moment of image capture all have an impact on dataset collection. Various stages of the plant's development were captured in the photographs(*S. D. Khirade*. et.al 2015)

Pre-Processing of Images

The visual appeal of plant photographs can be improved by preprocessing. Image pre processing has a considerable positive impact on the efficiency of feature extraction and the outcomes of picture analysis. Several activities, including artefact removal, image intensity stabilisation, and glamour reduction, are completed at this stage. Pre-processing images before they are computationally processed enhances the picture data. (*S. D. Khirade*. et.al 2015).

The noise in photos or other artefacts is reduced using a variety of preprocessing approaches. Using picture cropping, the necessary portion of the leaf image can be obtained by removing undesirable image components. This process also employs a smoothing filter to smooth the image. The image-enhancement technique raises the contrast of the image . As a result of the original image's large pixel size and lengthy processing time, picture scaling is used to reduce the original image's size into thumbnails, which reduces the original image's pixel size and speeds up the process overall.

Segmentation of Images

In order to diagnose plant diseases and extract information from photos, image segmentation is a crucial step. Finding and identifying sick plant leaves via image segmentation is the most important difficulty in the complicated environment. Separating symptom information from background information is the main goal of image segmentation. Using this method, an image is divided into a number of significant, non-overlapping portions. Regarding diseases, the infectious region of plant leaves is especially important. The crucial stage in determining the sickness picture and the kind of disease is segmenting the leaf disease image. The success or failure of subsequent image processing is directly influenced by image segmentation. In image processing, the threshold cannot be disregarded.

Feature Extraction

A crucial step in identifying plant diseases is the extraction of characteristics. Harvesting functionality for the detection of plant diseases has caused numerous issues (*S. D. Khirade*. Et.al 2015). For illness feature extraction, image features such textures, structure, colour, and motion-related attributes are essential . The characteristics of the leaf have a significant role in the precise classification of plant disease. Hu moments and Haralick texture characteristics are commonly used to abstract insightful information. Feature extraction is employed in many image processing applications. The features that are used in illness identification include texture edges , morphology and colour.

Some of the traits that are used to classify plant diseases.

Color : Colour values, Mean, Deviation
Shape : Eccentricity, major axis length, perimeter, Area, roundness and Minor axis length.
Texture: Energy, Entropy, Homogenity, Contrast, Correlation, Variance, Dissimilarity

Discovery and Cataloguing of Plant Diseases

Image classification is the process of assigning a class to photos, or the probability that a given input belongs to a certain class. Classification methods are comparable to detection methods in that they aim to identify and classify the pathology affecting the plant, rather than trying to discover just one disease among many diseases and symptoms. Similar to quantization, segmentation is typically the first stage in classification techniques. Next, various sorts of features are extracted and input into a classifier model of some kind (*S. D. Khirade*. Et.al 2015).

CLASSIFICATION

Classification is used to examine the sick zone in a photograph and determine the kind of disease infection in leaves. The following are some classifiers(Figure 5).

Classification NB

Gaussian Each attribute's continuous values are calculated by Naive Bayesian, and their distribution is determined by a Gaussian distribution, also known as the Normal Distribution. The regularity of the mean values may be estimated using the bell-shaped bell curve formed from the Gaussian distribution data.

$$p\left(\frac{x_i}{y}\right)=\frac{1}{\sqrt{2\Pi\sigma^2{}_y}}\exp\left(-\frac{(x_i-\mu_y)^2}{2\sigma^2 y}\right)\rightarrow \tag{1}$$

Figure 4. A straightforward plant disease detection system based on imaging methods. Images for the testing phase are from
(Phadikar S et.al2012)

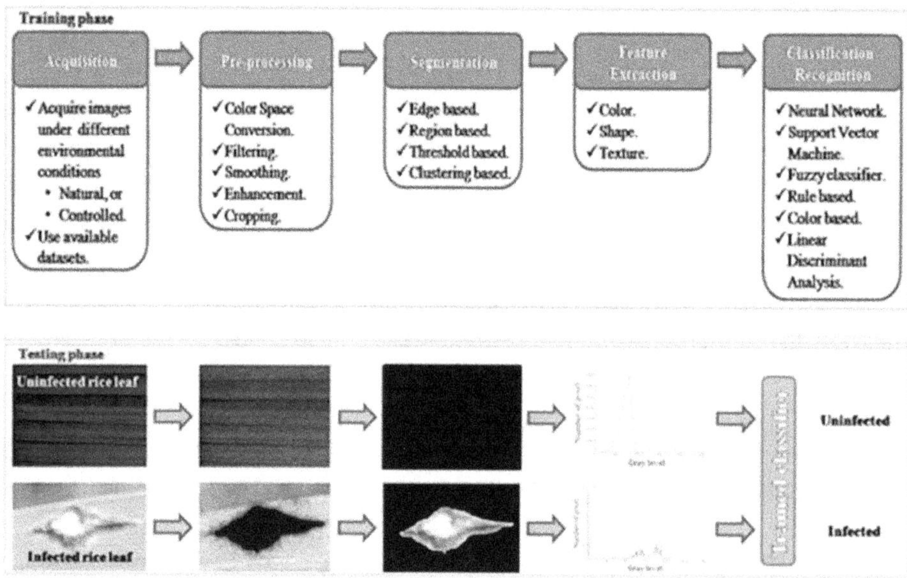

Figure 5. Classifiers are widely investigated in the realm of plant disease detection systems

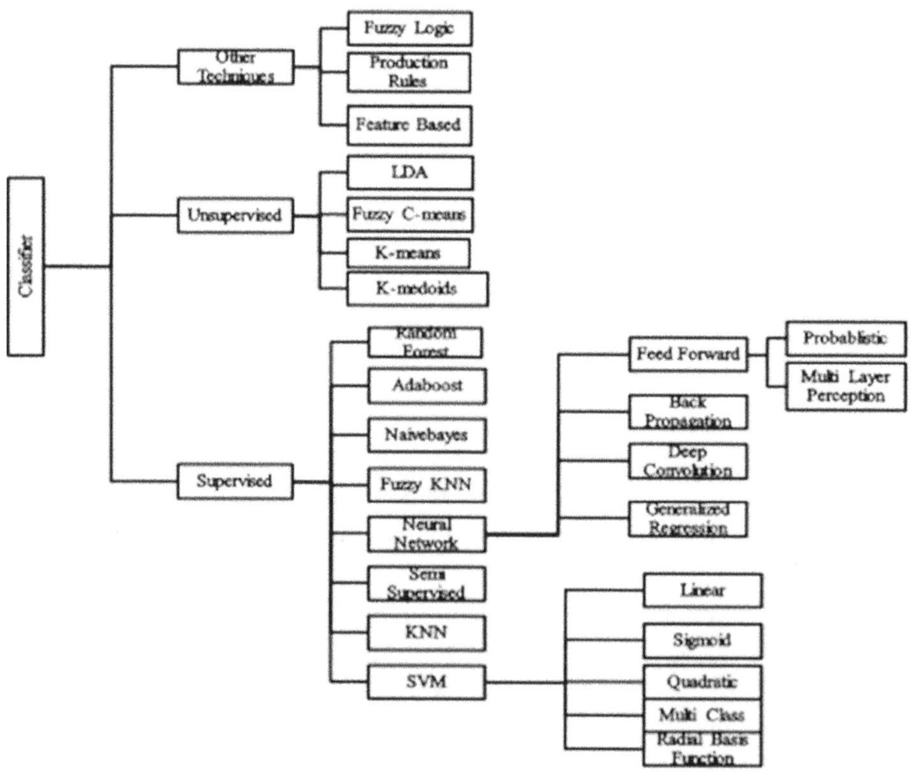

The premise of a Naive Bayes classifier is that the existence of one feature in a class has no bearing on the presence of any other features.When assessing the likelihood of a certain result, a Naive Bayes classifier would consider each of these qualities independently, even if they are connected to one another.A Naive Bayesian model is easy to build and works well with huge datasets. Despite its simplicity, it has been shown to outperform even the most advanced classification algorithms.

K-Means

It is an unsupervised learning strategy that handles clustering concerns. Data sets are divided into K clusters, so that each cluster's data points are homogenous and distinct from those in the other clusters.

Cluster formation procedure of K-means

- The K-means method chooses k centroids, or points, for each cluster.
- For every data point, the K cluster algorithm finds the cluster with the nearest centroids.
- The neighboring distance for each data point is now calculated using the newly created centroids that are based on the current cluster members. Till the centroids stay the same, this process is repeated.

The KNN Classifier

A non-parametric, controlled learning classifier, the k-closest neighbours computation (also known as KNN or k-NN) makes predictions or characterises the collection of a single piece of information based on its proximity. It may be used for classification or regression issues, but because it presumes that comparable points can be discovered nearby, it is typically employed as a classification technique (Figure 6). KNN classifier used for relapse and grouping difficulties, however regularly, KNN helps with characterization concerns. KNN uses a distribution tree method as the basis for its diagnostic. Non-parametric, or the structure that is formed from the characteristics of the collection, is employed when there is no creative imagination for the distribution of the data. When data sets don't follow fictitious mathematical theories, KNN is used to predict. KNN requires no preparation information to proceed. Because all of the training data can be used for testing, the Slow Learning (SL) method is known as such.

Classifier In DT

The most popular and favoured approach for classifying and making predictions is the DT classifier. A feature test is described by every internal node in DT, a test result is described by every branch, and a data label is held by every terminal node. Decision trees can swiftly produce understandable rules. Because of its simple flowchart estimation, a decision tree is an accessible and useful value-based strategy.

One of the most widely used machine learning algorithms today, the decision tree method uses supervised learning to classify events. It can classify both category and continuous dependent variables. Using this approach, the population is divided into two or more homogenous groups based on the most essential traits or independent variables.

Figure 6. Classification using KNN

RF Classification

One of the supervised algorithm is RF that can handle classification and regression techniques. A forest is a managed group of trees, and a strong forest is one with a lot of trees. It also finds the dataset-based decision tree approach, just as the decision trees. It obtained the forecast results from every branch of the tree before selecting the better course of action. It is known as an ensemble approach, and it performs better than a single DT classifier since average performance reduces over-fitting.

The phrase “Random Forest” refers to a grouping of decision trees.Each tree is assigned a class, and the trees “vote” to assign a class to a new object depending on its characteristics. Each tree is assigned a class, and the tree “votes” for that class in order to categorise a new item according to its characteristics. The categorization with the most votes is chosen by the forest (from among all the trees in the forest).

- If the training set contains N instances, a random sample of N cases is selected. This sample will be used to train the tree.
- A quantity If there are M input variables, the M tree grows to the fullest extent feasible. Nothing has been pruned.

Classifier MLP

MLP is a regression-based concept. Using this technique, nonlinear learners transform and transform a set of input data. Deviations from the contribution data, which is a property that is linearly unrelated. A hidden layer is created from the input data layer. The MLP Classifier only employs one hidden layer, failing which it functions as an artificial neural network. Even the use of many concealed layers serves the objective of classification well.

AdaBoost and Gradient Boost

When handling enormous amounts of data to create predictions with great accuracy, boosting techniques like the AdaBoosting and the Gradient Boosting Algorithms are utilised. Boosting is an ensemble learning approach that increases robustness by combining the predictive strength of numerous base estimators.

Linear Regression

Consider how you would organise a set of random wood logs in ascending weight order to comprehend how linear regression functions. The disadvantage is that it's impossible to weigh every log. By examining the log's height and girth (visual inspection) and organising them according to a combination of these observable factors, you must estimate its weight. This is how machine learning's linear regression works.

By fitting the independent and dependent variables to a line, a relationship between them is created. The linear equation Y=a*X+b represents this line, which is referred to as the regression line.

In this equation, the dependent variable is Y, the slope is a, and the independent variable is X, Intercept is b.

The coefficients a and b are obtained limiting the square of the distance between the information fragments and the amount of contrast of the regression line.

Classifier SVM

SVM analysis for classification and regression can be helpful. Inside the dual classes of data, SVM determines the hyper-plane with the increased margin. Support vectors are the name given to the hyper-plane vectors. By taking into account crucial circumstances, SVM may create a margin of hyper-plane that totally divides the hyperplane vector into dual classes that don't intersect. However, this is not always the case, so this classifier will look for support vector hyperplanes that have larger margins and fewer classification errors.

If you plot the raw data as dotss in an n-layered space (where n is the number of items you have), you can use the SVM technique to rank the data. Because each feature's value is linked to a specific coordinate, the data can then be easily classified. Classifier lines can be used to divide the data into groups and then plot them on a graph. By keeping two classes as far apart as possible, it is designed to maximize classification boundaries.

Figure 7 shows how SVM was used to extract the pertinent data point from this region, which is referred to as the hyperplane. The neighbouring points of the hyper-plane can be seen from either the dual sides of the plane or from the support vectors. The SVM can be successfully trained by creating a fixed margin between these vectors that must be as large as possible.

Figure 7. Classification using SVM

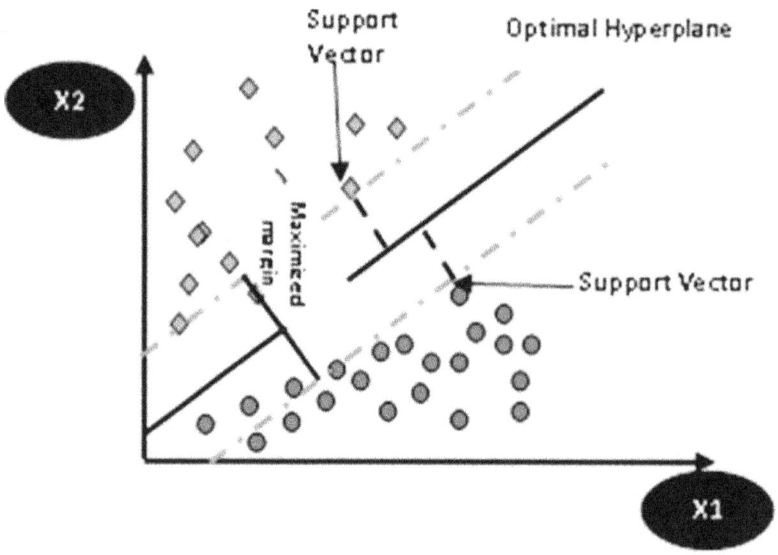

Deep Learning

A combination of AI and ML is called deep learning, replicates how people learn certain types of information. Along with statistics and predictive modelling, deep learning is a crucial part of data science. Deep learning accelerates and simplifies the process of obtaining, analysing, and interpreting enormous volumes of data for data scientists. Profound learning is frequently viewed as a means to automate predictive testing at its most fundamental level. Instead of standard AI calculations, which are linear, deep learning calculations are piled inside an order of increasing intricacy and consideration.

CNN (Convolutional Neural Network)

Convolution neural networks, a subclass of deep feed forward neural networks, allow for the processing of multidimensional input. CNN's goal is to condense images into a form that is simpler to comprehend while maintaining the information required for accurate prediction.

For CNN, a variety of architectures are available, including AlexNet, GoogleNet, and VGGNet. Researchers in several areas of computer science have been quite interested in it as a result of its expansion . It has been applied to agriculture to categorise plant diseases.

The CNN model consists of an input layer, a convolution layer, a pooling layer, a fully connected layer, and an output layer, as shown in figure 8. The photos are used to precisely categorise the plant disease. The photos are used as input to accurately classify the plant illness. The features from the pictures are extracted using the layer known as convolution. The pooling layer uses the recovered highlights to process the element values. The convolution and pooling layer can be stretched further to extract more features, depending on the complexity of the picture. The fully linked layer makes use of the output from earlier layers, combining them into a single vector that may be utilized as an input for the next layer. Finally, the output layer identifies the plant illness.

Figure 8. Plant disease classification through CNN
(Aditya Khamparia 2019)

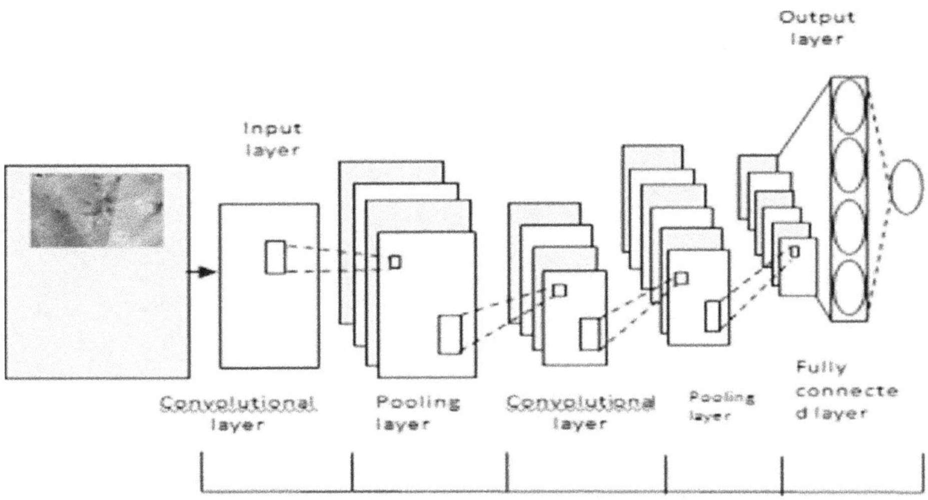

Artificial Neural Network (ANN)

ANNs, or artificial neural networks, are computer programmes that mimic how the human brain can perceive and process statistics. An ANN, like the human brain, is made up of connected processing units known as neurons in a directed graph (Jain, Mao, and Mohiuddin, 1996).

Two types of ANNs are feedforward ANNs, where the behavior of one layer does not affect the behavior of another layer, and feedback ANNs, where network circuits are required to allow signals to flow in both directions. (*Sladojevic, S* etal, 2016).

An artificial neuron, as depicted in figure 9, has one output but multiple inputs that can accept any value between 0 and 1. For each info the neuron has weight and a general inclination (*Sladojevic, S* etal, 2016).

Different forms of NNs include the RNN(Recurrent Neural Networ), CNN(Convolutional Neural Network), MLP(Multilayer Perceptron), and others. The Multilayer Perceptron network is the most widely used Neural Network (MLP). An input layer, one or more hidden layers, and an output layer are the three node levels that make up a typical MLP network. An input layer, one or more hidden layers, and an output layer are the different node levels.

A conventional MLP network has three levels of nodes ie., an input layer, one or more hidden layers, and an output layer. Every node is assigned a weight and a threshold, and they are all linked to one another. Only when each node's output exceeds the set threshold is data moved from one layer to the next. The weight reveals the relative importance of two linked nodes in a network (Jain, Mao, Mohiuddin, 1996). Recognising patterns in the data supplied into the network is an important part of the training process for an ANN.

The actual and intended outputs of the ANN are compared (using the Cost function) during this supervised learning phase, and the training procedure is repeated numerous times (using the periods) up to there is minute to no difference. The cost function minimization method, back propagation, means moving from the output layer back to the input layer and adjusting the weights of the network nodes until the difference between the actual and planned output is as small as possible. (Mao, Jain, Mohiuddin, 1996).

Figure 9. Classification using ANN

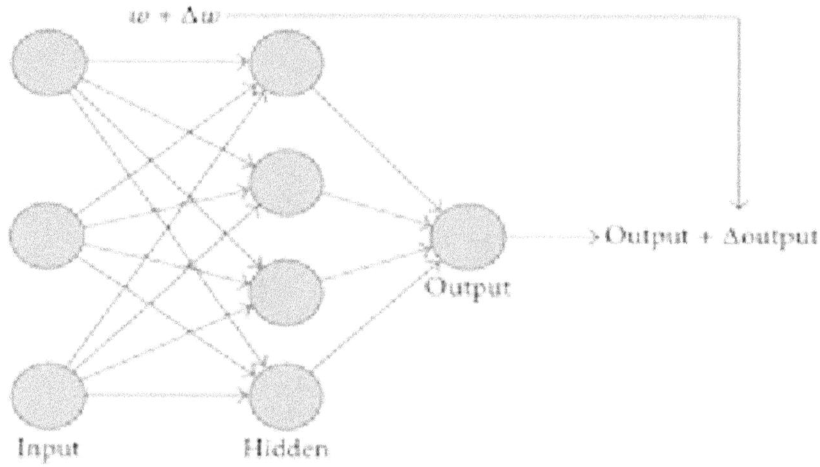

CASE STUDY

Dataset

This study's dataset consists of 10,000 tomato leaf pictures separated into ten groups. Each category has the same number of images. There are no missing photographs in the collection, and the diseased plant images are all 256x256 pixels. The collection, known as the plant village dataset (figure 10), is available on the public website Mendeley data 1.

LIBRARIES USED

- **CV2** : To address computer vision issues like reading or resizing an image, An open source library for computer vision and machine learning is called CV2.
- **Numpy**: The Python programming language's Numpy package aids in the management of large, multidimensional arrays and matrices. Additionally, this library makes it possible to use mathematical operations on arrays or matrices that have been built.
- **Keras** : An open source neural network library called Keras was created to make it possible to quickly develop deep neural networks within the Python interface.
- **Sklearn** : SVM, KNN,RF and other classification & regression methods are all included in the open source machine learning library known as Sklearn. The creator of this machine learning library was
- **Matplotlib**: Python's Matplotlib charting toolkit is used to make both static and animated visualisations.

Figure 10. Different tomato leaf diseases
"Link for the dataset - https://data.mendeley.com/datasets/tywbtsjrjv/1"

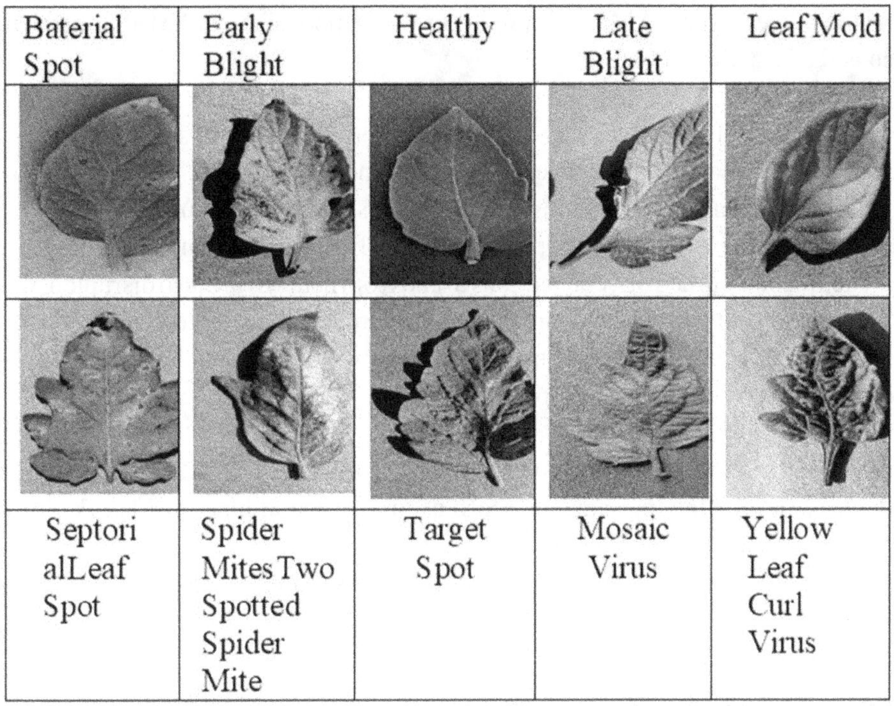

PREPROCESSING

Any machine learning model's pre-processing stage is crucial, since it should impact the performance and outcomes of the models selected. The following procedures were used in this report to ensure that the models gave the best results possible.

1. Read pictures - The root directory that was supplied was read using cv2.imread().
2. Resize photos. To modify the dimensions of the images, cv2.resize() was utilised. The system's RAM was unable to perform operations on all 10k 256x256 pixels images, the CNN and KNN models, therefore the images were reduced to the size 150x150.
3. Using np.array, we convert the photos into an array form after reading and resizing them ()
4. Using LabelBinarizer, Each plant image's labels are then assigned a unique value ()
5. A train set and a test set with a 75:25 ratio each are created from the plant village dataset.

CLASSIFICATION MODEL CONFIGURATION

CNN

The CNN architecture used in this work to identify plant illnesses is shown below:

- the Initial block comprises of 32 3 x 3 filters of convolutional layer,
- and while the ReLU activation function served as the activation function.
- To finish the process, apply a Max Pooling layer with a pool size of (3,3) after batch normalisation and then add a 25% dropout layer.

Batch normalisation, which is typically performed after each layer to standardise the output of the preceding layer and to speed up the convergence of the neural network, each individual layer in the network was allowed to learn independently (zhu, Marques, Garbin, 2020).

To avoid overfitting the model, the drop layer arbitrarily closes certain parts of the neurons. When certain parts of neurons are turned off, connections to and from neurons are also disrupted, which improves model learning and prevents the model from generalizing to test data. (Zhu,Garbin, Marques, 2020).

The second network chunk has two conv layers with 64 3 x 3 filters, a ReLU activation function, and batch normalisation. Then, a 25% dropout layer and a Max Pooling layer with a pool size of (2,2) were added.

The last block of the network has two conv layers of 128 3 x 3 filters, ReLU activation function and batch normalization. Following the same approach as in the second block, a Max Pooling layer of size (2, 2) and an abandonment layer of 25% were added.

The last convolution layer's three-dimensional output was smoothed down to one dimension before being given to the fully connected layers for classification, which was done before the model's final fully connected layer was built.

To enhance the model's performance for the classification of the features returned by the final convolution and pooling layer, a dense layer of 1024 neurons with ReLU activation function, batch normalisation, and a dropout layer of 5% dropout were added. The logits layer, the neural network's last layer, was used to return the values of the model's predictions(Yamashita et al., 2018). Since there are ten classes, a dense layer with 10 neurons and a Softmax activation technique was employed. We utilised a multi-label classification model in this layer, hence Softmax activation was employed.

The CNN Model's Variables

Batch size – 256 ; Epochs – 25 ; Learning rate: 1e-3 (0.001) ; Optimizer: Adam

The learning rate was chosen at 0.001 because difficult issues, such photo classification, require a greater learning rate.

Even though they result in longer training times, smaller learning rates are frequently good at providing higher generalisation accuracies and are known to increase the model's overall accuracy (Martinez & Wilson, 2001).

Testing various batch sizes on the model led to the discovery of its batch size. The model obtained an accuracy of 94.9% in the first experiment with a batch size of 32. In the second trial, the batch size was changed to 128 and the model produced results with an accuracy of 93.1%. In the third iteration of the experiment, the batch size was raised to 256, and the model generated an accuracy of 98.5% (Rolnick et al. 2018). Examine how batch size affects the noise resistance of neural networks during training. The MNIST dataset is used for the experiment, and batch sizes range from 32 to 256. The study found that raising the batch size significantly improved the dataset's tolerance to noise.

One of the most extensively used Neural Network optimizers is the Adam optimizer, which is well known for generating excellent results when used for immaturity classification (Ault, 2020).

The training vs. validation accuracy and training vs. validation loss for the CNN model were both displayed. Figure 11 illustrates this relationship between training and validation accuracy and loss, demonstrating that the model is continually learning and is neither overfitting nor underfitting.

Underfitting is the term used in machine learning models to describe when a model performs poorly on both the training and validation/testing datasets. Overfitting is the term used to describe when a machine learning model performs very well on a training dataset but poorly on a validation or testing dataset. A good fit model, according to Zhang, Zhang, and Jiang (2019), is one that can adequately generalise the validation dataset and learn from the training dataset.

Figure 11. Validation vs training accuracy and validation vs training loss

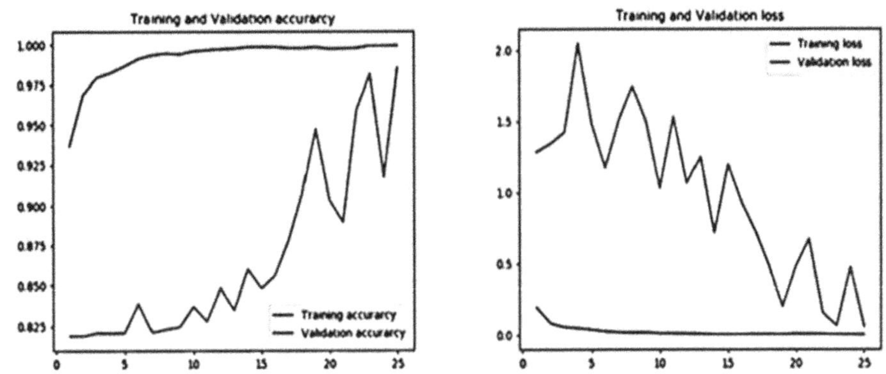

K-NEAREST NEIGHBOR

The performance of the model is determined by the assessment of "k" in the KNN. As previously stated, selecting a small "k" value increases the likelihood that the model will overfit, whereas selecting a large "k" value increases the likelihood that the model will consistently classify the majority of classes. All numbers between 1 and 25 were considered in order to get the optimum value of "k." Through the use of the Matplotlib Library, two graphs were placed. The first graph illustrates the connection between the "k" value and the model's accuracy. The second graph demonstrates the connection between error and the model's "k" value.

Parameters In The KNN Model : K - 21

As shown in the two graphs above, the best "k" value is chosen, and the experiment uses a "k" value with a low error rate and high accuracy. Figure 12 shows that the model's accuracy at k = 21 is 83.6%, and the error from Figure 3 may be considered to be the least significant with a value of 0.163. Therefore, for this investigation, the value of "k" was fixed to 21, and the Euclidean distance was employed to determine the distance.

Figure 12. Plotting accuracy versus various K values and error rate against various K values

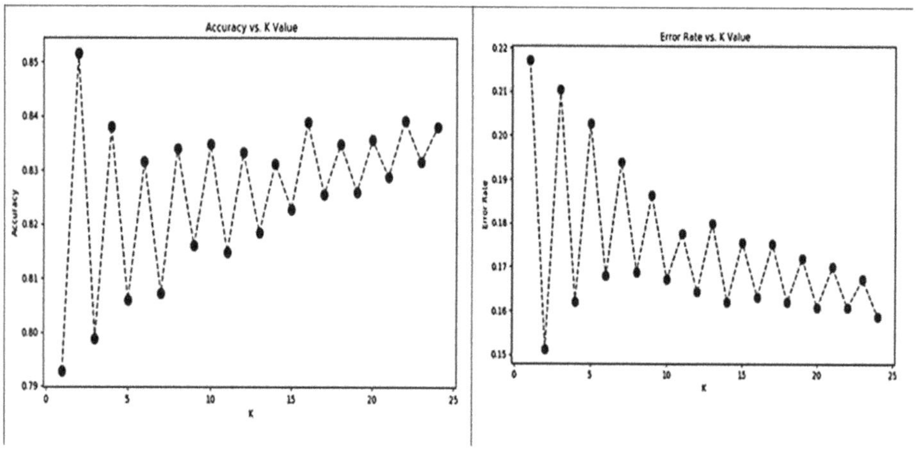

RESULT

The results below illustrate the precision, accuracy, F1-Score, and recall of the CNN and KNN models. To understand the idea of the aforementioned assessment metrics, it is essential to understand what True-negatives (TN), True-positives (TP), False-negatives (FN), and False-positives (FP) are.

Convolution Neural Network Evaluation Metrics

Precision(%): 93 Accuracy(%) : 98.5 F1-Score(%): 93 Recall(%): 93

K-Nearest Neighbor

Precision(%): 90 Accuracy(%) : 83.6 F1-Score(%): 86 Recall(%): 84

For instance, the confusion matrix for a tomato leaf image with the illness "Early blight" appears as depicted in Figure 13. One of the 10 categories/diseases is represented by the confusion matrix below.

Figure 13. 'Early blight' confusion matrix

		Actual disease	
		Early blight	Not Early blight
Predicted Disease	Early blight	TP	FP
	Not Early blight	FN	TN

The result is referred to as a True Positive (TP) outcome if the model properly recognised the plant as harbouring the illness. If the model correctly predicts that the plant does not have the disease, the outcome is referred to as a True Negative (TN) outcome.

A false positive (FP) result is one in which the model predicts that the plant will appear to be ill but in reality is not. When the model predicts inaccurately that the plant won't become infected by the pathogen, false negative (FN) results are generated.

Precision

To calculate this statistic, divide the total number of correctly predicted forecasts (True positives) by the total number of successful predictions (True positives plus False positives) (M and M.N, 2015). This statistic displays a classifier's accuracy and offers an answer to the question of how many of the samples predicted by the classifier to be "Early blight" were indeed "Early blight." Both the CNN and KNN models have accuracy scores of 93% and 90%, respectively. This shows that, when compared to the KNN model, the CNN model returns much more relevant results than irrelevant ones.

$$PRECISION = \frac{TP}{TP + FP}$$

Accuracy

By dividing the overall sample size by the total number of accurate predictions, this statistic is generated. (True positive and True negative outcomes) (Liu et al., 2014). The CNN model outperformed the KNN model in terms of accuracy, attaining 98.5% versus 83.6%.

Since both labels in the dataset used in this study have the same number of samples, accuracy may be used as a relevant metric to compare the two models. However, in situations where the dataset is unequal, accuracy cannot be demonstrated to be advantageous due to the Accuracy Paradox.

$$ACCURACY = \frac{TN + TP}{TN + FP + TP + FN}$$

F1-Score

Comparing the F1-Scores of the two models is not essential because the CNN model outperformed the KNN model in terms of accuracy and recall. A different metric called the F1-Score may be used to decide if a model with high accuracy but bad recall is better than one with low precision but high recall. Even in the F1-score measure, the CNN model outperforms the KNN model, with an F1-Score of 93% for CNN and an F1-Score of 86% for KNN.

$$F1SCORE = 2 * \frac{precision * recall}{precision + recall}$$

Recall

By the sum of True Positives and False Negatives (M and M.N, 2015), divide the overall number of accurate predictions (True positives). This statistic is used to demonstrate the model's sensitivity. It calculates the percentage of samples in the dataset that were correctly identified by the classifier as belonging to the "Early blight" category in its predictions. In comparison to the KNN model, the CNN model returns more relevant results from the plant village dataset, with a recall of 93% vs 84%.

$$RECALL = \frac{TP}{TP + FN}$$

CONCLUSION

The core objective of using AI methods in farming is to identify diseases from plant leaves. The four steps in the process of creating a model for plant disease identification are reprocessing, segmentation, attribute extraction, and classification. A digital camera is used in any approach based on image processing that uses digital images. The intended outcomes are achieved using algorithms for image processing. In the usecase four evaluation criteria, like precision, accuracy, f1 and recall scores, are taken into consideration in KNN,CNN to determine the effectiveness of the applied algorithm.

REFERENCES

Al-Sadi, A. (2017). Impact of Plant Diseases on Human Health. *International Journal of Nutrition, Pharmacology, Neurological Diseases*, *7*(2), 21–22. doi:10.4103/ijnpnd.ijnpnd_24_17

Ali, H., Lali, M. I., Nawaz, M. Z., Sharif, M., & Saleem, B. A. (2017). Symptom based automated detection of citrus diseases using color histogram and textural descriptors. *Computers and Electronics in Agriculture*, *138*, 92–104. doi:10.1016/j.compag.2017.04.008

Ault, R. (2020). *Optimization Study of an Image Classification Deep Neural Network.*

Bashish, D. A. (2010). A Fremework for Detection and Classification of Plant Leaf and Stem Diseases. *International Conference on Signal and Image Processing*, (pp. 113-118). Semantic Scholar.

Bhange, M. (2015), 'Smart Farming: Pomegranate Disease Detection Using Image Processing', *Second International Symposium on Computer Vision and the Internet*, Volume 58, (pp. 280-288). IEEE. 10.1016/j.procs.2015.08.022

Food and Agriculture Organization of the United Nations (Ed.). (2017). *The future of food and agriculture: trends and challenges.* Food and Agriculture Organization of the United Nation.

Franke, G., & Menz, G. (2007). Multi-temporal wheat disease detection by multi-spectral remote sensing. *Precision Agriculture*, *8*(3), 161–172. doi:10.100711119-007-9036-y

Garbin, C., Zhu, X., & Marques, O. (2020). Dropout vs. batch normalization: An empirical study of their impact to deep learning. *Multimedia Tools and Applications*, *79*(19), 12777–12815. doi:10.100711042-019-08453-9

Gavhale, M., & Gawande, U. (2014). An Overview of the Research on Plant Leaves Disease detection using Image Processing Techniques. *IOSR Journal of Computer Engineering*, *16*(1), 10–16. doi:10.9790/0661-16151016

Sujawata, G. & Chouhanb, J. (2021). Application of Artificial Intelligence in detection of diseases in plants: A Survey. *Turkish Journal of Computer and Mathematics Education, 12(*3), 3301-3305

He, K., Zhang, X., Ren, S., & Sun, J. Deep residual learning for image recognition, In *Proceedings of the IEEE conference on computer vision and pattern recognition*, (pp. 770–778). IEEE.

Huang, G., Liu, Z., Van Der Maaten, L., & Weinberger, K. Q. Densely connected convolutional networks, In *Proceedings of the IEEE conference on computer vision and pattern recognition,* (pp. 4700–4708). IEEE.

Hughes, D. & Salathé, M. (2015). *An open access repository of images on plant health to enable the development of mobile disease diagnostics*. arXiv, 1–13

Jean, W. (2009). *Extension plant pathologist, georgia plant disease loss estimates.* CAES. www.caes.uga.edu/publications

Kamdar, J. H., Jasani, M. D., & Jasani, J. D. (2021). *11 Artificial intelligence for plant disease detection: past, present, and future, Internet of Things and Machine Learning in Agriculture: Technological Impacts and Challenges* (pp. 223–238). De Gruyter. doi:10.1515/9783110691276-011

Khamparia, A., Singh, A., Luhach, A., Pandey, B., & Pandey, D. (2019). Classification and Identification of Primitive Kharif Crops using Supervised Deep Convolutional Networks. *Sustainable Computing: Informatics and Systems.* doi:10.1016/j.suscom.2019.07.003

Khirade, S. D., & Patil, A. B. (2015). Plant Disease Detection Using Image Processing. International Conference on Computing Communication Control and Automation, (pp.. 768-771). IEEE. 10.1109/ICCUBEA.2015.153

Kiani, E. (2017). Identification of plant disease infection using soft-computing: Application to modern botany. *9th International Conference on Theory an d Application of Soft Computing, Computing with Words and Perception*, (pp. 893-900). IEEE. 10.1016/j.procs.2017.11.323

Liu, Y., Zhou, Y., Wen, S., & Tang, C. (2014). A Strategy on Selecting Performance Metrics for Classifier Evaluation. *International Journal of Mobile Computing and Multimedia Communications*, *6*(4), 20–35. doi:10.4018/IJMCMC.2014100102

M., H. & M., S. (2015) A Review on Evaluation Metrics for Data Classifica- tion Evaluations. *International Journal of Data Mining & Knowledge Man- agement Process, 5*(2), pp. 01–11. . doi:10.5121/ijdkp.2015.5201

Manu, B.N. (2022). *Plant Leaf Disease Detection and Classification using Multiclass.*

Ngugi, L. C., Abelwahab, M., & Abo-Zahhad, M. (2021). Recent advances in image processing techniques for automated leaf pest and disease recognition – A review. *Information Processing in Agriculture*, *8*(1), 27–51. doi:10.1016/j.inpa.2020.04.004

Nilay Ganatra and Atul Patel, "A Survey on Diseases Detection and Classification of Agriculture Products using Image Processing and Machine Learning", International Journal of Computer Applications (0975 – 8887) Volume 180 – No.13, January 2018

Phadikar, S., Sil, J., & Das, A. K. (2012). Classification of rice leaf diseases based on morphological changes. *Int J Inf Electron Eng*, *2*(3), 460–463.

Pujari, J. D. (2015) Image Processing Based Detection of Fungal Diseases In Plants. *International Conference on Information and Communication Technologies,* (pp. 1802- 1808). Springer. 10.1016/j.procs.2015.02.137

Rolnick, D. (2018) Deep Learning is Robust to Massive Label Noise. *Open Review*. https://openreview.net/forum?id=B1p461b0W

Saradhambal, G. (2018). Plant Disease Detection and its Solution using Image Classification. *International Journal of Pure and Applied Mathematics*, *119*(14), 879–884.

Sharma, V., Verma, A., & Goel, N. (2020). Classification Techniques for Plant Disease Detection. International Journal of Recent Technology and Engineering (IJRTE), *8*(6).

Simonyan, K., & Zisserman, A. (2015). Very deep convolutional networks for large-scale image recognition. *The International Conference on Learning Representations*, (pp. 1–14). ACM.

Singh, D., Jain, N., Jain, P., Kayal, P., Kumawat, S., & Batra, N. (2020). PlantDoc: a dataset for visual plant disease detection, In *Proceedings of the 7th ACM IKDD CoDS and 25th COMAD*, (pp. 249–253). ACM. 10.1145/3371158.3371196

Singh, V. (2016). Detection of Plant Leaf Diseases Using Image Segmentation and Soft Computing Techniques. *Information Processing in Agriculture*, *8*, 252–277.

Sladojevic, S., Arsenovic, M., Anderla, A., Culibrk, D., & Stefanovic, D. (2016). Deep Neural Networks Based Recognition of Plant Diseases by Leaf Image Classification. In *Computational Intelligence and Neuroscience*. doi:10.1155/2016/3289801

Szegedy, C., Vanhoucke, V., Ioffe, S., Shlens, J., & Wojna, Z. Rethinking the inception architecture for computer vision, In *Proceedings of the IEEE conference on computer vision and pattern recognition,* (pp. 2818–2826). IEEE. 10.1109/CVPR.2016.308

Vijay, K. & Vani, K. (2018). Agricultural Robot: Leaf Disease Detection and Monitoring the Field Condition Using Machine Learning and Image Processing. *International Journal of Computational Intelligence Research, 14*(7), pp. 551-561

Wilson, D. R., & Martinez, & T. R. (2001). The need for small learning rates on large problems. IJCNN'01. *International Joint Conference on Neural Net- works. Proceedings*. IEEE. 10.1109/IJCNN.2001.939002

Yamashita, R., Nishio, M., Do, R. K. G., & Togashi, K. (2018). Convolutional neural networks: An overview and application in radiology. *Insights Into Imaging*, *9*(4), 611–629. doi:10.100713244-018-0639-9 PMID:29934920

Yialouris, A. B., & Sideridis, A. B. (1996). An expert system for tomato diseases. *Computers and Electronics in Agriculture*, *14*(1), 61–76. doi:10.1016/0168-1699(95)00037-2

Compilation of References

Aasen, H., & Bolten, A. (2018). Multi-temporal high-resolution imaging spectroscopy with hyperspectral 2D imagers – From theory to applicaton. *Remote Sensing of Environment, 205*, 374–389. doi:10.1016/j.rse.2017.10.043

Aasen, H., Honkavaara, E., Lucieer, A., & Zarco-Tejada, P. (2018). Quantitative remote sensing at ultra-high resolution with UAV spectroscopy: A review of sensor technology, measurement procedures, and data correction workflows. *Remote Sensing (Basel), 10*(7), 1091. doi:10.3390/rs10071091

Abadi, M., Barham, P., Chen, J., Chen, Z., Davis, A., Dean, J., Devin, M., Ghemawat, S., Irving, G., Isard, M., Kudlur, M., Levenberg, J., Monga, R., Moore, S., Murray, D. G., Steiner, B., Tucker, P., Vasudevan, V., Warden, P., & Zheng, X. (2016). TensorFlow: A system for large-scale machine learning. In *Proceedings of the 12th USENIX Symposium on Operating Systems Design and Implementation (OSDI 16) Savannah*, GA, USA. 10.48550/arXiv.1605.08695

Abbasi, R., Martinez, P., & Ahmad, R. (2022). The digitization of agricultural industry – a systematic literature review on agriculture 4.0. *Smart Agricultural Technology, 2*, 100042. doi:10.1016/j.atech.2022.100042

Abbaszadeh, P., Gavahi, K., Alipour, A., Deb, P., & Moradkhani, H. (2022). Bayesian multi-modeling of deep neural nets for probabilistic crop yield prediction. *Agricultural and Forest Meteorology, 314*, 108773. doi:10.1016/j.agrformet.2021.108773

Adarsh, P., Rathi, P., & Kumar, M. (2020). YOLO v3-Tiny: Object Detection and Recognition using one stage improved model. *2020 6th International Conference on Advanced Computing and Communication Systems (ICACCS)* (pp. 687–694). IEEE. 10.1109/ICACCS48705.2020.9074315

Adeux, G., Vieren, E., Carlesi, S., Bàrberi, P., Munier-Jolain, N., & Cordeau, S. (2019). Mitigating crop yield losses through weed diversity. *Nature Sustainability, 2*(11), 1018–1026. doi:10.103841893-019-0415-y

Aditya Shastry, K., Sanjay, H. A., & Sajini, M. C. (2022). Decision Tree Based Crop Yield Prediction Using Agro-climatic Parameters. *Lecture Notes in Electrical Engineering, 789*, 87–94. doi:10.1007/978-981-16-1338-8_8

Aditya, S., Natu., K., & S., C. (2016). Adoption and Utilization of Drones for Advanced Precision Farming: A Review. *International Journal on Recent and Innovation Trends in Computing and Communication, 4*, 563 - 565.

Agarwal, D., Sweta, & Bachan, P. (2023). Machine learning approach for the classification of wheat grains. *Smart Agricultural Technology, 3*, 100136. doi:10.1016/j.atech.2022.100136

Agarwal, S., & Tarar, S. (2021). A hybrid approach for crop yield prediction using machine learning and deep learning algorithms. *Journal of Physics: Conference Series, 1714*(1), 012012. doi:10.1088/1742-6596/1714/1/012012

Aggarwal, S., & Kumar, A. (2019). A Smart Irrigation System to Automate Irrigation Process Using IOT and Artificial Neural Network. *2nd International Conference on Signal Processing and Communication, ICSPC 2019 - Proceedings*, (pp. 310–314). IEEE. 10.1109/ICSPC46172.2019.8976631

Aghighi, H., Azadbakht, M., Ashourloo, D., Shahrabi, H. S., & Radiom, S. (2018, December). Machine learning regression techniques for the silage maize yield prediction using time-series images of landsat 8 OLI. *IEEE Journal of Selected Topics in Applied Earth Observations and Remote Sensing*, *11*(12), 4563–4577. doi:10.1109/JSTARS.2018.2823361

Agrawal, N., Verma, K., & Kumar, T. (2021). Soil Fertility Assessment using Web-based Spatial Decision Support System. *2021 IEEE International Conference on Emerging Trends in Industry 4.0, ETI 4.0 2021*. IEEE. 10.1109/ETI4.051663.2021.9619240

Ahmad, A., Saraswat, D., & El Gamal, A. (2023). A survey on using deep learning techniques for plant disease diagnosis and recommendations for development of appropriate tools. *Smart Agricultural Technology*, *3*, 100083. doi:10.1016/j.atech.2022.100083

Ahmad, I., Yang, Y., Yue, Y., Ye, C., Hassan, M., Cheng, X., Wu, Y., & Zhang, Y. (2022). Deep learning based detector YOLOv5 for identifying insect pests. *Applied Sciences (Basel, Switzerland)*, *12*(19), 10167. doi:10.3390/app121910167

Ahmed, I., & Yadav, P. K. (2022). Plant disease detection using machine learning approaches. *Expert Systems: International Journal of Knowledge Engineering and Neural Networks*, 13136.

Akaike, H. (1974). A new look at the statistical model identification. *IEEE Transactions on Automatic Control*, *19*(6), 716–723. doi:10.1109/TAC.1974.1100705

Al Sadeque, Z., & Bui, F. M. (2020). A deep learning approach to predict weather data using cascaded LSTM network. *2020 IEEE Canadian Conference on Electrical and Computer Engineering (CCECE)*, (pp. 1–5). IEEE. 10.1109/CCECE47787.2020.9255716

Aladhadh, S., Habib, S., Islam, M., Aloraini, M., Aladhadh, M., & Al-Rawashdeh, H. S. (2022). An Efficient Pest Detection Framework with a Medium-Scale Benchmark to Increase the Agricultural Productivity. *Sensors (Basel)*, *22*(24), 1–17. doi:10.339022249749 PMID:36560117

Alebele, Y., Wang, W., Yu, W., Zhang, X., Yao, X., Tian, Y., Zhu, Y., Cao, W., & Cheng, T. (2021). Estimation of crop yield from combined optical and SAR imagery using Gaussian kernel regression. *IEEE Journal of Selected Topics in Applied Earth Observations and Remote Sensing*, *14*, 10520–10534. doi:10.1109/JSTARS.2021.3118707

Alghazzawi, D., Bamasaq, O., Bhatia, S., Kumar, A., Dadheech, P., & Albeshri, A. (2021). Congestion Control in Cognitive IoT-Based WSN Network for Smart Agriculture. *IEEE Access : Practical Innovations, Open Solutions*, *9*, 151401–151420. doi:10.1109/ACCESS.2021.3124791

Ali, A. M., Abouelghar, M., Belal, A. A., Saleh, N., Yones, M., Selim, A. I., Amin, M. E. S., Elwesemy, A., Kucher, D. E., Maginan, S., & Savin, I. (2022). Crop yield prediction using multi sensors remote sensing (Review Article). *Egyptian Journal of Remote Sensing and Space Science*, *25*(3), 711–716. doi:10.1016/j.ejrs.2022.04.006

Ali, H., Lali, M. I., Nawaz, M. Z., Sharif, M., & Saleem, B. A. (2017). Symptom based automated detection of citrus diseases using color histogram and textural descriptors. *Computers and Electronics in Agriculture*, *138*, 92–104. doi:10.1016/j.compag.2017.04.008

Aliparo, J., Santos, C. D., Sese, C. M., Ver Trinidad, J., Beano, M. G., Capuno, M. E. A., Medina, O., Sigue, A. L., & Soriano, A. (2022). IoT-based Assessment and Monitoring of NPK Content and Fertility Condition of Soil. *IEEE Region 10 Annual International Conference, Proceedings/TENCON, 2022-Novem*. IEEE. 10.1109/TENCON55691.2022.9978040

Alli, A. A., Kassim, K., Mutwalibi, N., Hamid, H., & Ibrahim, L. (2021). Secure Fog-Cloud of Things: Architectures, Opportunities and Challenges. In M. Ahmed & P. Haskell-Dowland (Eds.), *Secure Edge Computing* (1st ed., pp. 3–20). CRC Press. doi:10.1201/9781003028635-2

Al-Naji, A., Fakhri, A. B., Gharghan, S. K., & Chahl, J. (2021). Soil color analysis based on a RGB camera and an artificial neural network towards smart irrigation: A pilot study. *Heliyon, 7*(1), e06078. doi:10.1016/j.heliyon.2021.e06078 PMID:33537493

Al-Sadi, A. (2017). Impact of Plant Diseases on Human Health. *International Journal of Nutrition, Pharmacology, Neurological Diseases, 7*(2), 21–22. doi:10.4103/ijnpnd.ijnpnd_24_17

Alsamhi, S. H., Afghah, F., Sahal, R., Hawbani, A., Al-qaness, M. A. A., Lee, B., & Guizani, M. (2021). Green internet of things using UAVs in B5G networks: A review of applications and strategies. *Ad Hoc Networks, 117*, 102505. doi:10.1016/j.adhoc.2021.102505

Alston, J. M., & Pardey, P. G. (2014). Agriculture in the global economy. *The Journal of Economic Perspectives, 28*(1), 121–146. doi:10.1257/jep.28.1.121

Altalak, M., Ammad uddin, M., Alajmi, A., & Rizg, A. (2022). Smart Agriculture Applications Using Deep Learning Technologies: A Survey. *Applied Sciences (Basel, Switzerland), 12*(12), 5919. doi:10.3390/app12125919

Al-Tarawneh, L., Mehyar, A., Alasasaf, S. E., & Al-Mariat, M. (2022). Environmental Tracking System using IoT Based WSN: Smart Agriculture. *2022 4th IEEE Middle East and North Africa COMMunications Conference, MENACOMM 2022*, (pp. 147–152). IEEE. 10.1109/MENACOMM57252.2022.9998269

Anand, R., Sethi, D., Sharma, K., & Gambhir, P. (2019). Soil Moisture and Atmosphere Components Detection System Using IoT and Machine Learning. *Proceedings of the 2nd International Conference on Smart Systems and Inventive Technology, ICSSIT 2019*, (pp. 842–847). IEEE. 10.1109/ICSSIT46314.2019.8987754

Anusha, N., Chaithanya, M. S., & Reddy, G. J. (2019). Weather prediction using multi linear regression algorithm. *IOP Conference Series. Materials Science and Engineering, 590*(1), 012034. doi:10.1088/1757-899X/590/1/012034

Arjaria, S., Sahu, R., Agrawal, S., Khare, S., Agarwal, Y., & Chaubey, G. (2021). Hand Gesture Identification System Using Convolutional Neural Networks. In *Proceedings of the International Conference on Artificial Intelligence and Machine Vision (AIMV)*, Gandhinagar, India. 10.1109/AIMV53313.2021.9670906

Arsov, M., Zdravevski, E., Lameski, P., Corizzo, R., Koteli, N., Gramatikov, S., Mitreski, K., & Trajkovik, V. (2021, February 10). Multi-Horizon Air Pollution Forecasting with Deep Neural Networks. *Sensors (Basel), 21*(4), 1235. doi:10.339021041235 PMID:33578633

Aruna Devi, M., Suresh, D., Jeyakumar, D., Swamydoss, D., & Lilly Florence, M. (2022). Agriculture Crop Selection and Yield Prediction using Machine Learning Algorithms. *Proceedings of the 2nd International Conference on Artificial Intelligence and Smart Energy, ICAIS 2022*, (pp. 510–517). IEEE. 10.1109/ICAIS53314.2022.9742846

Ashwinkumar, S., Rajagopal, S., Manimaran, V., & Jegajothi, B. (2022). Automated plant leaf disease detection and classification using optimal MobileNet based convolutional neural networks. *Materials Today: Proceedings, 51*, 480–487. doi:10.1016/j.matpr.2021.05.584

Ault, R. (2020). *Optimization Study of an Image Classification Deep Neural Network.*

Ayaz, M., Ammad-Uddin, M., Sharif, Z., Mansour, A., & Aggoune, E. H. M. (2019). Internet-of-Things (IoT)-based smart agriculture: Toward making the fields talk. *IEEE Access : Practical Innovations, Open Solutions, 7*, 129551–129583. doi:10.1109/ACCESS.2019.2932609

Ayoub Shaikh, T., Rasool, T., & Rasheed Lone, F. (2022). Towards leveraging the role of machine learning and artificial intelligence in precision agriculture and smart farming. *Computers and Electronics in Agriculture, 198*, 107119. doi:10.1016/j.compag.2022.107119

Azfar Firdaus Azlah, M., Suan Chua, L., Razan Rahmad, F., Izana Abdullah, F., & Rafidah Wan Alwi, S. (2019). Review on Techniques for Plant Leaf Classification and Recognition. *Computers*, *8*(77), 77. doi:10.3390/computers8040077

Babu, B. S., Kamalakannan, J., Meenatchi, N., M, S. K. S., S, K., & Boopathi, S. (2023). Economic impacts and reliability evaluation of battery by adopting Electric Vehicle. *IEEE Explore*, (pp. 1–6). IEEE. doi:10.1109/ICPECTS56089.2022.10046786

Balakrishna, K., Mohammed, F., Ullas, C. R., Hema, C. M., & Sonakshi, S. K. (2021). Application of IoT and machine learning in crop protection against animal intrusion, (pp. 221–238). Machine Learning Techniques for Sustainable Agriculture. doi:10.4018/978-1-7998-7488-6.ch010

Barbedo, J. G. (2018). Factors influencing the use of deep learning for plant disease recognition. *Biosystems Engineering*, *172*, 84–91. doi:10.1016/j.biosystemseng.2018.05.013

Barbedo, J. G. A. (2019). A review on the use of unmanned aerial vehicles and imaging sensors for monitoring and assessing plant stresses. *Drones (Basel)*, *3*(2), 40. doi:10.3390/drones3020040

Bashish, D. A. (2010). A Fremework for Detection and Classification of Plant Leaf and Stem Diseases. *International Conference on Signal and Image Processing*, (pp. 113-118). Semantic Scholar.

Belavadi, S. V., Rajagopal, S. R. R., & Mohan, R. (2020). Air Quality Forecasting using LSTM RNN and Wireless Sensor Networks. *Procedia Computer Science*, *170*, 241–248. doi:10.1016/j.procs.2020.03.036

Bendig, J., Bolten, A., & Bareth, G. (2012). Introducing a low-cost mini-UAV for thermal-and multispectral-imaging. *The International Archives of the Photogrammetry, Remote Sensing and Spatial Information Sciences*, *39*(B1), 345–349. doi:10.5194/isprsarchives-XXXIX-B1-345-2012

Bhange, M. (2015), 'Smart Farming: Pomegranate Disease Detection Using Image Processing', *Second International Symposium on Computer Vision and the Internet*, Volume 58, (pp. 280-288). IEEE. 10.1016/j.procs.2015.08.022

Bhansali, S., Shah, P., Shah, J., Vyas, P., & Thakre, P. (2022). Healthy harvest: Crop prediction and disease detection system. *7th International conference for Convergence in Technology (I2CT),* (pp. 1–5). IEEE Publications. 10.1109/I2CT54291.2022.9825446

Bini, D., Pamela, D., & Prince, S. (2020). Machine Vision and Machine Learning for Intelligent Agrobots: A review. *ICDCS 2020 - 2020 5th International Conference on Devices, Circuits and Systems*, (pp. 12–16). IEEE.

Birrell, S. J., Sudduth, K. A., & Borgelt, S. C. (1996). Comparison of sensors and techniques for crop yield mapping. *Computers and Electronics in Agriculture*, *14*(2–3), 215–233. doi:10.1016/0168-1699(95)00049-6

Bodapati, N., Himavaishnavi, J., Rohitha, V., Jagadeeswari, D. L., & Bhavana, P. (2022). Analyzing crop yield using machine learning. *IEEE International Conference on Electronics and Renewable Systems (ICEARS),* (pp. 1–8). IEEE. 10.1109/ICEARS53579.2022.9752242

Boechel, T., Policarpo, L. M., Ramos, G. de O., da Rosa Righi, R., & Singh, D. (2022). Prediction of Harvest Time of Apple Trees: An RNN-Based Approach. *Algorithms 2022, 15*(3), 95.

Bojanowski, J. S., Sikora, S., Musiał, J. P., Woźniak, E., Dąbrowska-Zielińska, K., Slesiński, P., Milewski, T., & Łączyński, A. (2022). Integration of Sentinel-3 and MODIS Vegetation Indices with ERA-5 Agro-Meteorological Indicators for Operational Crop Yield Forecasting. *Remote Sensing, 14*(5), 1238.

Boopathi, S., Arigela, S. H., Raman, R., Indhumathi, C., Kavitha, V., & Bhatt, B. C. (2023). Prominent Rule Control-based Internet of Things: Poultry Farm Management System. *IEEE Explore*, (pp. 1–6). IEEE. doi:10.1109/ICPECTS56089.2022.10047039

Boopathi, S., Siva Kumar, P. K., & Meena, R. S. J., S. I., P., S. K., & Sudhakar, M. (2023). Sustainable Developments of Modern Soil-Less Agro-Cultivation Systems. In Human Agro-Energy Optimization for Business and Industry (pp. 69–87). IGI Global. doi:10.4018/978-1-6684-4118-3.ch004

Boopathi, S. (2022a). An Extensive Review on Sustainable Developments of Dry and Near-Dry Electrical Discharge Machining Processes. *Journal of Manufacturing Science and Engineering*, *144*(5), 50801. doi:10.1115/1.4052527

Boopathi, S. (2022b). Cryogenically treated and untreated stainless steel grade 317 in sustainable wire electrical discharge machining process: A comparative study. *Environmental Science and Pollution Research International*, 1–10. doi:10.100711356-022-22843-x PMID:36057706

Boopathi, S. (2022c). Experimental investigation and multi-objective optimization of cryogenic Friction-stir-welding of AA2014 and AZ31B alloys using MOORA technique. *Materials Today. Communications*, *33*, 104937. doi:10.1016/j.mtcomm.2022.104937

Boopathi, S., Lewise, K. A. S., Subbiah, R., & Sivaraman, G. (2021). Near-dry wire-cut electrical discharge machining process using water-air-mist dielectric fluid: An experimental study. *Materials Today: Proceedings*, *49*(5), 1885–1890. doi:10.1016/j.matpr.2021.08.077

Boopathi, S., Thillaivanan, A., Pandian, M., Subbiah, R., & Shanmugam, P. (2021). Friction stir processing of boron carbide reinforced aluminium surface (Al-B4C) composite: Mechanical characteristics analysis. *Materials Today: Proceedings*, *50*(5), 2430–2435. doi:10.1016/j.matpr.2021.10.261

Bose, P., Kasabov, N. K., Bruzzone, L., & Hartono, R. N. (2016, November). Spiking neural networks for crop yield estimation based on spatiotemporal analysis of image time series. *IEEE Transactions on Geoscience and Remote Sensing*, *54*(11), 6563–6573. doi:10.1109/TGRS.2016.2586602

Boulent, J., Ge, Y., & Zhu, H. (2019). Deep Learning Techniques for Crop Classification: A Review. *Computers and Electronics in Agriculture*, *162*, 219–236. doi:10.1016/j.compag.2019.04.014

Box, G. E., Jenkins, G. M., & Reinsel, G. (1970). Time series analysis: forecasting and control Holden-day San Francisco. *BoxTime Series Analysis: Forecasting and Control Holden Day1970.*

Božnar, M. Z., Mlakar, P., Grašič, B., Gršić, Z., Hettrich, S., Mancini, F., Patryl, L., & Thiessen, K. M. (2022, June 1). Modelling air pollution around nuclear power plants: Validation of dispersion models using tracer data. *Journal of Radiological Protection*, *42*(2), 020519. doi:10.1088/1361-6498/ac7a6f PMID:35775740

Burema, H., & Filin, A. (2016). *Aerial Farm Robot System for Crop Dusting, Planting, Fertilizing and Other Field Jobs.* United States Patent US9382003B2.

Burugari, V. K., Selvaraj, P., & Palaniappan, K. (2021). Smart agriculture using WSN and IoT. In *Artificial Intelligence and IoT-Based Technologies for Sustainable Farming and Smart Agriculture* (pp. 192–212). IGI Global. doi:10.4018/978-1-7998-1722-2.ch012

Cai, G., Chen, B. M., & Lee, T. H. (2010). An overview on development of miniature unmanned rotorcraft systems. *Frontiers of Electrical and Electronic Engineering in China*, *5*(1), 1–14. doi:10.100711460-009-0065-3

Cedric, L. S., Adoni, W. Y. H., Aworka, R., Zoueu, J. T., Mutombo, F. K., Krichen, M., & Kimpolo, C. L. M. (2022). *Crops yield prediction based on machine learning models: Case of West African countries*. Smart Agricultural Technology.

Chakraborty, S., Paul, S., & Rahat-Uz-Zaman, M. (2021). Prediction of apple leaf diseases using multiclass support vector machine. *International Conference on Robotics, Electrical and Signal Processing Techniques*, (pp. 147–151). IEEE. 10.1109/ICREST51555.2021.9331132

Chandraprabha, M., & Dhanaraj, R. K. (2021). Soil based prediction for crop yield using predictive analytics. *3rd IEEE International Conference on Advances in Computing, Communication Control. and Networking (ICAC3N),* (pp. 265–270). IEEE. 10.1109/ICAC3N53548.2021.9725758

Chauhan, B. S. (2020). Grand challenges in weed management. *Frontiers in Agronomy*, *1*, 3. doi:10.3389/fagro.2019.00003

Chen, P., Niu, A., Liu, D., Jiang, W., & Ma, B. (2018). Time series forecasting of temperatures using SARIMA: An example from Nanjing. *IOP Conference Series. Materials Science and Engineering*, *394*, 052024. doi:10.1088/1757-899X/394/5/052024

Chen, Y., Cui, S., Chen, P., Yuan, Q., Kang, P., & Zhu, L. (2021, March 11). An LSTM-based neural network method of particulate pollution forecast in China. *Environmental Research Letters*, *16*(4), 044006. doi:10.1088/1748-9326/abe1f5

Cheruvu, B., Latha, S. B., Nikhil, M., Mahajan, H., & Prashanth, K. (2023). Smart Farming System using NPK Sensor. *9th International Conference on Advanced Computing and Communication Systems (ICACCS)*, Coimbatore, India. 10.1109/ICACCS57279.2023.10112795

Cheung, Y.-W., & Lai, K. S. (1995). Lag order and critical values of the augmented Dickey–Fuller test. *Journal of Business & Economic Statistics*, *13*(3), 277–280. doi:10.1080/07350015.1995.10524601

Chiranjeevi, & Nadagoudar, R. B. (2018). Analysis of Soil Nutrients using Data Mining Techniques. International Journal of Recent Trends in Engineering and Research, 4(7), 103–107. doi:10.23883/IJRTER.2018.4363.PDT1C

Chlingaryan, A., Sukkarieh, S., & Whelan, B. (2018). Machine learning approaches for crop yield prediction and nitrogen status estimation in precision agriculture: A review. *Computers and Electronics in Agriculture*, *151*, 61–69. doi:10.1016/j.compag.2018.05.012

Cogato, A., Meggio, F., De Antoni Migliorati, M., & Marinello, F. (2019). Extreme weather events in agriculture: A systematic review. *Sustainability (Basel)*, *11*(9), 2547. doi:10.3390u11092547

Colomina, I., & Molina, P. (2014). Unmanned aerial systems for photogrammetry and remote sensing: A review. *ISPRS Journal of Photogrammetry and Remote Sensing*, *92*, 79–97. doi:10.1016/j.isprsjprs.2014.02.013

Cordova, C. H., Portocarrero, M. N. L., Salas, R., Torres, R., Rodrigues, P. C., & López-Gonzales, J. L. (2021, December 20). Air quality assessment and pollution forecasting using artificial neural networks in Metropolitan Lima-Peru. *Scientific Reports*, *11*(1), 24232. doi:10.103841598-021-03650-9 PMID:34930975

Cunha, R. L. F., & Silva, B. (2020). Estimating crop yields with remote sensing and deep learning. IEEE Latin American GRSS y ISPRS Remote Sensing Conference (LAGIRS), (pp. 273–278). IEEE.

Cunha, R. L. F., Silva, B., & Netto, M. A. S. (2018). A scalable machine learning system for pre-season agriculture yield forecast. *14th International Conference on e-Science (e-Science),* (pp. 423–430). IEEE Publications. 10.1109/eScience.2018.00131

Dalai, R., & Senapati, K. K. (2019). An Intelligent Vision based Pest Detection System Using RCNN based Deep Learning Mechanism. *2019 International Conference on Recent Advances in Energy-Efficient Computing and Communication, ICRAECC 2019*, (pp. 1-4). IEEE. 10.1109/ICRAECC43874.2019.8995072

Darmawan, H., Yuliana, M., & Hadi, M. Z. (2022). Realtime Weather Prediction System Using GRU with Daily Surface Observation Data from IoT Sensors. *2022 International Electronics Symposium (IES)*, (pp. 221–226). IEEE. 10.1109/IES55876.2022.9888468

Das, P. K., Meher, S., Panda, R., & Abraham, A. (2019). A review of automated methods for the detection of sickle cell disease. *IEEE Reviews in Biomedical Engineering*, *13*, 309–324. doi:10.1109/RBME.2019.2917780 PMID:31107662

David, R., Duke, J., Jain, A., Reddi, V. J., Jeffries, N., Li, J., Kreeger, N., Nappier, I., Natraj, M., Regev, S., Rhodes, R., Wang, T., & Warden, P. (n.d.). Tensorflow lite micro: Embedded machine learning on TinyML systems. https://doi.org// arXiv.2010.08678 doi:10.48550

De Castro, A. I., Peña, J. M., Torres-Sánchez, J., Jiménez-Brenes, F., & López-Granados, F. (2017). mapping Cynodon dactylon in vineyards using UAV images for site-specifc weed control. *Advances in Animal Biosciences*, *8*(2), 267–271. doi:10.1017/S2040470017000826

De Silva, M., & Brown, D. (2022, August). Plant Disease Detection using Deep Learning on Natural Environment Images. In *2022 International Conference on Artificial Intelligence, Big Data, Computing and Data Communication Systems (icABCD)* (pp. 1-5). IEEE. 10.1109/icABCD54961.2022.9855925

Deepika, P., & Arthi, B. (2022). Prediction of plant pest detection using improved mask FRCNN in cloud environment. *Measurement: Sensors*, *24*, 100549. doi:10.1016/j.measen.2022.100549

Deng, F., Mao, W., Zeng, Z., Zeng, H., & Wei, B. (2022). Multiple Diseases and Pests Detection Based on Federated Learning and Improved Faster R-CNN. *IEEE Transactions on Instrumentation and Measurement*, *71*, 1–11. doi:10.1109/ TIM.2022.3201937

Deng, L., Wang, Y., Han, Z., & Yu, R. (2018). Research on insect pest image detection and recognition based on bio-inspired methods. *Biosystems Engineering*, *169*, 139–148. doi:10.1016/j.biosystemseng.2018.02.008

Department of Agriculture & Farmers Welfare (DA&CFW), Ministry of Agriculture and Farmers Welfare, Government of India. (n.d.). *Farmer's Portal: One stop shop for Farmer. Download Mobile Application (DMA).* DAFW. https:// farmer.gov.in/mobileappsdownload.aspx

Department of Agriculture & Farmers Welfare. (n.d.). *MoA&FW*. Agricoop. https://agridashboard.dac.gov.in/

Devi, K. G., Kumar, C. S., & Kishore, B. (2022). A Survey on the Design of Autonomous and Semi Autonomous Pesticide Sprayer Robot. *El-Cezeri*, *9*(1), 371–381.

Devi, V. B., Prabavathi, R., Subha, P., & Meenaloshini, M. (2022). An Efficient and Robust Random Forest Algorithm for Crop Disease Detection. *2022 International Conference on Communication, Computing and Internet of Things, IC3IoT 2022 - Proceedings*. IEEE. 10.1109/IC3IOT53935.2022.9767937

Dhaliwal, G. S., Jindal, V., & Mohindru, B. (2015). Crop losses due to insect pests: Global and Indian scenario. *Indian Journal of Entomology*, *77*(2), 165–168. doi:10.5958/0974-8172.2015.00033.4

Diaz-Gonzalez, F. A., Vuelvas, J., Correa, C. A., Vallejo, V. E., & Patino, D. (2022). Machine learning and remote sensing techniques applied to estimate soil indicators – review. *Ecological Indicators*, *135*, 108517. doi:10.1016/j. ecolind.2021.108517

Directorate of Economics and Statistics, Ministry of Agriculture and Farmers Welfare. (n.d.). Area and Production Statistics (APS). https://aps.dac.gov.in/Home.aspx

Divakar, M. S., Elayidom, M. S., & Rajesh, R. (2022). Forecasting crop yield with deep learning based ensemble model. *Materials Today: Proceedings*, *58*, 256–259. doi:10.1016/j.matpr.2022.02.109

Domakonda, V. K., Farooq, S., Chinthamreddy, S., Puviarasi, R., Sudhakar, M., & Boopathi, S. (2023). Sustainable Developments of Hybrid Floating Solar Power Plants. In *Human Agro-Energy Optimization for Business and Industry* (pp. 148–167). IGI Global. doi:10.4018/978-1-6684-4118-3.ch008

Domingues, T., Brandão, T., & Ferreira, J. C. (2022). Machine Learning for Detection and Prediction of Crop Diseases and Pests: A Comprehensive Survey. *Agriculture*, *12*(9), 1–23. doi:10.3390/agriculture12091350

Duan, T., Chapman, S. C., Guo, Y., & Zheng, B. (2017). Dynamic monitoring of NDVI in wheat agronomy and breeding trials using an unmanned aerial vehicle. *Field Crops Research*, *210*, 71–80. doi:10.1016/j.fcr.2017.05.025

Dubey, S. R., & Jalal, A. S. (2013). Adapted approach for fruit disease identification using images. In *Image processing: Concepts, methodologies, tools, and applications* (pp. 1395–1409). IGI Global. doi:10.4018/978-1-4666-3994-2.ch069

Edwin, T. (2017) [Online]. Onion, tomato price spike: Season not the only reason. *The Hindu Business Online.* https://www.thehindubusinessline.com/economy/agri-business/onion-tomato-price-spike-season-not-the-only-reason/article9957255.ece#.

Ejohwomu, O. A., Shamsideen Oshodi, O., Oladokun, M., Bukoye, O. T., Emekwuru, N., Sotunbo, A., & Adenuga, O. (2022). Modelling and Forecasting Temporal PM2.5 Concentration Using Ensemble Machine Learning Methods. *Buildings*, *12*(1), 46. doi:10.3390/buildings12010046

El Hoummaidi, L., Larabi, A., & Alam, K. (2021). Using unmanned aerial systems and deep learning for agriculture mapping in Dubai. *Heliyon*, *7*(10), e08154. doi:10.1016/j.heliyon.2021.e08154 PMID:34703924

Elavarasan, D., & Vincent, P. M. D. (2020). Crop yield prediction using deep reinforcement learning model for sustainable agrarian applications. *IEEE Access : Practical Innovations, Open Solutions*, *8*, 86886–86901. doi:10.1109/ACCESS.2020.2992480

Ercolini, L., Grossi, N., & Silvestri, N. (2022). A Simple Method to Estimate Weed Control Threshold by Using RGB Images from Drones. *Applied Sciences 2022, Vol. 12, Page 11935, 12*(23), 11935.

Eswaran, H., Ahrens, R., & Rice, T. J. S. B. A. (2002). Soil classification: a global desk reference. CRC Press. doi:10.1201/9781420040364

Eunice, J., Popescu, D. E., Chowdary, M. K., & Hemanth, J. (2022). Deep Learning-Based Leaf Disease Detection in Crops Using Images for Agricultural Applications. *Agronomy (Basel)*, *12*(10), 2395. doi:10.3390/agronomy12102395

Fawakherji, M., Potena, C., Bloisi, D. D., Imperoli, M., Pretto, A., & Nardi, D. (2019). UAV image based crop and weed distribution estimation on embedded GPU boards. *Communications in Computer and Information Science*, *1089*, 100–108. doi:10.1007/978-3-030-29930-9_10

Food and Agriculture Organization of the United Nations (Ed.). (2017). *The future of food and agriculture: trends and challenges*. Food and Agriculture Organization of the United Nation.

Fountas, S., Anastasiou, K., & Katsoulas, N. (2020). Forecasting Crop Yield Using Deep Learning: A Systematic Review. *Agronomy (Basel)*, *10*(4), 571. doi:10.3390/agronomy10040571

Franke, G., & Menz, G. (2007). Multi-temporal wheat disease detection by multi-spectral remote sensing. *Precision Agriculture*, *8*(3), 161–172. doi:10.100711119-007-9036-y

Franzen, D. W., Casey, F. X. M., & Derby, N. (2018, May). Yield Mapping and Use of Yield Map Data. *Ndsu Extension.* https://extensionpublications.unl.edu/assets/pdf/ec2004.pdf

Fu, J., Li, P., & Wang, Y., Phu Do, X., Bok Choi -, S., Sun, F., Ma, W., Li, H., & Wang, S. (2018). Research on water-fertilizer integrated technology based on neural network prediction and fuzzy control. *IOP Conference Series. Earth and Environmental Science*, *170*(3), 032168.

Garbin, C., Zhu, X., & Marques, O. (2020). Dropout vs. batch normalization: An empirical study of their impact to deep learning. *Multimedia Tools and Applications*, *79*(19), 12777–12815. doi:10.100711042-019-08453-9

García, L., Parra, L., Jimenez, J. M., Lloret, J., & Lorenz, P. (2020). IoT-Based Smart Irrigation Systems: An Overview on the Recent Trends on Sensors and IoT Systems for Irrigation in Precision Agriculture. *Sensors (Basel)*, *20*(4), 1042. doi:10.339020041042 PMID:32075172

Garcia-Ruiz, F. J., Wulfsohn, D., & Rasmussen, J. (2015). Sugar beet (Beta vulgaris L.) and thistle (Cirsium arvensis L.) discrimination based on feld spectral data. *Biosystems Engineering*, *139*, 1–15. doi:10.1016/j.biosystemseng.2015.07.012

Garg, A., & Rani, A. (2020). Agricultural Monitoring and Management using Deep Learning: A Review. *Computers and Electronics in Agriculture*, *173*, 105374. doi:10.1016/j.compag.2020.105374

Gautam, P. V., Kushwaha, H. L., Kumar, A., & Kumar, D. (2019). Mechatronics application in precision sowing: A review. *International Journal of Current Microbiology and Applied Sciences*, *8*(4), 1793–1807. doi:10.20546/ijcmas.2019.804.208

Gavhale, M., & Gawande, U. (2014). An Overview of the Research on Plant Leaves Disease detection using Image Processing Techniques. *IOSR Journal of Computer Engineering*, *16*(1), 10–16. doi:10.9790/0661-16151016

Geetha, M. C. S. (2015). A Survey on Data Mining Techniques in Agriculture. *International Journal of Innovative Research in Computer and Communication Engineering.*, *3*(2), 887–892.

Geetha, V., Punitha, A., Abarna, M., Akshaya, M., Illakiya, S., & Janani, A. P. (2020). An Effective Crop Prediction Using Random Forest Algorithm. *2020 International Conference on System, Computation, Automation and Networking, ICSCAN 2020*. IEEE. 10.1109/ICSCAN49426.2020.9262311

Ghodeswar, U., Balagotra, T., Borale, A., Pawar, K., Parate, T., & Mahajan, S. (2022). Greeniot: A Smart IoT Gateway for Connected Farming. *IEEE 7th International conference for Convergence in Technology (I2CT)*. IEEE. 10.1109/I2CT54291.2022.9824907

Giordano, S., Seitanidis, I., Ojo, M., Adami, D., & Vignoli, F. (2018). IoT solutions for crop protection against wild animal attacks. In *2018 IEEE International Conference on Environmental Engineering (EE)* (pp. 1-5). IEEE. 10.1109/EE1.2018.8385275

Girshick, R., Donahue, J., Darrell, T., & Malik, J. (2014). Rich feature hierarchies for accurate object detection and semantic segmentation. *Proceedings of the IEEE conference on computer vision and pattern recognition*, (pp. 580–587). IEEE. 10.1109/CVPR.2014.81

Giua, C., Materia, V. C., & Camanzi, L. (2022). Smart farming technologies adoption: Which factors play a role in the digital transition? *Technology in Society*, *68*, 101869. doi:10.1016/j.techsoc.2022.101869

Go, S. H., Lee, D. H., Na, S. il, & Park, J. H. (2022). Analysis of Growth Characteristics of Kimchi Cabbage Using Drone-Based Cabbage Surface Model Image. *Agriculture, 12*(2), 216.

Goel, L., & Mishra, A. (2022). A survey of recent deep learning algorithms used in Smart farming. *IEEE Region 10 Symposium (TENSYMP),* (pp. 1–6). IEEE. 10.1109/TENSYMP54529.2022.9864477

Goel, L., & Nagpal, J. (2022). A Systematic Review of Recent Machine Learning Techniques for Plant Disease Identification and Classification. *IETE Technical Review.*

Guo, Q., He, Z., Li, S., Li, X., Meng, J., Hou, Z., Liu, J., & Chen, Y. (2020). Air Pollution Forecasting Using Artificial and Wavelet Neural Networks with Meteorological Conditions. *Aerosol and Air Quality Research*, *20*(6), 1429–1439. doi:10.4209/aaqr.2020.03.0097

Gupta, H. P., Chopade, S., & Dutta, T. (2022). Computational Intelligence in Agriculture. *Emerging Computing Paradigms*, 125–142.

Gupta, M. V, S. K. B., B, K., Narapureddy, H. R., Surapaneni, N., & Varma, K. (2022). Various crop yield prediction techniques using machine learning algorithms. *IEEE Second International Conference on Artificial Intelligence and Smart Energy (ICAIS)*, (pp. 273–279). IEEE. 10.1109/ICAIS53314.2022.9742903

Gupta, R., Sharma, A. K., Garg, O., Modi, K., Kasim, S., Baharum, Z., Mahdin, H., & Mostafa, S. A. (2021). WB-CPI: Weather based crop prediction in India using big data analytics. *IEEE Access : Practical Innovations, Open Solutions*, *9*, 137869–137885. doi:10.1109/ACCESS.2021.3117247

Hafeez, A., Husain, M. A., Singh, S. P., Chauhan, A., Khan, M. T., Kumar, N., Chauhan, A., & Soni, S. K. (2022). Implementation of drone technology for farm monitoring & pesticide spraying: A review. *Information Processing in Agriculture*.

Hamadani, H., Rashid, S. M., Parrah, J. D., Khan, A. A., Dar, K. A., Ganie, A. A., Gazal, A., Dar, R. A., & Ali, A. (2021). Traditional Farming Practices and Its Consequences. *Microbiota and Biofertilizers*, *2*, 119–128. doi:10.1007/978-3-030-61010-4_6

Han, Y., Mi, L., Shen, L., Cai, C. S., Liu, Y., Li, K., & Xu, G. (2022). A short-term wind speed prediction method utilizing novel hybrid deep learning algorithms to correct numerical weather forecasting. *Applied Energy*, *312*, 118777. doi:10.1016/j.apenergy.2022.118777

Haq, K. R., & Harigovindan, V. P. (2022). Water quality prediction for smart aquaculture using hybrid deep learning models. *IEEE Access : Practical Innovations, Open Solutions*, *10*, 60078–60098. doi:10.1109/ACCESS.2022.3180482

Haq, M. A. (2021). SMOTEDNN: A Novel Model for Air Pollution Forecasting and AQI Classification. *Computers, Materials & Continua*, *71*(1), 1403–1425. doi:10.32604/cmc.2022.021968

Haque, F. F., Abdelgawad, A., Yanambaka, V. P., & Yelamarthi, K. (2020). Crop yield prediction using deep neural network. *6th World Forum on Internet of Things (WF-IoT)*, (pp. 1–4). IEEE Publications. 10.1109/WF-IoT48130.2020.9221298

Hara, P., Piekutowska, M., & Niedbała, G. (2021). Selection of Independent Variables for Crop Yield Prediction Using Artificial Neural Network Models with Remote Sensing Data. *Land 2021, 10*(6), 609.

Haribalaji, V., Boopathi, S., & Asif, M. M. (2021). Optimization of friction stir welding process to join dissimilar AA2014 and AA7075 aluminum alloys. *Materials Today: Proceedings*, *50*, 2227–2234. doi:10.1016/j.matpr.2021.09.499

Harikaran, M., Boopathi, S., Gokulakannan, S., & Poonguzhali, M. (2023). Study on the Source of E-Waste Management and Disposal Methods. In *Sustainable Approaches and Strategies for E-Waste Management and Utilization* (pp. 39–60). IGI Global. doi:10.4018/978-1-6684-7573-7.ch003

Harsha Vardhini, P. A., Koteswaramma, N., & Murali Chandra Babu, K. (2019). IoT-based Raspberry Pi Crop Vandalism Prevention system. [IJITEE]. *International Journal of Innovative Technology and Exploring Engineering*, *9*(1), 3188–3192. doi:10.35940/ijitee.A9156.119119

Harvey, C. A., Saborio-Rodríguez, M., Martinez-Rodríguez, M. R., Viguera, B., Chain-Guadarrama, A., Vignola, R., & Alpizar, F. (2018). Climate change impacts and adaptation among smallholder farmers in Central America. *Agriculture & Food Security*, *7*(1), 1–20. doi:10.118640066-018-0209-x

Hasan, N., Nath, N. C., & Rasel, R. I. (2015). A support vector regression model for forecasting rainfall. *2015 2nd international conference on electrical information and communication technologies (EICT)*, (pp. 554–559). IEEE. 10.1109/EICT.2015.7392014

Haseeb, K., Din, I. U., Almogren, A., & Islam, N. (2020). An energy efficient and secure IoT-based WSN framework: An application to smart agriculture. *Sensors (Basel)*, *20*(7), 2081. doi:10.339020072081 PMID:32272801

Hatcher, W. G., Qian, C., Gao, W., Liang, F., Hua, K., & Yu, W. (2021). Towards Efficient and Intelligent Internet of Things Search Engine. *IEEE Access : Practical Innovations, Open Solutions*, *9*, 15778–15795. doi:10.1109/ACCESS.2021.3052759

He, B., Jia, B., Zhao, Y., Wang, X., Wei, M., & Dietzel, R. (2022). Estimate soil moisture of maize by combining support vector machine and chaotic whale optimization algorithm. *Agricultural Water Management*, *267*, 107618. doi:10.1016/j.agwat.2022.107618

He, K., Zhang, X., Ren, S., & Sun, J. Deep residual learning for image recognition, In *Proceedings of the IEEE conference on computer vision and pattern recognition*, (pp. 770–778). IEEE.

Herwitz, S. R., Johnson, L. F., Dunagan, S. E., Higgins, R. G., Sullivan, D. V., Zheng, J., Slye, R. E. (2004) Imaging from an unmanned aerial vehicle: agricultural surveillance and decision support. *Computers and electronics in agriculture, 44*(1).

Hornberg, A. (2017). Front Matter. Handbook of Machine and Computer Vision.

Hossain, M. D., Kashem, M. A., & Mustary, S. (2023). IoT Based Smart Soil Fertilizer Monitoring and ML Based Crop Recommendation System. *International Conference on Electrical, Computer and Communication Engineering (ECCE)*, (pp. 1-6). IEEE. 10.1109/ECCE57851.2023.10100744

Huang, G., Liu, Z., Van Der Maaten, L., & Weinberger, K. Q. Densely connected convolutional networks, In *Proceedings of the IEEE conference on computer vision and pattern recognition,* (pp. 4700–4708). IEEE.

Huang, H., Huang, S., & Du, Q. (2023). Evaluation of soil heavy metal pollution based on K-means and SVM. *International Journal of Environmental Science and Technology*, *2023*, 1–10. doi:10.100713762-022-04727-2

Huang, J.-R., Sun, J.-Y., Liao, H.-J., & Liu, X.-D. (2015a). Detection of brown planthopper infestation based on SPAD and spectral data from rice under different rates of nitrogen fertilizer. *Precision Agriculture*, *16*(2), 148–163. doi:10.100711119-014-9367-4

Huang, Y., Hoffmann, W. C., Lan, Y., Wu, W., & Fritz, B. K. (2009). Development of a spray system for an unmanned aerial vehicle platform. *Applied Engineering in Agriculture*, *25*(6), 803–809. doi:10.13031/2013.29229

Hughes, D. & Salathé, M. (2015). *An open access repository of images on plant health to enable the development of mobile disease diagnostics*. arXiv, 1–13

Hughes, D. P., & Salathé, M. (2015). An open access repository of images on plant health to enable the development of mobile disease diagnostics through machine learning and crowdsourcing. In ArXiv.

Hunter, J. E. III, Gannon, T. W., Richardson, R. J., Yelverton, F. H., & Leon, R. G. (2020). Integration of remote-weed mapping and an autonomous spraying unmanned aerial vehicle for site-specific weed management. *Pest Management Science*, *76*(4), 1386–1392. doi:10.1002/ps.5651 PMID:31622004

Indrapandi, A., & Rizwana, S. (2021). Smart Agriculture Monitoring System based Data Aggregation Process in IoT and WSN using Multiple Sensor Nodes. *Turkish Online Journal of Qualitative Inquiry, 12*(7).

Iniyaa, K. K., Divya, J. K., Devdharshini, S., & Sangeethapriya, R. (2021). Crop protection from animals using deep learning. *International Journal of Progressive Research in Science and Engineering*, *2*(3), 41–44. https://journals.grdpublications.com/index.php/ijprse/article/view/240

Iniyan, S., Akhil Varma, V., & Teja Naidu, C. (2023). Crop yield prediction using machine learning techniques. *Advances in Engineering Software*, *175*, 103326. doi:10.1016/j.advengsoft.2022.103326

Iost Filho, F. H., Heldens, W. B., Kong, Z., & de Lange, E. S. (2020). Drones: Innovative technology for use in precision pest management. *Journal of Economic Entomology, 113*(1), 1–25. doi:10.1093/jee/toz268 PMID:31811713

Irias Tejeda, A. J., & Castro Castro, R. (2019). Algorithm of weed detection in crops by computational vision. *CONIELECOMP 2019 - 2019 International Conference on Electronics, Communications and Computers*, (pp. 124–128). IEEE. 10.1109/CONIELECOMP.2019.8673182

Isaev, E., Ajikeev, B., Shamyrkanov, U., Kalnur, K., Maisalbek, K., & Sidle, R. C. (2022). Impact of Climate Change and Air Pollution Forecasting Using Machine Learning Techniques in Bishkek. *Aerosol and Air Quality Research, 22*(3), 210336. doi:10.4209/aaqr.210336

Ishimwe, R., Abutaleb, K., & Ahmed, F. (2014). Applications of Thermal Imaging in Agriculture—A Review. *Advances in Remote Sensing, 3*(3), 128–140. doi:10.4236/ars.2014.33011

Islam, M. S., & Dey, G. K. (2019). Precision agriculture: Renewable energy based smart crop field monitoring and management system using WSN via IoT. *2019 International Conference on Sustainable Technologies for Industry 4.0, STI 2019*, (pp. 1–6). IEEE. 10.1109/STI47673.2019.9068017

Ismail, N., & Malik, O. A. (2022). Real-time visual inspection system for grading fruits using computer vision and deep learning techniques. *Information Processing in Agriculture, 9*(1), 24–37. doi:10.1016/j.inpa.2021.01.005

Jackson, R. D. (1984). *Remote Sensing of Vegetation Characteristics for Farm Management*. SPIE. doi:10.1117/12.966243

Jagtap, S. T., Phasinam, K., Kassanuk, T., Jha, S. S., Ghosh, T., & Thakar, C. M. (2022). Towards application of various machine learning techniques in agriculture. *Materials Today: Proceedings, 51*, 793–797. doi:10.1016/j.matpr.2021.06.236

Janardhana, K., Anushkannan, N. K., Dinakaran, K. P., Puse, R. K., & Boopathi, S. (2023). *Experimental Investigation on Microhardness, Surface Roughness, and White Layer Thickness of Dry EDM*. Engineering Research Express. doi:10.1088/2631-8695/acce8f

Jannoura, R., Brinkmann, K., Uteau, D., Bruns, C., & Joergensen, R. G. (2015). Monitoring of crop biomass using true colour aerial photographs taken from a remote controlled hexacopter. *Biosystems Engineering, 129*, 341–351. doi:10.1016/j.biosystemseng.2014.11.007

Javidan, S. M., Banakar, A., Vakilian, K. A., & Ampatzidis, Y. (2023). Diagnosis of grape leaf diseases using automatic K-means clustering and machine learning. *Smart Agricultural Technology, 3*, 100081. doi:10.1016/j.atech.2022.100081

Jean, W. (2009). *Extension plant pathologist, georgia plant disease loss estimates*. CAES. www.caes.uga.edu/publications

Jeevanantham, Y. A., A, S., V, V., J, S. I., Boopathi, S., & Kumar, D. P. (2023). Implementation of Internet-of Things (IoT) in Soil Irrigation System. *IEEE Explore*, 1–5. doi:10.1109/ICPECTS56089.2022.10047185

Jiao, L., Dong, S., Zhang, S., Xie, C., & Wang, H. (2020). AF-RCNN: An anchor-free convolutional neural network for multi-categories agricultural pest detection. *Computers and Electronics in Agriculture, 174*, 105522. doi:10.1016/j.compag.2020.105522

Jiao, L., Xie, C., Chen, P., Du, J., Li, R., & Zhang, J. (2022). Adaptive feature fusion pyramid network for multi-classes agricultural pest detection. *Computers and Electronics in Agriculture, 195*, 106827. doi:10.1016/j.compag.2022.106827

Jiao, L., Zhang, F., Liu, F., Yang, S., Li, L., Feng, Z., & Qu, R. (2019). A Survey of Deep Learning-Based Object Detection. *IEEE Access : Practical Innovations, Open Solutions, 7*, 128837–128868. doi:10.1109/ACCESS.2019.2939201

Jin, X., Zhang, J., Kong, J., Su, T., & Bai, Y. (2022). A Reversible Automatic Selection Normalization (RASN) Deep Network for Predicting in the Smart Agriculture System. *Agronomy 2022, 12*(3), 591.

Jocher, G., Stoken, A., & Borovec, J., Christopher, S., Changyu, L., & Rai, P. (2020). ultralytics/yolov5: v3.1 - Bug Fixes and Performance Improvements. *ultralytics/yolov5: v3.1 - Bug Fixes and Performance Improvements.* Zenodo. doi:10.5281/zenodo.4154370

Kabra, T. S., Kardile, A. V., Deeksha, M. G., Mane, D. B., Bhosale, P. R., & Belekar, A. M. (2017). Design, Development & Optimization of a Quad-Copter for Agricultural Applications. *International Research Journal of Engineering and Technology, 04*(7).

Kadam, A. A., Peddi, P., & Gujar, S. N. (n.d.). *Design Of Deep Cnn Approach For Optimal Water Control Using Iot Based Wsn For Smart Agriculture Applications.*

Kadiyala, A., & Kumar, A. (2014). Vector time series models for prediction of air quality inside a public transportation bus using available software. *Environmental Progress & Sustainable Energy, 33*(2), 1069–1073. doi:10.1002/ep.11959

Kakani, V., Nguyen, V. H., Kumar, B. P., Kim, H., & Pasupuleti, V. R. (2020). A critical review on computer vision and artificial intelligence in food industry. *Journal of Agriculture and Food Research, 2*, 100033. doi:10.1016/j.jafr.2020.100033

Kalaiarasi, E., & Anbarasi, A. (2022). Multi-parametric multiple kernel deep neural network for crop yield prediction. *Materials Today: Proceedings, 62*(7), 4635–4642. doi:10.1016/j.matpr.2022.03.115

Kale, S. D., Khandagale, S. V., Gaikwad, S. S., Narve, S. S., & Gangal, P. V. (2015). Agriculture Drone for Spraying Fertilizer and Pesticides. Published in *International journal of Advanced Research in Computer science and Software engineering, 5*(12), 804- 807.

Kalimuthu, M., Vaishnavi, P., & Kishore, M. (2020). Crop prediction using machine learning. *Third IEEE International Conference on Smart Systems and Inventive Technology (ICSSIT),* (pp. 926–932). IEEE. 10.1109/ICSSIT48917.2020.9214190

Kalinaki, K., Thilakarathne, N. N., Mubarak, H. R., Malik, O. A., & Abdullatif, M. (2023). Cybersafe Capabilities and Utilities for Smart Cities. In *Cybersecurity for Smart Cities* (pp. 71–86). Springer. doi:10.1007/978-3-031-24946-4_6

Kamdar, J. H., Jasani, M. D., & Jasani, J. D. (2021). *11 Artificial intelligence for plant disease detection: past, present, and future, Internet of Things and Machine Learning in Agriculture: Technological Impacts and Challenges* (pp. 223–238). De Gruyter. doi:10.1515/9783110691276-011

Kamilaris, A., & Prenafeta-Boldú, F. X. (2018). A review of the use of convolutional neural networks in agriculture. *Journal of Agricultural Science, 156*(3), 312–322. doi:10.1017/S0021859618000436

Kannan, E., Trabelsi, Y., Boopathi, S., & Alagesan, S. (2022). Influences of cryogenically treated work material on near-dry wire-cut electrical discharge machining process. *Surface Topography : Metrology and Properties, 10*(1), 15027. doi:10.1088/2051-672X/ac53e1

Karlapudy, A. M., Kommasani, R. R., Malla, K. B., & Vasam, S. P. (2019). IoT-based agricultural fields protection from animals. [IJRASET]. *International Journal for Research in Applied Science and Engineering Technology, 7*(5), 3086–3090. doi:10.22214/ijraset.2019.5509

Kawai, T., & Mineno, H. (2020). Evaluation environment using edge computing for artificial intelligence-based irrigation system. *Proceedings - 2020 16th International Conference on Mobility, Sensing and Networking, MSN 2020.* IEEE. 10.1109/MSN50589.2020.00046

Kaya, K., & Gündüz Öğüdücü, U. (2020, February 25). Deep Flexible Sequential (DFS) Model for Air Pollution Forecasting. *Scientific Reports, 10*(1), 3346. doi:10.103841598-020-60102-6 PMID:32098977

Kedari, S., Lohagaonkar, P., Nimbokar, M., Palve, G., & Yevale, P. (2016). Quadcopter-A Smarter Way of Pesticide Spraying. *Imperial Journal of Interdisciplinary Research, 2*(6).

Kerkech, M., Hafiane, A., & Canals, R. (2018). Deep leaning approach with colorimetric spaces and vegetation indices for vine diseases detection in UAV images. *Computers and Electronics in Agriculture*, *155*, 237–243. doi:10.1016/j.compag.2018.10.006

Khalifeh, A., Al-Qammaz, A., Darabkh, K. A., Abualigah, L., Khasawneh, A. M., & Zinonos, Z. (2021). An AI Based Irrigation and Weather Forecasting System utilizing LoRaWAN and Cloud Computing Technologies. *Proceedings of the 2021 IEEE Conference of Russian Young Researchers in Electrical and Electronic Engineering, ElConRus 2021*, (pp. 443–448). IEEE. 10.1109/ElConRus51938.2021.9396431

Khamparia, A., Singh, A., Luhach, A., Pandey, B., & Pandey, D. (2019). Classification and Identification of Primitive Kharif Crops using Supervised Deep Convolutional Networks. *Sustainable Computing: Informatics and Systems.* doi:10.1016/j.suscom.2019.07.003

Khan, M. A., Akram, T., Sharif, M., Alhaisoni, M., Saba, T., & Nawaz, N. (2021). A probabilistic segmentation and entropy-rank correlation-based feature selection approach for the recognition of fruit diseases. *EURASIP Journal on Image and Video Processing*, *2021*(1), 1–28. doi:10.118613640-021-00558-2

Khan, R., Ali, I., Zakarya, M., Ahmad, M., Imran, M., & Shoaib, M. (2018). Technology-Assisted Decision Support System for Efficient Water Utilization: A Real-Time Testbed for Irrigation Using Wireless Sensor Networks. *IEEE Access : Practical Innovations, Open Solutions*, *6*, 25686–25697. doi:10.1109/ACCESS.2018.2836185

Khirade, S. D., & Patil, A. B. (2015). Plant Disease Detection Using Image Processing. International Conference on Computing Communication Control and Automation, (pp.. 768-771). IEEE. 10.1109/ICCUBEA.2015.153

Kiani, E. (2017). Identification of plant disease infection using soft-computing: Application to modern botany. *9th International Conference on Theory an d Application of Soft Computing, Computing with Words and Perception*, (pp. 893-900). IEEE. 10.1016/j.procs.2017.11.323

Kogekar, A. P., Nayak, R., & Pati, U. C. (2021). A CNN-GRU-SVR based Deep Hybrid Model for Water Quality Forecasting of the River Ganga. *2021 International Conference on Artificial Intelligence and Machine Vision (AIMV)* (pp. 1-6). IEEE. 10.1109/AIMV53313.2021.9670916

Kok, Z. H., Mohamed Shariff, A. R., Alfatni, M. S. M., & Khairunniza-Bejo, S. (2021). Support Vector Machine in Precision Agriculture: A review. *Computers and Electronics in Agriculture*, *191*, 106546. doi:10.1016/j.compag.2021.106546

Korche, M., Tokse, S., Shirbhate, S., Thakre, V., & Jolhe, S. P. (2021). Smart Crop Protection System. [IJLES]. *International Journal of Latest Engineering Science*, *4*(4), 101–106. https://www.ijlesjournal.org/2021/volume-4%20issue-4/ijles-v4i4p101.pdf

Koshelev, I., Savinov, M., Menshchikov, A., & Somov, A. (2023). Drone-Aided Detection of Weeds: Transfer Learning for Embedded Image Processing. *IEEE Journal of Selected Topics in Applied Earth Observations and Remote Sensing*, *16*, 102–111. doi:10.1109/JSTARS.2022.3224657

Kothandaraman, D., Praveena, N., Varadarajkumar, K., Madhav Rao, B., Dhabliya, D., Satla, S., & Abera, W. (2022, June 27). Intelligent Forecasting of Air Quality and Pollution Prediction Using Machine Learning. *Adsorption Science and Technology*, *2022*, 1–15. doi:10.1155/2022/5086622

Kulkarni, P., Karwande, A., Kolhe, T., Kamble, S., Joshi, A., & Wyawahare, M. (2021). Plant disease detection using image processing and machine learning. *arXiv preprint arXiv:2106.10698.*

Kulkarni, P., Wyawahare, M., Karwande, A., Kolhe, T., Kamble, S., & Joshi, A. (2022). Agricultural Field Analysis Using Satellite Hyperspectral Data and Autoencoder. *Communications in Computer and Information Science, 1576 CCIS*, (pp. 363–375). IEEE.

Kumar Sahu, S., & Pandey, M. (2023). An optimal hybrid multiclass SVM for plant leaf disease detection using spatial Fuzzy C-Means model. *Expert Systems with Applications*, *214*, 118989. doi:10.1016/j.eswa.2022.118989

Kumar, M., Kumar, A., & Palaparthy, V. S. (2021). Soil sensors-based prediction system for plant diseases using exploratory data analysis and machine learning. In IEEE Sensors Journal, 21(16). doi:10.1109/JSEN.2020.3046295

Kumara, V., Mohanaprakash, T. A., Fairooz, S., Jamal, K., Babu, T., & B., S. (2023). Experimental Study on a Reliable Smart Hydroponics System. In *Human Agro-Energy Optimization for Business and Industry* (pp. 27–45). IGI Global. doi:10.4018/978-1-6684-4118-3.ch002

Kumar, A., Sarkar, S., & Pradhan, C. (2019). Recommendation system for crop identification and pest control technique in agriculture. *Proceedings of the 2019 IEEE International Conference on Communication and Signal Processing, ICCSP 2019*, (pp. 185–189). IEEE. 10.1109/ICCSP.2019.8698099

Kumar, K. B. P., Bhavitha, T., Karishma, S., Pavithra, M., & Kumar, M. P. (2019). IoT-based crop monitoring from animals. *Journal of Emerging Technologies and Innovative Research*, *6*(4), 173–177. https://www.jetir.org/view?paper=JETIRAK06049. doi:10.24247/jetir.6.4.30

Kumar, V., & Laxmi, V. (2022). Pests Detection Using Artificial Neural Network and Image Processing: A Review. *International Conference on Sustainable Computing and Data Communication Systems, ICSCDS 2022 - Proceedings*, (pp. 462–467). 10.1109/ICSCDS53736.2022.9760976

Kwiatkowski, D., Phillips, P. C., Schmidt, P., & Shin, Y. (1992). Testing the null hypothesis of stationarity against the alternative of a unit root: How sure are we that economic time series have a unit root? *Journal of Econometrics*, *54*(1-3), 159–178. doi:10.1016/0304-4076(92)90104-Y

Lahouar, A., & Slama, J. B. (2017). Hour-ahead wind power forecast based on random forests. *Renewable Energy*, *109*, 529–541. doi:10.1016/j.renene.2017.03.064

Lama, G. F. C., Errico, A., Francalanci, S., Solari, L., Preti, F., & Chirico, G. B. (2020). of flow resistance models based on field experiments in a partly vegetated reclamation channel. *Geosciences*, *10*(2), 47. doi:10.3390/geosciences10020047

Lawal, A., Rehman, S., Alhems, L. M., & Alam, M. M. (2021). Wind speed prediction using hybrid 1D CNN and BLSTM network. *IEEE Access : Practical Innovations, Open Solutions*, *9*, 156672–156679. doi:10.1109/ACCESS.2021.3129883

Lee, H., Moon, A., Moon, K., & Lee, Y. *Disease and pest prediction IoT system in orchard: A preliminary study*. In *Proceedings of the 2017 Ninth International Conference on Ubiquitous and Future Networks (ICUFN)*, Milan, Italy. 10.1109/ICUFN.2017.7993840

Li, G., Huang, Y., Chen, Z., Chesser, G. D., Purswell, J. L., Linhoss, J., & Zhao, Y. (2021). Practices and Applications of Convolutional Neural Network-Based Computer Vision Systems in Animal Farming: A Review. *Sensors, 21*(4), 1492.

Li, Z. Ding, L., & Xu, D. (2022). Exploring the potential role of environmental and multi-source satellite data in crop yield prediction across Northeast China. *Science of the Total Environment, 815*. doi:10.1016/j.scitotenv.2021.152880

Liang, W. C., Yang, Y. J., & Chao, C. M. (2019). Low-cost weed identification system using drones. *Proceedings - 2019 7th International Symposium on Computing and Networking Workshops, CANDARW 2019*, (pp. 260–263). IEEE. 10.1109/CANDARW.2019.00052

Li, D., Ahmed, F., Wu, N., & Sethi, A. I. (2022). YOLO-JD: A Deep Learning Network for Jute Diseases and Pests Detection from Images. *Plants*, *11*(7), 1–13. doi:10.3390/plants11070937 PMID:35406915

Li, G., Ma, Z., & Wang, H. (2012). Image recognition of grape downy mildew and grape powdery mildew based on support vector machine. In *Computer and Computing Technologies in Agriculture V: 5th IFIP TC 5/SIG 5.1 Conference*. Springer.

Li, K., Wang, J., Jalil, H., & Wang, H. (2023). A fast and lightweight detection algorithm for passion fruit pests based on improved YOLOv5. *Computers and Electronics in Agriculture, 204*, 107534. doi:10.1016/j.compag.2022.107534

Li, L., Zhang, S., & Wang, B. (2021). Plant disease detection and classification by deep learning—A review. *IEEE Access : Practical Innovations, Open Solutions, 9*, 56683–56698. doi:10.1109/ACCESS.2021.3069646

Limouni, T., Yaagoubi, R., Bouziane, K., Guissi, K., & Baali, E. H. (2022). Univariate and Multivariate LSTM Models for One Step and Multistep PV Power Forecasting. *International Journal of Renewable Energy Development, 11*(3), 815–828. doi:10.14710/ijred.2022.43953

Lin, T. L., Chang, H. Y., & Chen, K. H. (2020). The pest and disease identification in the growth of sweet peppers using faster r-cnN and mask r-CNn. *Journal of Internet Technology, 21*, 605–614. doi:10.3966/160792642020032102027

Lippi, M., Bonucci, N., Carpio, R. F., Contarini, M., Speranza, S., & Gasparri, A. (2021). A YOLO-based pest detection system for precision agriculture. *2021 29th Mediterranean Conference on Control and Automation. Med (New York, N.Y.), 2021*, 342–347. doi:10.1109/MED51440.2021.9480344

Lippi, M., Carpio, R. F., Contarini, M., Speranza, S., & Gasparri, A. (2022). A Data-Driven Monitoring System for the Early Pest Detection in the Precision Agriculture of Hazelnut Orchards. *IFAC-PapersOnLine, 55*(32), 42–47. doi:10.1016/j.ifacol.2022.11.112

Liu, B., Gan, H., Chen, D., & Shu, Z. (2023). Research on Fault Early Warning of Marine Diesel Engine Based on CNN-BiGRU. *Journal of Marine Science and Engineering, 11*(1), 56. doi:10.3390/jmse11010056

Liu, Q., Yan, Q., Tian, J., & Yuan, K. (2021). Key technologies and applications in intelligent agriculture. *Journal of Physics: Conference Series, 1757*(1), 012059. doi:10.1088/1742-6596/1757/1/012059

Liu, W., Anguelov, D., Erhan, D., Szegedy, C., Reed, S., Fu, C.-Y., & Berg, A. C. (2016). Ssd: Single shot multibox detector. *Computer Vision–ECCV 2016: 14th European Conference, Amsterdam, The Netherlands, October 11–14, 2016. Proceedings, 14*(Part I), 21–37. doi:10.1007/978-3-319-46448-0_2

Liu, Y., Zhou, Y., Wen, S., & Tang, C. (2014). A Strategy on Selecting Performance Metrics for Classifier Evaluation. *International Journal of Mobile Computing and Multimedia Communications, 6*(4), 20–35. doi:10.4018/IJMCMC.2014100102

Liu, Z., Bashir, R. N., Iqbal, S., Shahid, M. M. A., Tausif, M., & Umer, Q. (2022). Internet of things (IoT) and machine learning model of plant disease prediction–blister blight for tea plant. *IEEE Access : Practical Innovations, Open Solutions, 10*, 44934–44944. doi:10.1109/ACCESS.2022.3169147

Lottes, P., Behley, J., Milioto, A., & Stachniss, C. (2018). Fully convolutional networks with sequential information for robust crop and weed detection in precision farming. *IEEE Robotics and Automation Letters, 3*(4), 2870–2877. doi:10.1109/LRA.2018.2846289

Louargant, M., Jones, G., Faroux, R., Paoli, J. N., Maillot, T., Gée, C., & Villette, S. (2018). Unsupervised classification algorithm for early weed detection in row-crops by combining spatial and spectral information. *Remote Sensing (Basel), 10*(5), 1–18. doi:10.3390/rs10050761

Lu, B., Dao, P. D., Liu, J., He, Y., & Shang, J. (2020). Recent advances of hyperspectral imaging technology and applications in agriculture. *Remote Sensing (Basel), 12*(16), 2659. doi:10.3390/rs12162659

Lube, V., Noyan, M. A., Przybysz, A., Salama, K., & Blilou, I. (2022). MultipleXLab: A high-throughput portable live-imaging root phenotyping platform using deep learning and computer vision. *Plant Methods, 18*(1), 1–22. doi:10.118613007-022-00864-4 PMID:35346267

Lu, Y., Chen, D., Olaniyi, E., & Huang, Y. (2022). Generative adversarial networks (GANs) for image augmentation in agriculture: A systematic review. *Computers and Electronics in Agriculture*, *200*, 107208. doi:10.1016/j.compag.2022.107208

Lyu, Z., Jin, H., Zhen, T., Sun, F., & Xu, H. (2021). Small Object Recognition Algorithm of Grain Pests Based on SSD Feature Fusion. *IEEE Access : Practical Innovations, Open Solutions*, *9*, 43202–43213. doi:10.1109/ACCESS.2021.3066510

M., H. & M., S. (2015) A Review on Evaluation Metrics for Data Classifica- tion Evaluations. *International Journal of Data Mining & Knowledge Man- agement Process, 5*(2), pp. 01–11. . doi:10.5121/ijdkp.2015.5201

Madhavan, M. V., Thanh, D. N. H., Khamparia, A., Pande, S., Malik, R., & Gupta, D. (2021). Recognition and classification of pomegranate leaves diseases by image processing and machine learning techniques. *Computers, Materials & Continua*, *66*(3), 2939–2955. doi:10.32604/cmc.2021.012466

Maes, W. H., & Steppe, K. (2019). Perspectives for remote sensing with unmanned aerial vehicles in precision agriculture. *Trends in Plant Science*, *24*(2), 152–164. doi:10.1016/j.tplants.2018.11.007 PMID:30558964

Mahlein, A. K. (2016). Plant disease detection by imaging sensors—Parallels and specific demands for precision agriculture and plant phenotyping. *Plant Disease*, *100*(2), 241–251. doi:10.1094/PDIS-03-15-0340-FE PMID:30694129

Ma, J., Zhuang, D., & Gao, Y. (2018). A review of deep learning models for crop recognition. *Precision Agriculture*, *19*(4), 693–723. doi:10.100711119-017-9513-9

Makkithaya, K., & N. V. G. (2022). A federated learning-based crop yield prediction for agricultural production risk management. IEEE Delhi Section Conference (DELCON), 2022 (pp. 1–7). IEEE.

Malik, O. A., Faisal, M., & Hussein, B. R. (2021). Ensemble Deep Learning Models for Fine-grained Plant Species Identification. *2021 IEEE Asia-Pacific Conference on Computer Science and Data Engineering, CSDE 2021*. IEEE. 10.1109/CSDE53843.2021.9718387

Manfreda, S., McCabe, M. E., Miller, P. E., Lucas, R., Madrigal, V. P., & Mallinis, G. (2018). On the use of unmanned aerial systems for environmental monitoring. *Remote Sensing (Basel)*, *10*(4), 641. doi:10.3390/rs10040641

Manu, B.N. (2022). *Plant Leaf Disease Detection and Classification using Multiclass.*

Mariammal, G., Suruliandi, A., Raja, S. P., & Poongothai, E. (2021, October). Prediction of land suitability for crop cultivation based on soil and environmental characteristics using modified recursive feature elimination technique with various classifiers. *IEEE Transactions on Computational Social Systems*, *8*(5), 1132–1142. doi:10.1109/TCSS.2021.3074534

Martínez-Ferrer, L., Piles, M., & Camps-Valls, G. (2021, December). Crop yield estimation and interpretability with Gaussian processes. *IEEE Geoscience and Remote Sensing Letters*, *18*(12), 2043–2047. doi:10.1109/LGRS.2020.3016140

Mathivanan, S. K., & Jayagopal, P. (2022). Utilizing satellite and UAV data for crop yield prediction and monitoring through deep learning. *Acta Geophysica*, *70*(6), 2991–3004. doi:10.100711600-022-00911-7

Mehta, T., Manolkar, O., Multaikar, S., Patil, S., & Gawande, N. (2021). Intensification of agriculture using deep learning and machine learning: A survey. *10th IEEE International Conference on Internet of Everything, Microwave Engineering*, (pp. 1–6). IEEE. 10.1109/IEMECON53809.2021.9689196

Mekhalfi, M. L., Nicolò, C., Ianniello, I., Calamita, F., Goller, R., Barazzuol, M., & Melgani, F. (2020). Vision System for Automatic On-Tree Kiwifruit Counting and Yield Estimation. *Sensors 2020, 20*(15), 4214.

Meng, Y., Zhong, W., Liu, Y., Wang, M., & Lan, Y. (2022). Droplet Distribution of an Autonomous UAV-based Sprayer in Citrus Tree Canopy. *Journal of Physics: Conference Series*, *2203*(1), 012022. doi:10.1088/1742-6596/2203/1/012022

Michaelides, S., Paronis, D., Retalis, A., & Tymvios, F. (2017). Monitoring and Forecasting Air Pollution Levels by Exploiting Satellite, Ground-Based, and Synoptic Data, Elaborated with Regression Models. *Advances in Meteorology, 2017*, 1–17. doi:10.1155/2017/2954010

Ministry of Agriculture and Farmers Welfare. (n.d.). *Soil Health Card (SHC).* Soil Health. https://soilhealth.dac.gov.in/soil-health-map/MADHYA_PRADESH [Accessed: 26/3/2023]

Misbah, K., Laamrani, A., Chehbouni, A., Dhiba, D., & Ezzahar, J. (2021). Use of Hyperspectral Prisma Level-1 Data and ISDA Soil Fertility Map for Soil Macronutrient Availability Quantification In A Moroccan Agricultural Land. *International Geoscience and Remote Sensing Symposium (IGARSS)*, (pp. 7051–7054). 10.1109/IGARSS47720.2021.9553190

Mishra, D., Khan, A., Tiwari, R., & Upadhay, S. (2018). *Automated Irrigation System-IoT Based Approach. 3rd International Conference on Internet of Things: Smart Innovation and Usages (IoT-SIU)*, Bhimtal, India. 10.1109/IoT-SIU.2018.8519886

Mogili, U. R., & Deepak, B. B. V. L. (2018). Review on application of drone systems in precision agriculture. *Procedia Computer Science*, *133*, 502–509. doi:10.1016/j.procs.2018.07.063

Mohanty, S. P., Hughes, D. P., & Salathé, M. (2018). Using Deep Learning for Image-Based Plant Disease Detection. *Frontiers in Plant Science*, *9*, 1420. doi:10.3389/fpls.2018.01420 PMID:27713752

Mohanty, S. P., Hughes, D. P., & Salathé, M. (2019). Deep Learning for Plant Disease Detection: A Comprehensive Review. *Computers and Electronics in Agriculture*, *162*, 139–153. doi:10.1016/j.compag.2019.03.016

Mohi-Alden, K., Omid, M., Soltani Firouz, M., & Nasiri, A. (2022). Design and evaluation of an intelligent sorting system for bell pepper using deep convolutional neural networks. *Journal of Food Science*, *87*(1), 289–301. doi:10.1111/1750-3841.15995 PMID:34940977

Mondal, A., & Banerjee, S. (2021). Effective crop prediction using deep learning. *IEEE International Conference on Smart Generation Computing, Communication and Networking (SMART GENCON).* IEEE. 10.1109/SMARTGENCON51891.2021.9645872

Mondal, A., Kundu, S., Mukhopadhyay, A., & ... (2012). Rainfall trend analysis by Mann-Kendall test: A case study of north-eastern part of Cuttack district, Orissa. *International Journal of Geology. Earth and Environmental Sciences*, *2*, 70–78.

Mopidevi, S., Singitham, V., Thippani, B., Shamanthula, R., & Satya Phanindra Vallabhaneni, N. S. (2022). Plant growth and yield prediction using ML and DL algorithms. *IEEE International Conference on Electronics and Renewable Systems (ICEARS),* (pp. 1470–1477). IEEE. 10.1109/ICEARS53579.2022.9751906

Morellos, A., Pantazi, X. E., Moshou, D., Alexandridis, T., Whetton, R., Tziotzios, G., Wiebensohn, J., Bill, R., & Mouazen, A. M. (2016). Machine learning based prediction of soil total nitrogen, organic carbon and moisture content by using VIS-NIR spectroscopy. *Biosystems Engineering*, *152*, 104–116. doi:10.1016/j.biosystemseng.2016.04.018

Morey, N. S., Mehere, P. N., & Hedaoo, K. (2017). Agriculture Drone for Fertilizers and Pesticides Spraying. *International Journal for engineering applications and Technology.*

Muhammad, A., Amjad, A., & Khan, I. (2021). Crop yield prediction using deep learning: A review. *Computers and Electronics in Agriculture*, *184*, 106024. doi:10.1016/j.compag.2020.106024

Muhammed, F. (2022). Computer vision-based plant disease identification system: A review. *Computer*, *1*(1), 59–78.

Mulla, D. J. (2013). Twenty five years of remote sensing in precision agriculture: Key advances and remaining knowledge gaps. *Biosystems Engineering*, *114*(4), 358–371. doi:10.1016/j.biosystemseng.2012.08.009

Munje, R. K., Warungase, P., Sarode, M., Lohar, V., & Mehare, A. (2022). Developing a Prototype of Smart Crop Protection System from Animals. In A. Pandian & V. Balaji (Eds.), *Intelligent Agriculture: The Role of IoT*. Big Data and AI. doi:10.5281/zenodo.6874633

Mustafi, S., Ghosh, P., Roy, K., Dan, S., Mukherjee, K., & Nath Mandal, S. (2021). Drones for Intelligent Agricultural Management. In P. Krause & F. Xhafa (Eds.), *IoT-based Intelligent Modelling for Environmental and Ecological Engineering. Lecture Notes on Data Engineering and Communications Technologies* (Vol. 67). Springer. doi:10.1007/978-3-030-71172-6_4

Mustak, A., Ma, H., Song, L., & Jin, Y. (2022). *Embedded Design of Automatic Pesticide Spraying Robot Control System.*

Nancy, P., Pallathadka, H., Naved, M., Kaliyaperumal, K., Arumugam, K., & Garchar, V. (2022). Deep learning and machine learning based efficient framework for image based plant disease classification and detection. *IEEE International Conference on Advanced Computing Technologies and Applications (ICACTA),* (pp. 1–6). IEEE. 10.1109/ICACTA54488.2022.9753623

Nanda, I., Chadalavada, S., Swathi, M., & Khatua, L. (2021). Implementation of IIoT based smart crop protection and irrigation system. In A. Pandian & V. Balaji (Eds.), *Intelligent Agriculture: The Role of IoT* (pp. 1–13). Big Data and AI. doi:10.4018/IJIA.2021010101

Nandal, V., & Dahiya, S. (2021). Energy Efficient Data Aggregation Protocol Based Iot-WSN Framework For Smart Agriculture. *Webology, 18*(6), 4032–4045.

Nansen, C. (2016). The potential and prospects of proximal remote sensing of arthropod pests. *Pest Management Science, 72*(4), 653–659. doi:10.1002/ps.4209 PMID:26663253

Nansen, C., & Elliott, N. (2016). Remote sensing and reflectance profiling in entomology. *Annual Review of Entomology, 61*(1), 139–158. doi:10.1146/annurev-ento-010715-023834 PMID:26982438

Narasimha Murthy, K. V., Saravana, R., & Vijaya Kumar, K. (2018). Modeling and forecasting rainfall patterns of southwest monsoons in North–East India as a SARIMA process. *Meteorology and Atmospheric Physics, 130*(1), 99–106. doi:10.100700703-017-0504-2

Narayan Mishra, K., Kumar, S., Patel, N. R., Shripathi Rao, M., Singh, A., Subba Reddy, N., & Acharya, D. U. (2022). Crop prediction using machine learning. *Journal of Physics: Conference Series, 2161*(1), 012033. doi:10.1088/1742-6596/2161/1/012033

Narayana, B. V. V. S., Ravi, K. S., & Ramesh, N. V. K. (2018). A review on advanced crop field monitoring system in agriculture field through top notch sensors. *Journal of Advanced Research in Dynamical and Control Systems, 10*(6), 1572–1578.

Nasiri, A., Omid, M., Taheri-Garavand, A., & Jafari, A. (2022). Deep learning-based precision agriculture through weed recognition in sugar beet fields. *Sustainable Computing: Informatics and Systems, 35*, 100759. doi:10.1016/j.suscom.2022.100759

National Food Security Mission (NFSM). (n.d.). *Paper on Rice*. Farmer. farmer.gov.in. https://farmer.gov.in/imagedefault/pestanddiseasescrops/rice.pdf (Accessed: 19/2/2023)

Navaneetha, P., Ramiya Devi, R., Vennila, S., & Manikandan, P. (2020). IoT-Based Crop Protection System against Birds and Wild Animal Attacks. *International Journal of innovative research in Technology (IJIRT).* http://www.ijirt.org/master/publishedpaper/IJIRT149166_PAPER.pdf

Nawar, S., Corstanje, R., Halcro, G., Mulla, D., & Mouazen, A. M. (2017). Delineation of soil management zones for variable-rate fertilization: A review. *Advances in Agronomy, 143*, 175–245. doi:10.1016/bs.agron.2017.01.003

Ngugi, L. C., Abelwahab, M., & Abo-Zahhad, M. (2021). Recent advances in image processing techniques for automated leaf pest and disease recognition – A review. *Information Processing in Agriculture, 8*(1), 27–51. doi:10.1016/j.inpa.2020.04.004

Nilay Ganatra and Atul Patel, "A Survey on Diseases Detection and Classification of Agriculture Products using Image Processing and Machine Learning", International Journal of Computer Applications (0975 – 8887) Volume 180 – No.13, January 2018

Nisha, G., & Megala, J. (2015). Wireless sensor Network based automated irrigation and crop field monitoring system. *Sixth International Conference on Advanced Computing, ICoAC 2014*, (pp. 189–194). IEEE. 10.1109/ICoAC.2014.7229707

Oliphant, A., Thenkabail, P., & Teluguntla, P. (2022). Global food-security-support-analysis data at 30-m resolution (GFSAD30) cropland-extent products—Download Analysis. In Open-File Report. US Geological Survey. doi:10.3133/ofr20221001

Oo, Z. Z., & Sabai, P. H. (2020). Time Series Prediction Based on Facebook Prophet: A Case Study, Temperature Forecasting in Myintkyina. *International Journal of Applied Mathematics Electronics and Computers, 8*(4), 263–267. doi:10.18100/ijamec.816894

Ouhame, S., & Hadi, Y. (2019). Multivariate workload prediction using Vector Autoregressive and Stacked LSTM models. *Proceedings of the New Challenges in Data Sciences: Acts of the Second Conference of the Moroccan Classification Society*, (pp. 1–7). ACM. 10.1145/3314074.3314084

Ouhami, M., Hafiane, A., Es-Saady, Y., El Hajji, M., & Canals, R. (2021). Computer vision, IoT and data fusion for crop disease detection using machine learning: A survey and ongoing research. *Remote Sensing (Basel), 13*(13), 2486. doi:10.3390/rs13132486

Palaniappan, M., Tirlangi, S., Mohamed, M. J. S., Moorthy, R. M. S., Valeti, S. V., & Boopathi, S. (2023). Fused deposition modelling of polylactic acid (PLA)-based polymer composites: A case study. In Development, Properties, and Industrial Applications of 3D Printed Polymer Composites (pp. 66–85). IGI Global. doi:10.4018/978-1-6684-6009-2.ch005

Pande, S. M., Ramesh, P. K., Anmol, A., Aishwarya, B. R., Rohilla, K., & Shaurya, K. (2021). *Crop Recommender System Using Machine Learning Approach. 5th International Conference on Computing Methodologies and Communication (ICCMC)*, Erode, India. 10.1109/ICCMC51019.2021.9418351

Pandiri, D. N. K., Murugan, R., & Goel, T. (n.d.). *Indian Regions Soil Image Database (IRSID): A dataset for classification of Indian soils.* IEEE Dataport. https://ieee-dataport.org/documents/indian-regions-soil-image-database-irsid-dataset-classification-indian-soils

Pandithurai, O., Aishwarya, S., Aparna, B., & Kavitha, K. (2017). Agro-tech: A digital model for monitoring soil and crops using internet of things (IoT). *ICONSTEM 2017 - Proceedings: 3rd IEEE International Conference on Science Technology, Engineering and Management*. IEEE. 10.1109/ICONSTEM.2017.8261306

Pang, L., Men, S., Yan, L., & Xiao, J. (2020). Rapid vitality estimation and prediction of corn seeds based on spectra and images using deep learning and hyperspectral imaging techniques. *IEEE Access : Practical Innovations, Open Solutions, 8*, 123026–123036. doi:10.1109/ACCESS.2020.3006495

Panpatte, D. G. (2018). *Artificial Intelligence in Agriculture: An Emerging Era of Research. Intutional Science*. CANADA.

Pan, S. J., & Yang, Q. (2010). A survey on transfer learning. *IEEE Transactions on Knowledge and Data Engineering*, *22*(10), 1345–1359. doi:10.1109/TKDE.2009.191

Pant, J., Pant, R. P., Kumar Singh, M., Pratap Singh, D., & Pant, H. (2021). Analysis of agricultural crop yield prediction using statistical techniques of machine learning. *Materials Today*, *46*(20), 10922–10926. doi:10.1016/j.matpr.2021.01.948

Pantazi, X. E., Moshou, D., Alexandridis, T., Whetton, R. L., & Mouazen, A. M. (2016). Wheat yield prediction using machine learning and advanced sensing techniques. *Computers and Electronics in Agriculture*, *121*, 57–65. doi:10.1016/j.compag.2015.11.018

Parraga-Alava, J., Cusme, K., Loor, A., & Santander, E. (2019). RoCoLe: A robusta coffee leaf images dataset for evaluation of machine learning based methods in plant diseases recognition. *Data in Brief*, *25*, 104414. doi:10.1016/j.dib.2019.104414 PMID:31516934

Patel, D., & Bhatt, N. (2021). Improved accuracy of pest detection using augmentation approach with Faster R-CNN. *IOP Conference Series. Materials Science and Engineering*, *1042*(1), 012020. doi:10.1088/1757-899X/1042/1/012020

Pathan, M., Patel, N., Yagnik, H., & Shah, M. (2020). Artificial cognition for applications in smart agriculture: A comprehensive review. *Artificial Intelligence in Agriculture*, *4*, 81–95. doi:10.1016/j.aiia.2020.06.001

Patrício, D. I., & Rieder, R. (2018). Computer vision and artificial intelligence in precision agriculture for grain crops: A systematic review. *Computers and Electronics in Agriculture*, *153*, 69–81. doi:10.1016/j.compag.2018.08.001

Pavani, S., & Augusta Sophy Beulet, P. (2022). Prediction of Jowar Crop Yield Using K-Nearest Neighbor and Support Vector Machine Algorithms. *Lecture Notes in Electrical Engineering*, *792*, 497–503. doi:10.1007/978-981-16-4625-6_49

Penchalaiah, N., Pavithra, D., Bhargavi, P., & Madhuri, D. P. (2020). Smart irrigation and crop protection from wild animals. *Journal of Engineering Sciences (JES)*. https://jespublication.com/upload/2020-1104126.pdf

Peng, T., Zhang, C., Zhou, J., & Nazir, M. S. (2021). An integrated framework of Bi-directional long-short term memory (BiLSTM) based on sine cosine algorithm for hourly solar radiation forecasting. *Energy*, *221*, 119887. doi:10.1016/j.energy.2021.119887

Phadikar, S., Sil, J., & Das, A. K. (2012). Classification of rice leaf diseases based on morphological changes. *Int J Inf Electron Eng*, *2*(3), 460–463.

Phute, M., Madake, J., & Bhatlawande, S. (2021). Precision Farming Based Soil Fertility Assessment Techniques. *ACCESS 2021 - Proceedings of 2021 2nd International Conference on Advances in Computing, Communication, Embedded and Secure Systems*, (pp. 170–172). IEEE. 10.1109/ACCESS51619.2021.9563284

Population Division of the United Nations Department of Economic and Social Affairs. (2017). *World Population Prospects: The 2017 Revision*. UN. https://www.un.org/en/desa/world-population-projected-reach-98-billion-2050-and-112-billion-2100, (Accessed: 17/2/2021)

Potdar, R. P., Shirolkar, M. M., Verma, A. J., More, P. S., & Kulkarni, A. (2021). *Determination of soil nutrients (NPK) using optical methods: a mini review*, *44*(12), 1826–1839. Taylor & Francis. doi:10.1080/01904167.2021.1884702

Prajapati, H. B., Shah, J. P., & Dabhi, V. K. (2017). Detection and classification of rice plant diseases. *Intelligent Decision Technologies*, *11*(3), 357–373. doi:10.3233/IDT-170301

Priyanka, R. R., Mahesh, M., Pallavi, S. S., Jayapala, G., & Pooja, M. R. (2018). Crop protection by an alert-based system using deep learning concept. International Journal of Scientific Research in Computer Science. *Engineering and Information Technology*, *6*(6), 33–38. doi:10.26438/ijsrcse/v6i6.4749

Pujari, J. D. (2015) Image Processing Based Detection of Fungal Diseases In Plants. *International Conference on Information and Communication Technologies,* (pp. 1802- 1808). Springer. 10.1016/j.procs.2015.02.137

Puno, J. C., Sybingco, E., Dadios, E., Valenzuela, I., & Cuello, J. (2017). Determination of soil nutrients and pH level using image processing and artificial neural network. *HNICEM 2017 - 9th International Conference on Humanoid, Nanotechnology, Information Technology, Communication and Control, Environment and Management.* IEEE. 10.1109/HNICEM.2017.8269472

Puri, D., Kumar, A., Virmani, J., & Kriti. (2022). Classification of leaves of medicinal plants using laws' texture features. *International Journal of Information Technology : an Official Journal of Bharati Vidyapeeth's Institute of Computer Applications and Management, 14*(2), 931–942. doi:10.100741870-019-00353-3

Qiang, Z., He, L., & Dai, F. (2019). Identification of plant leaf diseases based on inception V3 transfer learning and fine-tuning. In *Smart City and Informatization: 7th International Conference, iSCI 2019.* Springer.

Qiao, M. He, X., Cheng, Xijie, Li, Panle, Luo, Haotian, Zhang, Lehan, & Tian, Z. (2021). Crop yield prediction from multi-spectral, multi-temporal remotely sensed imagery using recurrent 3D convolutional neural networks. *International Journal of Applied Earth Observation and Geoinformation, 102,* 102436. doi:10.1016/j.jag.2021.102436

Qiao, M., He, X., Cheng, X., Li, P., Luo, H., Tian, Z., & Guo, H. (2021). Exploiting hierarchical features for crop yield prediction based on 3-D convolutional neural networks and multikernel Gaussian process. *IEEE Journal of Selected Topics in Applied Earth Observations and Remote Sensing, 14,* 4476–4489. doi:10.1109/JSTARS.2021.3073149

Qing, X., & Niu, Y. (2018). Hourly day-ahead solar irradiance prediction using weather forecasts by LSTM. *Energy, 148,* 461–468. doi:10.1016/j.energy.2018.01.177

Radhika, Y., & Shashi, M. (2009). Atmospheric temperature prediction using support vector machines. *International journal of computer theory and engineering, 1,* 55. doi:10.7763/IJCTE.2009.V1.9

Raja, S. P., Sawicka, B., Stamenkovic, Z., & Mariammal, G. (2022). Crop prediction based on characteristics of the agricultural environment using various feature selection techniques and classifiers. *IEEE Access : Practical Innovations, Open Solutions, 10,* 23625–23641. doi:10.1109/ACCESS.2022.3154350

Ranganathan, J., Waite, R., & Searchinger, T. (2018). *How to Sustainably Feed 10 Billion People by 2050, in 21 Charts.* WRI. https://www.wri.org/insights/how-sustainably-feed-10-billion-people-2050-21-charts

Rao, A. K., Nagwanshi, K. K., & Pathak, S. (2022). Empirical Study on Energy-Efficient IoT-Based WSN Routing Protocols for Smart Agriculture System. In *Lecture Notes in Networks and Systems* (Vol. 392, pp. 259–271). Springer. doi:10.1007/978-981-19-0619-0_23

Rasheed, N., Khan, S. A., Hassan, A., & Safdar, S. (2021). A decision support framework for national crop production planning. *IEEE Access : Practical Innovations, Open Solutions, 9,* 133402–133415. doi:10.1109/ACCESS.2021.3115801

Rashid, M., Bari, B. S., Yusup, Y., Kamaruddin, M. A., & Khan, N. (2021). A Comprehensive Review of Crop Yield Prediction Using Machine Learning Approaches with Special Emphasis on Palm Oil Yield Prediction. *IEEE Access : Practical Innovations, Open Solutions, 9,* 63406–63439. doi:10.1109/ACCESS.2021.3075159

Raspberry Pi Foundation. (n.d.). Raspberry Pi. https://www.raspberrypi.org/

Raval, M. S., Chaudhary, S., & Adinarayana, J. (2022). Computer Vision and Machine *Learning in Agriculture,* 97–126.

Ray, R. K., & S. K. (2022). Das and S. Chakravarty. Smart Crop Recommender System-à Machine Learning Approach. *12th IEEE International Conference on Cloud Computing, Data Science & Engineering (Confluence),* (pp. 494–499). IEEE.

Reddy, D. J., & Kumar, M. R. (2021). Crop yield prediction using machine learning algorithm. *5th IEEE International Conference on Intelligent Computing and Control. Systems (ICICCS),* (pp. 1466–1470). IEEE. 10.1109/ICICCS51141.2021.9432236

Reddy, K. N., Huang, Y., Lee, M. A., Nandula, V. K., Fletcher, R. S., Thomson, S. J., & Zhao, F. (2014). Glyphosate-resistant and glyphosate-susceptible Palmer amaranth(Amaranthus palmeri S. Wats.): Hyperspectral refectance properties of plants and potential for classifcation. *Pest Management Science, 70*(12), 1910–1917. doi:10.1002/ps.3755 PMID:24497403

Reddy, M. A., Reddy, B. M., Mukund, C. S., Venneti, K., Preethi, D. M. D., & Boopathi, S. (2023). Social Health Protection During the COVID-Pandemic Using IoT. In *The COVID-19 Pandemic and the Digitalization of Diplomacy* (pp. 204–235). IGI Global. doi:10.4018/978-1-7998-8394-4.ch009

Reddy, M. C. Mohan, Raju, K., Kodi, K., Anapalli, B., & Pulla, M. (2020). Smart Crop Protection System from Living Objects and Fire Using Arduino. Science. *Technology and Development Journal, 23*(2), 287–292. doi:10.32508tdj.v23i2.1081

Redij, J., Shitap, P., Singh, S., Zagade, D., & Chougule, S. (2022). Smart Crop Protection System from Animals. In A. Pandian & V. Balaji (Eds.), *Intelligent Agriculture: The Role of IoT, Big Data and AI.* IJCRT. https://www.ijcrt.org/papers/IJCRTO020033.pdf

Redmon, J., Divvala, S., Girshick, R., & Farhadi, A. (2016). You only look once: Unified, real-time object detection. *Proceedings of the IEEE conference on computer vision and pattern recognition,* (pp. 779–788). IEEE. 10.1109/CVPR.2016.91

Redmon, J., & Farhadi, A. (2017). YOLO9000: better, faster, stronger. *Proceedings of the IEEE conference on computer vision and pattern recognition,* (pp. 7263–7271). IEEE. 10.48550/arXiv.1612.08242

Reinecke, M., & Prinsloo, T. (2017, July). The influence of drone monitoring on crop health and harvest size. *International Conference on Next Generation Computing Applications.* IEEE. 10.1109/NEXTCOMP.2017.8016168

Ren, S., He, K., Girshick, R., & Sun, J. (2015). Faster r-cnn: Towards real-time object detection with region proposal networks. *Advances in Neural Information Processing Systems, 28,* 1–14. doi:10.48550/arXiv.1506.01497

Ren, X., Li, X., Ren, K., Song, J., Xu, Z., Deng, K., & Wang, X. (2021). Deep learning-based weather prediction: A survey. *Big Data Research, 23,* 100178. doi:10.1016/j.bdr.2020.100178

Repositório Digipathos: Home. (n.d.). DigiPathos. https://www.digipathos-rep.cnptia.embrapa.br/

Rolnick, D. (2018) Deep Learning is Robust to Massive Label Noise. *Open Review.* https://openreview.net/forum?id=B1p461b0W

Romeijn, L. J. (2008). Agriculture Crop Management and Production Improved by Satellite Remote Sensing Technology and Geographic Information Systems (GIS). *Satellite Imaging Corporation, Houston, Texas.* https://www.prweb.com/releases/2008/06/prweb1044924.htm

S., P. K., Sampath, B., R., S. K., Babu, B. H., & N., A. (2022). Hydroponics, Aeroponics, and Aquaponics Technologies in Modern Agricultural Cultivation. In *Trends, Paradigms, and Advances in Mechatronics Engineering* (pp. 223–241). IGI Global. doi:10.4018/978-1-6684-5887-7.ch012

Sabat, N. K., Nayak, R., Pati, U. C., & Das, S. K. (2022). *A Comparative Analysis of Univariate Deep Learning-based Time-series Models for Temperature Forecasting of the Bhubaneshwar.* IEEE. doi:10.1109/iSSSC56467.2022.10051494

Sabat, N. K., Pati, U. C., & Das, S. K. (2021). A Short Survey on Real-Time Object Detection and Its Challenges. *Advances in Systems, Control and Automations: Select Proceedings of ETAEERE 2020,* (pp. 93–100). Springer. 10.1007/978-981-15-8685-9_9

Saeed, Khaki, & Lizhi, W. (2019). ISSN=1664-462X. Crop yield prediction using deep neural networks. *Frontiers in Plant Science*, 10.

Safonova, A., Guirado, E., Maglinets, Y., Alcaraz-Segura, D., & Tabik, S. (2021). Olive Tree Biovolume from UAV Multi-Resolution Image Segmentation with Mask R-CNN. *Sensors 2021, 21*(5).

Sagana, C., Keerthika, P., Thangatamilan, M., Kamali, R., Nanthini, K., & Maghathani, S. (2022). Identification of Suitable Crop Based on Weather Condition. *2022 International Conference on Computer Communication and Informatics, ICCCI 2022*. IEEE. 10.1109/ICCCI54379.2022.9740905

Sagarika, G. K., Prasad, S. K., & Kumar, S. M. (2020). Paddy Plant Disease Classification and Prediction Using Convolutional Neural Network. *International Conference on Recent Trends on Electronics, Information, Communication & Technology (RTEICT)*, Bangalore, India. 10.1109/RTEICT49044.2020.9315634

Saha, S., & Ahsan, S. M. M. (2021). Rice Disease Detection using Intensity Moments and Random Forest. *2021 International Conference on Information and Communication Technology for Sustainable Development, ICICT4SD 2021 - Proceedings*, (pp. 166–170). IEEE. 10.1109/ICICT4SD50815.2021.9396986

Sa, I., Chen, Z., Popovic, M., Khanna, R., Liebisch, F., Nieto, J., & Siegwart, R. (2018). WeedNet: Dense Semantic Weed Classification Using Multispectral Images and MAV for Smart Farming. *IEEE Robotics and Automation Letters*, *3*(1), 588–595. doi:10.1109/LRA.2017.2774979

Saini, P., & Nagpal, B. (2022). *Deep-LSTM Model for Wheat Crop Yield Prediction in India.*

Saleem, M. H., Potgieter, J., & Arif, K. M. (2021). *Automation in Agriculture by Machine and Deep Learning Techniques: A Review of Recent Developments. 22*, 2053–2091. Springer. doi:10.1007/s11119-021-09806-x

Saleem, M. H., Potgieter, J., & Arif, K. M. (2019). Plant disease detection and classification by deep learning. *Plants*, *8*(11), 468. doi:10.3390/plants8110468 PMID:31683734

Saleh, A. R. B. M., Venkatasubramanian, S., Paul, N. R. R., Maulana, F. I., Effendy, F., & Sharma, D. K. (2022). Real-time Monitoring System in IoT for Achieving Sustainability in the Agricultural Field. *International Conference on Edge Computing and Applications (ICECAA)*, Tamilnadu, India. 10.1109/ICECAA55415.2022.9936103

Salleh, M. A. M., Mahmoud, D. K., Karim, W. A. W. A., & Idris, A. (2011). Cationic and anionic dye adsorption by agricultural solid wastes: A comprehensive review. *Desalination*, *280*(1-3), 1–13. doi:10.1016/j.desal.2011.07.019

Salman, A. G., & Kanigoro, B. (2021). Visibility forecasting using autoregressive integrated moving average (ARIMA) models. *Procedia Computer Science*, *179*, 252–259. doi:10.1016/j.procs.2021.01.004

Salman, A., Semwal, A., Bhatt, U., & Thakkar, V. M. (2017, October 13). Leaf classification and identification using Canny Edge Detector and SVM classifier. *Proceedings of the International Conference on Inventive Systems and Control, ICISC 2017*. IEEE. 10.1109/ICISC.2017.8068597

Salnikov, V., Filin, A., & Burema, H. Hybrid Airship-Drone Farm Robot System for Crop Dusting, Planting, Fertilizing and Other Field Jobs, United States Patent US9852644B2, Dec. 26, 2017.

Samal, K. K., Babu, K. S., Das, S. K., & Acharaya, A. (2019). Time Series based Air Pollution Forecasting using SARIMA and Prophet Model. *Proceedings of the 2019 International Conference on Information Technology and Computer Communications*. IEEE. 10.1145/3355402.3355417

Samikannu, R., Koshariya, A. K., Poornima, E., Ramesh, S., Kumar, A., & Boopathi, S. (2023). Sustainable Development in Modern Aquaponics Cultivation Systems Using IoT Technologies. In *Human Agro-Energy Optimization for Business and Industry* (pp. 105–127). IGI Global. doi:10.4018/978-1-6684-4118-3.ch006

Sampath, B. C. S., & Myilsamy, S. (2022). Application of TOPSIS Optimization Technique in the Micro-Machining Process. In Trends, Paradigms, and Advances in Mechatronics Engineering (pp. 162–187). IGI Global. doi:10.4018/978-1-6684-5887-7.ch009

Sampath, B., Naveenkumar, N., Sampathkumar, P., Silambarasan, P., Venkadesh, A., & Sakthivel, M. (2021). Experimental comparative study of banana fiber composite with glass fiber composite material using Taguchi method. *Materials Today: Proceedings*, *49*(5), 1475–1480. doi:10.1016/j.matpr.2021.07.232

Sandeep, T., Manushree, B., Rahul, S., & Bharath, T. (2022). Crop protection from animals using CNN. *International Research Journal of Modernization in Engineering Technology and Science, 1*(1), 12-18. https://www.irjmets.com/uploadedfiles/paper/issue_3_march_2022/19639/final/fin_irjmets1646900628.pdf

Sandino, J., & Gonzalez, F. A Novel Approach for invasive weeds and vegetation surveys using UAS and artifcial intelligence. 2018 23rd Int Conf Methods Models Autom Robo. 2018.

Sanghvi, Y., Gupta, H., Doshi, H., Koli, D., Ansh, A., & Gupta, U. Comparison of Self organizing maps and Sammon's mapping on agricultural datasets for precision agriculture. In Proceedings of the 2015.International Conference on Innovations in Information, Embedded and Communication Systems (ICIIECS), Coimbatore, India, 19–20 March 2015; pp. 1–5. 10.1109/ICIIECS.2015.7193206

Santi, E., Pettinato, S., Paloscia, S., Dabboor, M., Notarnicola, C., Padovano, A., Greifeneder, F., & Cuozzo, G. (2018). Estimating soil moisture from C and X band SAR using machine learning algorithms and compact polarimetry. *International Geoscience and Remote Sensing Symposium (IGARSS)*. IEEE. 10.1109/IGARSS.2018.8518469

Santos, L., Santos, F., Mendes, J., Costa, P., Lima, J., Reis, R., & Shinde, P. (2020). Path planning aware of robot's center of mass for steep slope vineyards. *Robotica*, *38*(4), 684–698. doi:10.1017/S0263574719000961

Saradhambal, G. (2018). Plant Disease Detection and its Solution using Image Classification. *International Journal of Pure and Applied Mathematics*, *119*(14), 879–884.

Sarghini, F., & De Vivo, A. (2017). Analysis of preliminary design requirements of a heavy lift multi rotor drone for agricultural use. *Chemical Engineering Transactions*, *58*, 625–630.

Sarghini, F., & De Vivo, A. (2017). Interference analysis of a heavy lift multi rotor drone flow field and transported spraying system. [30]. *Chemical Engineering Transactions*, *58*, 631–636.

Scherrer, B., Sheppard, J., Jha, P., & Shaw, J. A. (2019). Hyperspectral imaging and neural networks to classify herbicide-resistant weeds. *Journal of Applied Remote Sensing*, *13*(4), 1. doi:10.1117/1.JRS.13.044516

Seireg, H. R., Omar, Y. M. K., El-Samie, F. E. A., El-Fishawy, A. S., & Elmahalawy, A. (2022). Ensemble machine learning techniques using computer simulation data for wild blueberry yield prediction. *IEEE Access : Practical Innovations, Open Solutions*, *10*, 64671–64687. doi:10.1109/ACCESS.2022.3181970

Selvakumar, S., Adithe, S., Isaac, J. S., Pradhan, R., Venkatesh, V., & Sampath, B. (2023). A Study of the Printed Circuit Board (PCB) E-Waste Recycling Process. In Sustainable Approaches and Strategies for E-Waste Management and Utilization (pp. 159–184). IGI Global.

Selvaraj, M. G., Vergara, A., Ruiz, H., Safari, N., Elayabalan, S., Ocimati, W., & Blomme, G. (2019). AI-powered banana diseases and pest detection. *Plant Methods*, *15*(1), 1–11. doi:10.118613007-019-0475-z

Senthil, T. S. R. Ohmsakthi vel, Puviyarasan, M., Babu, S. R., Surakasi, R., & Sampath, B. (2023). Industrial Robot-Integrated Fused Deposition Modelling for the 3D Printing Process. In Development, Properties, and Industrial Applications of 3D Printed Polymer Composites (pp. 188–210). IGI Global. doi:10.4018/978-1-6684-6009-2.ch011

Seru, L. K., Seru, & Gandham, S. M. (2022). *Crop recommendation system using machine learning algorithm.* Department of Computer Science and Engineering, School of computing, Sathyabama institute of science and technology.

Shafik, W., Mojtaba Matinkhah, S., Shokoor, F., & Nur Sanda, M. (2021). Internet of Things-Based Energy Efficiency Optimization Model in Fog Smart Cities. *JOIV : International Journal on Informatics Visualization, 5*(2), 105–112. doi:10.30630/joiv.5.2.373

Shafi, U., Mumtaz, R., García-Nieto, J., Hassan, S. A., Zaidi, S. A. R., & Iqbal, N. (2019). Precision agriculture techniques and practices: From considerations to applications. *Sensors (Basel), 19*(17), 3796. doi:10.339019173796 PMID:31480709

Shank, D. B., Hoogenboom, G., & McClendon, R. W. (2008). Dewpoint temperature prediction using artificial neural networks. *Journal of Applied Meteorology and Climatology, 47*(6), 1757–1769. doi:10.1175/2007JAMC1693.1

Shao, L., Zhu, F., & Li, X. (2014). Transfer learning for visual categorization: A survey. *IEEE Transactions on Neural Networks and Learning Systems, 26*(5), 1019–1034. doi:10.1109/TNNLS.2014.2330900 PMID:25014970

Sharma, P., & Rathi, Y. (2016). Efficient densit- based clustering using automatic parameter detection. *Proceedings of the International Congress on Information and Communication Technology.* Springer.

Sharma, S. K., Sharma, D. P., & Verma, J. K. (2021). Study on Machine-Learning Algorithms in Crop Yield Predictions specific to Indian Agricultural Contexts. *IEEE International Conference on Computational Performance Evaluation (ComPE),* (pp. 155–166). IEEE. 10.1109/ComPE53109.2021.9752260

Sharma, V., Verma, A., & Goel, N. (2020). Classification Techniques for Plant Disease Detection. Inte*rnational Journal of Recent Technology and Engineering (IJRTE), 8*(6).

Sharma, A., Jain, A., Gupta, P., & Chowdary, V. (2021). Machine Learning Applications for Precision Agriculture: A Comprehensive Review. *IEEE Access : Practical Innovations, Open Solutions, 9*, 4843–4873. doi:10.1109/ACCESS.2020.3048415

Shashi, V., Manoj, N., Hemanth, M., Namith, M. N., & Surekha, M. (2022). Smart Crop Protection System from Birds and Animals. In A. Pandian & V. Balaji (Eds.), *Intelligent Agriculture: The Role of IoT, Big Data and AI.* IJRTC. https://ijcrt.org/papers/IJCRT2207476.pdf

Shekhar, Y., Dagur, E., & Mishra, S. (2017). Intelligent IoT Based Automated Irrigation System. *International Journal of Applied Engineering Research: IJAER, 12*(18), 7306–7320.

Shevchenko, I., & Aliiev, E. (2022). Precise grading and sorting of sunflower plant materials in industrial facilities. *Journal of Central European Agriculture, 23*(2), 327–341. doi:10.5513/JCEA01/23.2.3345

Shivhare, N., Rahul, A. K., Dwivedi, S. B., & Dikshit, P. K. (2019). ARIMA based daily weather forecasting tool: A case study for Varanasi. *Mausam (New Delhi), 70*(1), 133–140. doi:10.54302/mausam.v70i1.179

Shokouhifar, M., & Ranjbarimesan, M. (2022). Multivariate time-series blood donation/demand forecasting for resilient supply chain management during COVID-19 pandemic. *Cleaner Logistics and Supply Chain, 5*, 100078. doi:10.1016/j.clscn.2022.100078

Shorten, C., & Khoshgoftaar, T. M. (2019). A survey on image data augmentation for deep learning. *Journal of Big Data, 6*(1), 1–48. doi:10.118640537-019-0197-0

Siddiqui, F., Gupta, S., Dubey, S., Murtuza, S., & Jain, A. (2020, January). Classification and diagnosis of invasive ductal carcinoma using deep learning. In *2020 10th International Conference on Cloud Computing, Data Science & Engineering (Confluence)* (pp. 242-247). IEEE. 10.1109/Confluence47617.2020.9058077

Simonyan, K., & Zisserman, A. (2015). Very deep convolutional networks for large-scale image recognition. *The International Conference on Learning Representations*, (pp. 1–14). ACM.

Singh, D., Jain, N., Jain, P., Kayal, P., Kumawat, S., & Batra, N. (2020). PlantDoc: A dataset for visual plant disease detection. *ACM International Conference Proceeding Series*, (pp. 249–253). ACM. 10.1145/3371158.3371196

Singh, V. (2016). Detection of Plant Leaf Diseases Using Image Segmentation and Soft Computing Techniques. *Information Processing in Agriculture*, *8*, 252–277.

Singh, V., & Misra, A. K. (2017). Detection of plant leaf diseases using image segmentation and soft computing techniques. *Information Processing in Agriculture*, *4*(1), 41–49. doi:10.1016/j.inpa.2016.10.005

Sivanantham, V., Sangeetha, V., & Alnuaim, A. A. Hatamleh, Wesam A., Anilkumar, C., Hatamleh, A. A., & Sweidan, D. (2022). Quantile correlative deep feedforward multilayer Perceptron for crop yield prediction. *Computers and Electrical Engineering*, *98*. doi:10.1016/j.compeleceng.2022.107696

Sladojevic, S., Arsenovic, M., Anderla, A., Culibrk, D., & Stefanovic, D. (2016). Deep Neural Networks Based Recognition of Plant Diseases by Leaf Image Classification. In *Computational Intelligence and Neuroscience*. doi:10.1155/2016/3289801

Sladojevic, S., Arsenovic, M., & Anderla, A. (2020). Deep Learning Techniques for Plant Disease Detection and Diagnosis: A Review. *Computers and Electronics in Agriculture*, *166*, 105073. doi:10.1016/j.compag.2019.105073

Small Farmers Agri-Business Consortium (SFAC). (n.d.). Department of Agriculture & Farmers Welfare Ministry of Agriculture & Farmers Welfare. Government of India. Weather Forecast. https://enam.gov.in/web/weather_forecast

Somasundaram, R. S., Nagamani, K., Florence, M. L., & Swamydoss, D. (2022). *Estimation and prediction of crop yielding rate using machine learning techniques.* 2022 International Conference on Computer Communication and Informatics, ICCCI, Coimbatore, India.

Soneji, H. (2020). Historical Weather Data for Indian Cities. *Historical Weather Data for Indian Cities. Kaggle.* doi:10.34740/KAGGLE/DSV/1129180

Spoorthi, S., & Manasa, V. K. (2017). Freyr drone: pesticide/fertilizers spraying drone – an agricultural approach. International Conference on Computing and Communications Technologies, (pp. 252–255). IEEE. 10.1109/ICCCT2.2017.7972289

Srikanth, N., Aishwarya, K., Kavita, H. M., Rashmi Reddy, K., & Soumya, D. B. (2019). Smart crop protection system from animals and fire using Arduino. *International Journal of Electrical, Electronics, and Computer Engineering*. https://www.technoarete.org/common_abstract/pdf/IJERECE/v6/i4/Ext_90718.pdf

Srinivasan, R., Shashikumar, B. N., & Singh, S. K. (2022). Mapping of soil nutrient variability and delineating site-specific management zones using fuzzy clustering analysis in eastern coastal region, India. *Photonirvachak (Dehra Dun)*, *50*(3), 533–547. doi:10.100712524-021-01473-9

Stenberg, B., Viscarra Rossel, R. A., Mouazen, A. M., & Wetterlind, J. (2010). Visible and Near Infrared Spectroscopy in Soil Science. In Advances in Agronomy, Vol. 107, (pp. 163–215). Academic Press. doi:10.1016/S0065-2113(10)07005-7

Stepanov, A., Dubrovin, K., & Sorokin, A. (2022). Function fitting for modeling seasonal normalized difference vegetation index time series and early forecasting of soybean yield. *The Crop Journal*, *10*(5), 1452–1459. doi:10.1016/j.cj.2021.12.013

Stöcker, C., Bennett, R., Nex, F., Gerke, M., & Zevenbergen, J. (2017). Review of the current state of UAV regulations. *Remote Sensing (Basel)*, *9*(5), 459. doi:10.3390/rs9050459

Subramanian, M., Shanmugavadivel, K., & Nandhini, P. S. (2022). On fine-tuning deep learning models using transfer learning and hyper-parameters optimization for disease identification in maize leaves. *Neural Computing & Applications*, *34*(16), 13951–13968. doi:10.100700521-022-07246-w

Sudheer, C. L., Kaveri, K., Kavya, M., Nath, G. K., Javeed, P. M., & Kumar, B. N. (2023). Crop Protection and Monitoring from Animal Attacks by Using IoT Solutions. In A. Pandian & V. Balaji (Eds.), *Intelligent Agriculture: The Role of IoT, Big Data and AI*. IJCRT. https://www.ijcrt.org/papers/IJCRT2303875.pdf

Sujatha, R., Chatterjee, J. M., Jhanjhi, N., & Brohi, S. N. (2021). Performance of deep learning vs machine learning in plant leaf disease detection. *Microprocessors and Microsystems*, *80*, 103615. doi:10.1016/j.micpro.2020.103615

Sujawata, G. & Chouhanb, J. (2021). Application of Artificial Intelligence in detection of diseases in plants: A Survey. *Turkish Journal of Computer and Mathematics Education, 12(*3), 3301-3305

Sumana, P. B., Sanjana, R., Sharanya, M., & Harish, N. J. (2021). Smart crop protection system from wild animals and birds using IoT. *International Journal of Advanced Research, Ideas and Innovations in Technology (IJARIIT), 7*(2), 2232-2237. https://www.ijariit.com/manuscripts/v7i4/V7I4-1724.pdf

Sun, J., Lai, Z., Di, L., Sun, Z., Tao, J., & Shen, Y. (2020). Multilevel deep learning network for county-level corn yield estimation in the U. S. Corn Belt. *IEEE Journal of Selected Topics in Applied Earth Observations and Remote Sensing*, *13*, 5048–5060. doi:10.1109/JSTARS.2020.3019046

Suomalainen, J., Anders, N., Iqbal, S., Roerink, G., Franke, J., Wenting, P., Hünniger, D., Bartholomeus, H., Becker, R., & Kooistra, L. (2014). A lightweight hyperspectral mapping system and photogrammetric processing chain for unmanned aerial vehicles. *Remote Sensing (Basel)*, *6*(11), 11013–11030. doi:10.3390/rs61111013

Svensgaard, J., Jensen, S. M., Westergaard, J. C., Nielsen, J., Christensen, S., & Rasmussen, J. (2019). Can reproducible comparisons of cereal genotypes be generated in feld experiments based on AV imagery using RGB cameras? *European Journal of Agronomy*, *106*, 49–57. doi:10.1016/j.eja.2019.03.006

Swarnakantha, S., Chathurika, B., Weragoda, K. V., Bowatte, W. M. I. K., Thalawala, E. V., & Bandara, M. M. U. L. (2022). Decision-making platform for SMART plantation agriculture using machine learning and image processing. *7th International conference for Convergence in Technology (I2CT),* (pp. 1–6). IEEE Publications. 10.1109/I2CT54291.2022.9825063

Szegedy, C., Vanhoucke, V., Ioffe, S., Shlens, J., & Wojna, Z. Rethinking the inception architecture for computer vision, In *Proceedings of the IEEE conference on computer vision and pattern recognition,* (pp. 2818–2826). IEEE. 10.1109/CVPR.2016.308

Tagad, C. K., Kim, H. U., Aiyer, R. C., More, P., Kim, T., Moh, S. H., Kulkarni, A., & Sabharwal, S. G. (2013). A sensitive hydrogen peroxide optical sensor based on polysaccharide stabilized silver nanoparticles. *RSC Advances*, *3*(45), 22940–22943. doi:10.1039/c3ra44547j

Tahaseen, M., & Moparthi, N. R. (2021). An assessment of the machine learning algorithms used in agriculture. *5th International Conference on Electronics, Communication and Aerospace Technology (ICECA),* (pp. 1579–1584). IEEE Publications. 10.1109/ICECA52323.2021.9675918

Tahmasebi, M., Gohari, M., & Emami, A. (2022). An autonomous pesticide sprayer robot with a color-based vision system. *International Journal of Robotics and Control Systems*, *2*(1), 115–123. doi:10.31763/ijrcs.v2i1.480

Talaviya, T., Shah, D., Patel, N., Yagnik, H., & Shah, M. (2020). Implementation of artificial intelligence in agriculture for optimisation of irrigation and application of pesticides and herbicides. *Artificial Intelligence in Agriculture*, *4*, 58–73. doi:10.1016/j.aiia.2020.04.002

Tang, Z., Chen, Z., Qi, F., Zhang, L., & Chen, S. (2021). Pest-YOLO: Deep Image Mining and Multi-Feature Fusion for Real-Time Agriculture Pest Detection. *2021 IEEE International Conference on Data Mining (ICDM). 2021-December*, pp. 1348–1353. IEEE. 10.1109/ICDM51629.2021.00169

Tariq, A., Yan, J., Gagnon, A. S., Riaz Khan, M., & Mumtaz, F. (2022). Mapping of cropland, cropping patterns and crop types by combining optical remote sensing images with decision tree classifier and random forest. *T and F Online*. Http://Www.Tandfonline.Com/Action/JournalInformation?Show=aimsScope&journalCode=tgsi20#.VsXpLiCLRhE/

Technology Innovation Hub for IoT and IoE by IIT-Bombay. (n.d.). *Agriculture*. TIH. https://www.tih.iitb.ac.in/technology-verticals/agriculture/ [Accessed: 5/1/2023].

Tejoyadav, M., Nayak, R., & Pati, U. C. (2022). A Comparative Analysis of Multivariate Statistical Time Series Models for Water Quality Forecasting of the River Ganga. In *Ambient Intelligence in Health Care: Proceedings of ICAIHC 2022* (pp. 429–441). Springer. 10.1007/978-981-19-6068-0_41

Teng, Y., Zhang, J., Dong, S., Zheng, S., & Liu, L. (2022). MSR-RCNN: A Multi-Class Crop Pest Detection Network Based on a Multi-Scale Super-Resolution Feature Enhancement Module. *Frontiers in Plant Science*, *13*, 1–12. doi:10.3389/fpls.2022.810546 PMID:35310676

Thai-Nghe, N., Tri, N. T., & Hoa, N. H. (2022). Deep Learning for Rice Leaf Disease Detection in Smart Agriculture. In *Artificial Intelligence in Data and Big Data Processing* (pp. 659–670). Springer International Publishing. doi:10.1007/978-3-030-97610-1_52

Thilakarathne, N. N., Bakar, M. S. A., Abas, P. E., & Yassin, H. (2022). A cloud enabled crop recommendation platform for machine learning-driven precision farming. *Sensors, 22*(16), 6299.

Tian, H., Wang, T., Liu, Y., Qiao, X., & Li, Y. (2020). Computer vision technology in agricultural automation—A review. *Information Processing in Agriculture*, *7*(1), 1–19. doi:10.1016/j.inpa.2019.09.006

Tiwari, O., Goyal, V., Kumar, P., & Vij, S. (2018). An experimental set up for utilizing convolutional neural network in automated weed detection. Proc 2019 4th Int Conf Internet Things Smart Innov Usages. IEEE.

Toharudin, T., Pontoh, R. S., Caraka, R. E., Zahroh, S., Lee, Y., & Chen, R. C. (2020). Employing long short-term memory and Facebook prophet model in air temperature forecasting. *Communications in Statistics. Simulation and Computation*, 1–24. doi:10.1080/03610918.2020.1854302

Traore, B. B., Kamsu-Foguem, B., & Tangara, F. (2017). Data mining techniques on satellite images for discovery of risk areas. *Expert Systems with Applications*, *72*, 443–456. doi:10.1016/j.eswa.2016.10.010

Tripodi, P., Nicastro, N., Pane, C., Cammarano, D., Tripodi, P., Nicastro, N., Pane, C., & Cammarano, D. (2022). Digital applications and artificial intelligence in agriculture toward next-generation plant phenotyping. *Crop & Pasture Science*, *74*(6), 597–614. doi:10.1071/CP21387

Trojovský, P., Dhasarathan, V., & Boopathi, S. (2023). Experimental investigations on cryogenic friction-stir welding of similar ZE42 magnesium alloys. *Alexandria Engineering Journal*, *66*(1), 1–14. doi:10.1016/j.aej.2022.12.007

Tsouros, D. C., Bibi, S., & Sarigiannidis, P. G. (2019). A Review on UAV-Based Applications for Precision Agriculture. *Information, 10*(11), 349. doi:10.3390/info10110349

Tyagi, S., Pandey, D. K., Putrevu, D., Srivastava, P. K., & Misra, A. (2021). Machine Learning Based Soil Moisture Retrieval Algorithm and Validation at Selected Agricultural Sites Over India Using CYGNSS Data. *International Geoscience and Remote Sensing Symposium (IGARSS)*, (pp. 6335–6338). IEEE. 10.1109/IGARSS47720.2021.9555095

Uddin, M. S., & Bansal, J. C. (2021). *Introduction to computer vision and machine learning applications in agriculture*, 1–8.

Uddin, M. S., & Bansal, J. C. (Eds.). (2022). *Computer Vision and Machine Learning in Agriculture, 2.* doi:10.1007/978-981-16-9991-7

Udutalapally, V., Mohanty, S. P., Pallagani, V., & Khandelwal, V. (2021). sCrop: A novel device for sustainable automatic disease prediction, crop selection, and irrigation in Internet-of-agro-things for Smart agriculture. In IEEE Sensors Journal, 21(16), 17525–17538. doi:10.1109/JSEN.2020.3032438

Umamaheswari, S., Arjun, R., & Meganathan, D. (2018, July 2). Weed Detection in Farm Crops using Parallel Image Processing. *2018 Conference on Information and Communication Technology, CICT 2018.* IEEE. 10.1109/INFOCOMTECH.2018.8722369

United Nations. (2017). *World population projected to reach 9.8 billion in 2050, and 11.2 billion in 2100.* United Nations. https://www.un.org/en/desa/world-population-projected-reach-98-billion-2050-and-112-billion-2100/

Uryasheva, A., Kalashnikova, A., Shadrin, D., Evteeva, K., Moskovtsev, E., & Rodichenko, N. (2022). Computer vision-based platform for apple leaves segmentation in field conditions to support digital phenotyping. *Computers and Electronics in Agriculture, 201*, 107269. doi:10.1016/j.compag.2022.107269

Van Dijk, M., Morley, T., Rau, M. L., & Saghai, Y. (2021). A meta-analysis of projected global food demand and population at risk of hunger for the period 2010–2050. *Nature Food 2021 2:7, 2*(7), 494–501. doi:10.1038/s43016-021-00322-9

van Klompenburg, T., Kassahun, A., & Catal, C. (2020). Crop yield prediction using machine learning: A systematic literature review. *Computers and Electronics in Agriculture, 177.* doi:10.1016/j.compag.2020.105709

Vanitha, S. K. R., & Boopathi, S. (2023). Artificial Intelligence Techniques in Water Purification and Utilization. In *Human Agro-Energy Optimization for Business and Industry* (pp. 202–218). IGI Global., doi:10.4018/978-1-6684-4118-3.ch010

Varshney, D., Babukhanwala, B., Khan, J., Saxena, D., & Singh, A. K. (2022). Plant Disease Detection Using Machine Learning Techniques. *2022 3rd International Conference for Emerging Technology, INCET 2022.* IEEE.

Vashisht, S., Kumar, P., & Trivedi, M. C. (2022). Improvised extreme learning Machine for crop yield prediction. *3rd IEEE International Conference on Intelligent Engineering and Management (ICIEM),* (pp. 754–757). IEEE. 10.1109/ICIEM54221.2022.9853054

Vennila, T., Karuna, M. S., Srivastava, B. K., Venugopal, J., Surakasi, R., & B., S. (2023). New Strategies in Treatment and Enzymatic Processes. In *Human Agro-Energy Optimization for Business and Industry* (pp. 219–240). IGI Global. doi:10.4018/978-1-6684-4118-3.ch011

Verma, S., Tripathi, S., Singh, A., Ojha, M., & Saxena, R. R. (2021). Insect Detection and Identification using YOLO Algorithms on Soybean Crop. *IEEE Region 10 Annual International Conference, Proceedings/TENCON, 2021-December*, (pp. 272–277). IEEE. 10.1109/TENCON54134.2021.9707354

Vihari, M. M., & Nelakuditi, U. R. (2018). IoT based unmanned aerial vehicle system for agriculture applications, in: International Conference on Smart Systems and Inventive Technology, (pp. 26–28). IEEE. 10.1109/ICSSIT.2018.8748794

Vijay, K. & Vani, K. (2018). Agricultural Robot: Leaf Disease Detection and Monitoring the Field Condition Using Machine Learning and Image Processing. *International Journal of Computational Intelligence Research, 14*(7), pp. 551-561

Vikhram, B., Revathi, B., Shanmugapriya, R., Sowmiya, S., & Pragadeeswaran, G. (2017). Animal detection system in farm areas. [IJARCCE]. *International Journal of Advanced Research in Computer and Communication Engineering, 6*(1), 1–5. https://ijarcce.com/upload/2017/march-17/IJARCCE%20137.pdf

Vining, K. C. (1990). Effects of weather on agricultural crops and livestock: An overview. *The International Journal of Environmental Studies*, *36*(1-2), 27–39. doi:10.1080/00207239008710581

Vlachopoulos, O., Leblon, B., Wang, J., Haddadi, A., LaRocque, A., & Patterson, G. (2022). Evaluation of crop health status with UAS multispectral imagery. *IEEE Journal of Selected Topics in Applied Earth Observations and Remote Sensing*, *15*, 297–308. doi:10.1109/JSTARS.2021.3132228

Wakchaure, M., Patle, B. K., & Mahindrakar, A. K. (2023). Application of AI techniques and robotics in agriculture: A review. *Artificial Intelligence in the Life Sciences*, *3*, 100057. doi:10.1016/j.ailsci.2023.100057

Wang, X., Wang, S., Huang, J., Lv, S., & Li, X. (2022). Design of an autonomous navigation intelligent spraying robot. *IEEE 6th Information Technology and Mechatronics Engineering Conference, ITOEC 2022*, (pp. 496–499). IEEE.

Wang, J., Li, J., Wang, X., Wang, T., & Sun, Q. (2022). An quality prediction model based on CNN-BiNLSTM-attention. *Environment, Development and Sustainability*, 1–16. doi:10.100710668-021-02102-8

Wang, J., Lin, C., Ji, L., & Liang, A. (2012). A new automatic identification system of insect images at the order level. *Knowledge-Based Systems*, *33*, 102–110. doi:10.1016/j.knosys.2012.03.014

Wang, Q. J., Zhang, S. Y., Dong, S. F., Zhang, G. C., Yang, J., Li, R., & Wang, H. Q. (2020). Pest24: A large-scale very small object data set of agricultural pests for multi-target detection. *Computers and Electronics in Agriculture*, *175*, 105585. doi:10.1016/j.compag.2020.105585

Wang, R., Jiao, L., Xie, C., Chen, P., Du, J., & Li, R. (2021). S-RPN: Sampling-balanced region proposal network for small crop pest detection. *Computers and Electronics in Agriculture*, *187*, 106290. doi:10.1016/j.compag.2021.106290

Wang, Y., Liu, H., & Chen, Y. (2019). Deep Learning Approaches for Crop Yield Prediction and Nitrogen Status Estimation in Precision Agriculture. *IEEE Access : Practical Innovations, Open Solutions*, *7*, 92025–92036. doi:10.1109/ACCESS.2019.2922124

Wiesner-Hanks, T., Stewart, E. L., Kaczmar, N., Dechant, C., Wu, H., Nelson, R. J., Lipson, H., & Gore, M. A. (2018). Image set for deep learning: Field images of maize annotated with disease symptoms. *BMC Research Notes*, *11*(1), 1–3. doi:10.118613104-018-3548-6 PMID:29970178

Wilson, D. R., & Martinez, & T. R. (2001). The need for small learning rates on large problems. IJCNN'01. *International Joint Conference on Neural Net- works. Proceedings*. IEEE. 10.1109/IJCNN.2001.939002

World Bank. (n.d.). [Data set]. World Bank. https://data.worldbank.org/

Wspanialy, P., Brooks, J., & Moussa, M. (2020). *An image labeling tool and agricultural dataset for deep learning*. arXiv preprint arXiv:2004.03351.

Wu, K., Wu, J., Feng, L., Yang, B., Liang, R., Yang, S., & Zhao, R. (2021). An attention-based CNN-LSTM-BiLSTM model for short-term electric load forecasting in integrated energy system. *International Transactions on Electrical Energy Systems*, *31*(1), e12637. doi:10.1002/2050-7038.12637

Wu, X., Zhan, C., Lai, Y.-K., Cheng, M.-M., & Yang, J. (2019). Ip102: A large-scale benchmark dataset for insect pest recognition. *Proceedings of the IEEE/CVF conference on computer vision and pattern recognition*, (pp. 8787–8796). IEEE. 10.1109/CVPR.2019.00899

Wu, Z., Zhao, W., & Lv, Y. (2022). An ensemble LSTM-based AQI forecasting model with decomposition-reconstruction technique via CEEMDAN and fuzzy entropy. *Air Quality, Atmosphere & Health*, *15*(12), 2299–2311. doi:10.100711869-022-01252-6 PMID:36196368

Xie, C., Wang, R., Zhang, J., Chen, P., Dong, W., Li, R., Chen, T., & Chen, H. (2018). Multi-level learning features for automatic classification of field crop pests. *Computers and Electronics in Agriculture*, *152*, 233–241. doi:10.1016/j.compag.2018.07.014

Yadav, A. K., Jain, A., Morato, J. L., & Yadav, D. (2021). Retinal Blood Vessel Segmentation using Convolutional Neural Networks. In KDIR (pp. 292-298). doi:10.5220/0010719500003064

Yadav, D., Gupta, A., Jain, A., & Yadav, A. K. (2022). Plant leaf disease detection using CNN with transfer learning and XGBoost. *International Journal of Data Analysis Techniques and Strategies*, *14*(3), 244–265. doi:10.1504/IJDATS.2022.128273

Yallappa, D., Veerangouda, M., Maski, D., Palled, V., & Bheemanna, M. (2017, October) Development and evaluation of drone mounted sprayer for pesticide applications to crops. *IEEE Global Humanitarian Technology Conference (GHTC) 2017 IEEE* (pp. 1-7). IEEE. 10.1109/GHTC.2017.8239330

Yamashita, R., Nishio, M., Do, R. K. G., & Togashi, K. (2018). Convolutional neural networks: An overview and application in radiology. *Insights Into Imaging*, *9*(4), 611–629. doi:10.100713244-018-0639-9 PMID:29934920

Yamparla, R., Shaik, H. S., Guntaka, N. S. P., Marri, P., & Nallamothu, S. (2022). Crop yield prediction using random forest algorithm. *7th International Conference on Communication and Electronics Systems (ICCES)*, (pp. 1538–1543). IEEE. 10.1109/ICCES54183.2022.9835756

Yang, C., Everitt, J. H., Bradford, J. M., & Murden, D. (2009a). Comparison of airborne multispectral and hyperspectral imagery for estimating grain sorghum yield. *Transactions of the ASAE. American Society of Agricultural Engineers*, *52*, 641–649.

Yang, Z., Jianjun, L., Faqiri, H., Shafik, W., Talal Abdulrahman, A., Yusuf, M., & Sharawy, A. M. (2021). Green internet of things and big data application in smart cities development. *Complexity*, *2021*, 2021. doi:10.1155/2021/4922697

Yang, Z., Rao, M. N., Elliott, N. C., Kindler, S. D., & Popham, T. W. (2009). Differentiating stress induced by greenbugs and Russian wheat aphids in wheat using remote sensing. *Computers and Electronics in Agriculture*, *67*(1-2), 64–70. doi:10.1016/j.compag.2009.03.003

Yanliang, Z., Qi, L., & Wei, Z. (2017). Design and test of a six-rotor unmanned aerial vehicle (UAV) electrostatic spraying system for crop protection. *International Journal of Agricultural and Biological Engineering*, *10*(6), 68–76. doi:10.25165/j.ijabe.20171006.3460

Yialouris, A. B., & Sideridis, A. B. (1996). An expert system for tomato diseases. *Computers and Electronics in Agriculture*, *14*(1), 61–76. doi:10.1016/0168-1699(95)00037-2

Yoo, M. (2022). An Efficient Multivariate Autoscaling Framework Using Bi-LSTM for Cloud Computing. *Applied Sciences (Basel, Switzerland)*, *12*(7), 3523. doi:10.3390/app12073523

Yun, S., Xianfeng, W., Shanwen, Z., & Chuanlei, Z. (2015). PNN based crop disease recognition with leaf image features and meteorological data. *International Journal of Agricultural and Biological Engineering*, *8*(4), 60–68.

Yupapin, P., Trabelsi, Y., Nattappan, A., & Boopathi, S. (2023). Performance Improvement of Wire-Cut Electrical Discharge Machining Process Using Cryogenically Treated Super-Conductive State of Monel-K500 Alloy. *Iranian Journal of Science and Technology. Transaction of Mechanical Engineering*, *47*(1), 267–283. doi:10.100740997-022-00513-0

Zakaria, N. N., Othman, M., Sokkalingam, R., Daud, H., Abdullah, L., & Abdul Kadir, E. (2019, September 22). Markov Chain Model Development for Forecasting Air Pollution Index of Miri, Sarawak. *Sustainability (Basel)*, *11*(19), 5190. doi:10.3390u11195190

Zamani Joharestani, M., Cao, C., Ni, X., Bashir, B., & Talebiesfandarani, S. (2019). PM2. 5 prediction based on random forest, XGBoost, and deep learning using multisource remote sensing data. *Atmosphere (Basel)*, *10*(7), 373. doi:10.3390/atmos10070373

Zendehboudi, A., Baseer, M. A., & Saidur, R. (2018). Application of support vector machine models for forecasting solar and wind energy resources: A review. *Journal of Cleaner Production*, *199*, 272–285. doi:10.1016/j.jclepro.2018.07.164

Zhalehdoost, A., & Taleai, M. (2022). A Review of the Application of Machine Learning and Geospatial Analysis Methods in Air Pollution Prediction. *Pollution*, *8*(3), 904–933. doi:10.22059/poll.2022.336044.1300

Zhang, H., Yanne, P., & Liang, S. (2012). Plant species classification using leaf shape and texture. *Proceedings of the 2012 International Conference on Industrial Control and Electronics Engineering, ICICEE 2012*, (pp. 2025–2028). IEEE. 10.1109/ICICEE.2012.538

Zhang, S., Guo, B., Dong, A., He, J., Xu, Z., & Chen, S. X. (2017). Cautionary tales on air-quality improvement in Beijing. *Proceedings - Royal Society. Mathematical, Physical and Engineering Sciences*, *473*(2205), 20170457. doi:10.1098/rspa.2017.0457 PMID:28989318

Zhang, W., Huang, H., Sun, Y., & Wu, X. (2022). AgriPest-YOLO: A rapid light-trap agricultural pest detection method based on deep learning. *Frontiers in Plant Science*, *13*, 1–16. doi:10.3389/fpls.2022.1079384 PMID:36589124

Zhang, Y., Cai, W., Fan, S., Song, R., & Jin, J. (2022). Object Detection Based on YOLOv5 and GhostNet for Orchard Pests. *Information (Basel)*, *13*(11), 548. doi:10.3390/info13110548

Zhang, Y., Ma, B., Hu, Y., Li, C., & Li, Y. (2022). Accurate cotton diseases and pests detection in complex background based on an improved YOLOX model. *Computers and Electronics in Agriculture*, *203*, 107484. doi:10.1016/j.compag.2022.107484

Zhao, N., Zhou, L., Huang, T., Taha, M. F., He, Y., & Qiu, Z. (2022). Development of an automatic pest monitoring system using a deep learning model of DPeNet. *Measurement*, *203*, 111970. doi:10.1016/j.measurement.2022.111970

Zheng, H., & Wu, Y. (2019). A xgboost model with weather similarity analysis and feature engineering for short-term wind power forecasting. *Applied Sciences (Basel, Switzerland)*, *9*(15), 3019. doi:10.3390/app9153019

Zhou, J., Lu, X., Yang, R., Chen, H., Wang, Y., Zhang, Y., Huang, J., & Liu, F. (2022). Developing novel rice yield index using UAV remote sensing imagery fusion technology. *Drones, 6*(6), 151.

Zhou, X., Xu, J., Zeng, P., & Meng, X. (2019). Air pollutant concentration prediction based on GRU method. *Journal of Physics: Conference Series*, *1168*, 032058. doi:10.1088/1742-6596/1168/3/032058

Zia, H., Rehman, A., Harris, N. R., Fatima, S., & Khurram, M. (2021). An Experimental Comparison of IoT-Based and Traditional Irrigation Scheduling on a Flood-Irrigated Subtropical Lemon Farm. *Sensors (Basel)*, *21*(12), 4175. doi:10.339021124175 PMID:34204584

Ziliani, M. G., Altaf, M. U., Aragon, B., Houborg, R., Franz, T. E., Lu, Y., Sheffield, J., Hoteit, I., & McCabe, M. F. (2022). Early season prediction of within-field crop yield variability by assimilating CubeSat data into a crop model. *Agricultural and Forest Meteorology, 313*. doi:10.1016/j.agrformet.2021.108736

About the Contributors

Rajeev Kumar Gupta is working as an Assistant Professor at Pandit Deendayal Energy University (An Autonomous Institution) Gandhinagar, Gujarat. He has completed his M.Tech and PhD from MANIT, Bhopal and is a recipient of the Best Young Researcher Award by RSRI in 2019. He has published more than 30 referred articles in various book chapters, conferences and international repute peer-reviewed journals of Elsevier, Springer, IEEE etc. He has a total of more than ten years of teaching experience. He is a life member of some of the reputed societies like CSI India, IAENG (Hongkong) etc. He has organized several STTP/FDP and taken several expert lectures at various institutes. He has supervised 20 M.Tech thesis and around 40 B.Tech projects in various domains. He is a TPC member and reviewer of several International Conferences. His area of interest includes Machine Learning, Deep Learning and Cloud computing.

Arti Jain (SMIEEE) is working as Assistant Professor (Sr. Grade), Computer Science & Engineering and Information Technology, Jaypee Institute of Information Technology, Noida (Uttar Pradesh), India. She has academic experience of 20+ years. She is an editorial board member of IGI Global and SciencePG. She is an advisory committee member of International Conferences held in Maldives, Turkey, and UAE. She is a reviewer of reputed International Journals and guest-edited Special Issues. She is a TPC member of several International Conferences and has chaired Special Sessions. She is an invited speaker for various workshops, seminars, FDPs, and conferences affiliated with India, China, UK, and USA. She has published 02 books, and 35 research papers in indexed International Journals, book chapters, and conferences. Her research interests include Natural language processing, machine learning, data science, deep learning, social media analytics, and data mining.

John Wang is a professor in the Department of Information Management and Business Analytics at Montclair State University, USA. Having received a scholarship award, he came to the USA and completed his PhD in operations research from Temple University. Due to his extraordinary contributions beyond a tenured full professor, Dr. Wang has been honored with two special range adjustments in 2006 and 2009, respectively. He has published over 100 refereed papers and seventeen books. He has also developed several computer software programs based on his research findings. He serves as Editor-in-Chief for ten Scopus-indexed journals, such as Int. J. of Business Analytics, Int. J. of Information Systems and Supply Chain Management, Int. J. of Information Systems in the Service Sector, Int. J. of Applied Management, Int. J. of Information and Decision Sciences, Int. J. of Data Mining, Modelling and Management, etc. He is the Editor of Encyclopedia of Business Analytics and Optimization (five-volume), Data Warehousing and Mining: Concepts, Methodologies, Tools, and Applications (six-volume) and the Editor of the Encyclopedia of Data Warehousing and Mining, 1st (two-volume) and 2nd (four-volume). His long-term research goal is on the synergy of operations research, data mining and cybernetics.

Santosh Kumar Bharti received the BE degree in Computer Science and Engineering from Visveswaraiah Technological University, Karnataka, India in 2007 and an M.Tech degree in Computer Science and Engineering from Graphic Era University, Dehradun, India in 2012. He has completed his Ph.D. in the Department of Computer Science at the National Institute of Technology, Rourkela, India. Currently, he is working as an Assistant Professor at Pandit Deendayal Petroleum University, Gandhinagar. His research interest includes Opinion Mining and Sarcasm Detection.

Samir B. Patel obtained his Ph.D. Degree from Nirma University in Computer Engineering in the month of October, 2012. He has published more than 18 papers of National and International Repute. He is author of one book. Before joining PDEU, he has worked as Principal, GMFE, Sr. Associate Professor at Nirma University, Assistant Prof., Senior Lecturer at AESICS and before that as lecturer and programmer cum lecturer at CPICA and PDPICA. He has total 23 years of teaching and administrative experience. He is reviewer of various Journals of repute and is a life member of Computer Society of India and ISTE. His area of interest includes, Big Data Analytics, Data Mining and Parallel computing.

***

Siddhartha Arjaria is B.E., Mtech, and Ph.D in computer science and Engineering. He has over 14 years of experience of teaching in institutes like BITS pilani Goa Campus, IIITM Gwalior. Currently, He is working as an Assistant Professor - IT Department in Rajkiya Engineering College, Banda,U.P, India. He has published over 20 papers/Book chapters in various national/international journals/Conference Proceedings. He is a reviewer of many international journals/conferences. He has organized many FDPs/Workshops as Co-Ordinator/Convenor. His research interests include Machine learning, Data Analytics and text mining.

A. Arokiaraj Jovith is a distinguished faculty member in the Department of Networking and Communications at the College of Engineering and Technology, SRM Institute of Science and Technology, located in Kattankulathur, Tamil Nadu, India

Santosh Kumar Bharti has awarded his Ph.D. (CSE) from the National Institute of Technology Rourkela India in 2019. He has a remarkable journey of 10 years as an active researcher, teacher, and collaborator. Currently, he is working as an Assistant Professor in the Department of Computer Science and Engineering at the School of Technology of Pandit Deendayal Energy University, Gandhinagar Gujarat. He has actively contributed to the field of Natural Language Processing: Sentiment analysis, Sarcasm Detection, Text Summarization, etc. Currently, he is supervising 2 Ph.D. students in this domain. He has published more than 50 research papers in top venues of various well-known publishers such as IEEE, ACM, Elsevier, and Springer. Moreover, he has published a few books (Edited/Text Book) and book chapters as well to his credit. With his expertise, he has delivered more than 05 expert talks across India at various institutions. He has also served as a reviewer and a session chair at conferences of IEEE, ACM, Springer, etc. He is a senior IEEE member, professional ACM member and CSI Life member.

Sampath Boopathi completed his undergraduate in Mechanical Engineering and postgraduate in the field of Computer-Aided Design. He completed his Ph.D. from Anna University and his field of research includes Manufacturing and optimization. He published 60 more research articles in Internationally Peer-reviewed journals, one Patent grant, and three published patents.He has 16 more years of academic and research experiences in the various Engineering Colleges in Tamilnadu, India.

Pankaj Dadheech received his Ph.D. degree in Computer Science & Engineering from Suresh Gyan Vihar University (Accredited by NAAC with 'A' Grade), Jaipur, Rajasthan, India. He received his M.Tech. degree in Computer Science & Engineering from Rajasthan Technical University, Kota and he has received his B.E. in Computer Science & Engineering from University of Rajasthan, Jaipur. He has more than 18 years of experience in teaching. He is currently working as an Associate Professor & Dy. HOD in the Department of Computer Science & Engineering (NBA Accredited), Swami Keshvanand Institute of Technology, Management & Gramothan (SKIT), Jaipur, Rajasthan, India. He has published 25 Patents at Intellectual Property India, Office of the Controller General of Patents, Design and Trade Marks, Department of Industrial Policy and Promotion, Ministry of Commerce and Industry, Government of India. He has published 5 Australian Patents at Commissioner of Patents, Intellectual Property Australia, Australian Government. He has 61 publications in various International & National Journals, 55 papers in various National & International conferences. He has published 9 Books & 14 Book Chapters.

Santos Das is an associate Professor in the department of ECE, NIT Rourkela.

Prashant Dwivedi received the B.Tech degree in Electronics and Communication Engineering from the Uttar Pradesh Technical University, Lucknow, Uttar Pradesh, India, in 2015. Presently, he is pursuing his M.Tech degree in electronics and communication engineering (specialization: Signal and Image Processing) from the NIT Rourkela, Odisha, India. His current research interests include Artificial Intelligence and Object Detection.

Rajeev Kumar Gupta is working as a senior Assistant Professor, Pandit Deendayal Energy University (An Autonomous Institution) Gandhinagar, Gujarat. He has completed his M.Tech and PhD from MANIT, Bhopal and a recipient of the Best Young Researcher Award by RSRI in 2019. He has authored two books and published more than 70 referred articles in various book chapters, conferences and international repute peer-reviewed journals of Elsevier, Springer, IEEE etc. He has a total of more than ten years of teaching experience. He is an active member of IEEE and life member of some of the reputed societies like CSI India, IAENG (Hongkong) etc. He has organized several STTP/FDP and taken several expert lectures at various institutes. He has supervised 20 M.Tech thesis and around 40 B.Tech projects in various domains. Currently, he is supervising three PhD students in the domain of AI in agriculture and medical. He is a TPC member and reviewer of several International Conferences and Journals and also serve as a session chair at many conferences. His area of interest includes Machine Learning, Deep Learning and Cloud computing

Tar Janet Gutu, an Expert in Computing with over 10years experience. holds a Master of Science Degree in Computer Science with distinctions from University of Mysore-Manasagangothri in India and a Bachelor of Science Degree in Computer Science from Makerere University. Currently working as an Assistant Lecturer and Head of Department, Electronics and Computer Engineering at the School of

Engineering and Technology-Soroti University. As Head of Department, Lynn has demonstrated exceptional capabilities in financial management. My research interests are in areas of Artificial Intelligence, Image processing, Pattern Recognition, Internet of things (IOT), Cyber Security and Big data amongst others. While at Soroti, Lynn has participated in writing a number of research proposals including the "Soroti University ICT and Innovation Hub"

Daniel Lawrence is an researcher from Garuda Aerospace Private Ltd, Chennai. The author has published 34 Research articles and contributed to research in topic(s): Materials, UAV Development, Design and Optimization. The author has an h index of 6 and the publication(s) receiving 115 citation(s).

J. Agnishwar is an Industrial researcher from Garuda Aerospace Private limited, Chennai. The author contributed to research in UAV Design and Optimization.

Kalaiyarasi is a dedicated faculty member in the Department of Electronics and Communication Engineering at Panimalar Engineering College, situated in Chennai, Tamil Nadu, India

Kassim Kalinaki (MIEEE) is a passionate technologist, researcher, and educator with more than ten years of experience in industry and academia. He received his Diploma in Computer engineering from Kyambogo University, a BSc in computer science and engineering, and an MSc. Computer Science and Engineering from Bangladesh's Islamic University of Technology (IUT). Since 2014, He has been lecturing at the Islamic University in Uganda (IUIU), where he most recently served as the Head of Department Computer Science department (2019-2022). Currently, he's pursuing his Ph.D. in Computer Science at the School of Digital Science at Universiti Brunei Darussalam (UBD) since January 2022 and is slated to complete in August 2025. He's the founder and principal investigator of Borderline Research Laboratory (BRLab) and his areas of research include Ecological Informatics, Data Analytics, Computer Vision, ML/DL, Digital Image Processing, Cybersecurity, IoT/AIoMT, Remote Sensing, and Educational Technologies. He has authored and co-authored several published peer-reviewed articles in renowned journals and publishers, including in Springer, Elsevier, Taylor and Francis, Emerald and IEEE.

A.V.Senthil Kumar is working as a Director & Professor in the Department of Research and PG in Computer Applications, Hindusthan College of Arts and Science, Coimbatore since 05/03/2010. He has to his credit 11 Book Chapters, 265 papers in International and National Journals, 25 papers in International Conferences, 5 papers in National Conferences, and edited Nine books (IGI Global, USA). He is an Editor-in-Chief for various journals. Key Member for India, Machine Intelligence Research Lab (MIR Labs). He is an Editorial Board Member and Reviewer for various International Journals. He is also a Committee member for various International Conferences. He is a Life member of International Association of Engineers (IAENG), Systems Society of India (SSI), member of The Indian Science Congress Association, member of Internet Society (ISOC), International Association of Computer Science and Information Technology (IACSIT), Indian Association for Research in Computing Science (IARCS), and committee member for various International Conferences

Arvind Kumar is currently pursuing B.Tech. in Information Technology from Rajkiya Engineering College Banda Uttar Pradesh.

Ashok Kumar is a respected faculty member of the Department of Plant Pathology at the School of Agriculture, Lovely Professional University, located in Jalandhar, Punjab, India

Ashwani Maurya is currently pursuing B.Tech. in Information Technology from Rajkiya Engineering College Banda Uttar Pradesh.

Prathik Narayan is a dedicated and tech-savvy individual pursuing B-Tech in Computer Science and Engineering constantly on the search for opportunities that offer immense growth and personal development, proficient mainly in the domains of Data Science and Analysis, Project Management And Human Resource Management.

Rashmiranjan Nayak has received the BTech degree in Electronics and Telecommunication Engineering from the Biju Patnaik University of Technology, Rourkela, Odisha, India, in 2010, and the MTech degree in Electronics and Communication Engineering (Specialization: Communication and Networks) from the NIT Rourkela, Odisha, India, in 2016. Currently, he is pursuing his Ph.D. in the department of Electronics & Communication Engineering, NIT Rourkela. His current research interests include Time Series Forecasting, Artificial Intelligence, Video Anomaly Detection, Computer Vision, Internet of Things, and UWB Antennas for GPR. He is a graduate student member of IEEE.

Praveen Kumar Panth is currently pursuing B.Tech. in Information Technology from Rajkiya Engineering College Banda Uttar Pradesh.

Sreya U. Parvathi received her Ph.D. degree in Soil Science and Agricultural Chemistry at Kerala Agriculture University, India, in the year 2017. She has 04 years of teaching and research experience. She has published several research articles in reputed international, national journals. She has attended seminars, conferences and workshops. Her research interest includes organic farming, soil fertility and remote sensing.

Umesh C. Pati is a Full Professor at the Department of Electronics and Communication Engineering, National Institute of Technology (NIT), Rourkela. He has obtained his B.Tech. Degree in Electrical Engineering from National Institute of Technology (NIT), Rourkela, Odisha. He received both M.Tech. and Ph.D. degrees in Electrical Engineering with specialization in Instrumentation and Image Processing, respectively, from the Indian Institute of Technology (IIT), Kharagpur. He has also served as Head, Career Development Center, NIT Rourkela for three years. His current areas of interest are Internet of Things (IoT), Industrial Automation, Instrumentation Systems, Artificial Intelligence, Image/Video Processing, Computer Vision, and Medical Imaging. He has authored/edited two books and published more than 100 articles in the peer-reviewed international journals as well as conference proceedings. Dr. Pati has filed 2 Indian patents. He has served as a reviewer in a wide range of reputed international journals and conferences. He also has guest-edited special issues of Cognitive Neurodynamics and the International Journal of Signal and Imaging System Engineering. He has delivered many Keynote/Invited talks in India as well as abroad. He has visited countries like the USA, Italy, Austria, Singapore, Mauritius, etc., in connection with research collaboration and paper presentation. He was also an academic visitor to the Department of Electrical and Computer Engineering, San Diego State University, USA, and the Institute for Automation, University of Leoben, Austria. He is a Senior member of IEEE, Fellow of The Institution of Engineers (India), Fellow of The Institution of Electronics and Telecommunication Engineers (IETE), and life member of various professional bodies like MIR Labs (USA), The Indian Society for Technical Education, Instrument Society of India, Computer Society of India, and Odisha Bigyan Academy. His biography has been included in the 32nd edition of MARQUIS Who's Who in the World 2015.

R. Vijayakumar is an Industrial researcher from Garuda Aerospace Private limited, Chennai. The author contributed to research in topic(s): UAV Design and Optimization.

Mahesh Kumar S. V. currently serves as an Associate Professor at Amrita College of Engineering and Technology, Nagercoil, Tamil Nadu, India. He received his Ph.D. degree in Electronics and Communication Engineering at Pondicherry University, Puducherry in the year 2018. During his academic and research activities, he has been involved in several research works in digital image processing, machine learning research fields. His research works appeared in top-level journals and international conferences. He is an IEEE Member and served as an organizing chair for several scientific events. His research interest is in several topics mostly related to biomedical imaging and artificial intelligence.

Naba Sabat is currently a PhD research Scholar in the department of Electronics and Communication Engineering, National Institute of Technology Rourkela, Odisha, India. His research interest includes time series forecasting.

Dler Salih Hasan is a faculty member in the Department of Computer Science and Information Technology at the University of Salahaddin-Erbil, located in Iraq.

Wasswa Shafik (IEEE member, P.Eng) received a B.Sc. degree with honour rank in Mathematics and Computer Science in 2016 from Ndejje University, and an M.Sc. degree in Information Technology Engineering (MIT) in 2020, from the Computer Engineering Department, Yazd University, Islamic Republic of Iran. He is an associate researcher at the Computer Science department, Network interconnectivity Lab at Yazd University, Islamic Republic of Iran, and at Information Sciences, Prince Sultan University, Saudi Arabia. He is the author and co-author of more than 80 papers in IEEE/ACM/IET/CRC press/IGI-Global/Wiley/Springer/Elsevier journals, conference papers, books and book chapters. He received many awards; the best M.Sc. student award from Yazd University, Iran, 2020, Young scientific research award. His areas of interest are Anomaly Detection, Drones (UAVs), Machine/Deep Learning, AI-enabled IoT/IoMTs, IoT/IIoT/OT Security, Cyber Security and Privacy. Shafik is the chair/co-chair/program chair of some Scopus/EI conferences. Also, academic editor/ associate editor for set of indexed journals (Scopus journals' quartile ranking). He is the founder and lead investigator of Digital Connectivity Research Laboratory (DCR-Lab) since 2019.

Priyanka Sharma is born in Jaipur, Rajasthan, India, in 1991. She Received her B.Tech. degree from Rajasthan Technical University, Kota, Rajasthan India, in 2012 in Information Technology and M.Tech. degree from Rajasthan Technical University, Kota, Rajasthan India, in 2016 in Computer Science. She is working as an Assistant Professor in the department of Computer Science & Engineering, Swami Keshvanand Institute of Technology, Management & Gramothan, Jaipur, Rajasthan, India. She has guided various M.Tech. and UG scholars. She is member of IEEE, ACM and various professional societies. She has published more than 15 research papers in national and international journals/conferences, book chapters. Also, she has taken part in forums hosted by Infosys, TCS, Oracle, and IBM. Her area of interest is investigating the advancements in machine learning and deep learning applications. She has received accolades on a number of times in a variety of fields, including Designation as an Active Reviewer by some well-known journals. She was also given recognition by Infosys, India, for her remarkable performance in the Campus Link Program.

T. Sivakami is a faculty member in the Department of Electronics and Communication Engineering at Bharath Institute of Higher Education and Research, located in Chennai, Tamil Nadu, India

Aman Verma is currently pursuing B.Tech. in Information Technology from Rajkiya Engineering College Banda Uttar Pradesh.

Index

K

L

M

P

R

S

T

U

V

W

Y

www.ingramcontent.com/pod-product-compliance
Lightning Source LLC
LaVergne TN
LVHW060423300926
849162LV00006B/1509